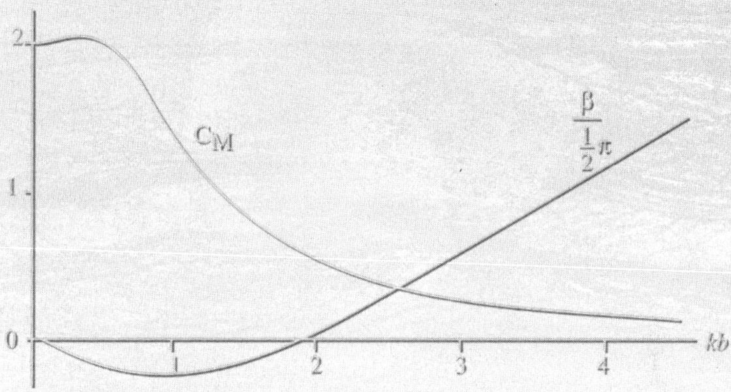

$$P\left(i\frac{\partial}{\partial t},-i\frac{\partial}{\partial x},-i\frac{\partial}{\partial y},-i\frac{\partial}{\partial z}\right)\phi \;=\; e^{-i\omega_0 t}f(\mathbf{r}-\mathbf{U}t)$$

Sir James Lighthill and Modern Fluid Mechanics

Sir James Lighthill (1924 - 1998)

Sir James Lighthill and Modern Fluid Mechanics

$$\rho(\mathbf{x}) - \rho_0 \sim \frac{1}{4\pi a_0^2} \int \frac{(x_i - y_i)(x_j - y_j)}{|\mathbf{x}-\mathbf{y}|^3} a_0^{-2} \frac{\partial^2}{\partial t^2} T_{ij}\left(y, t - \frac{|\mathbf{x}-\mathbf{y}|}{a_0}\right) d\mathbf{y}$$

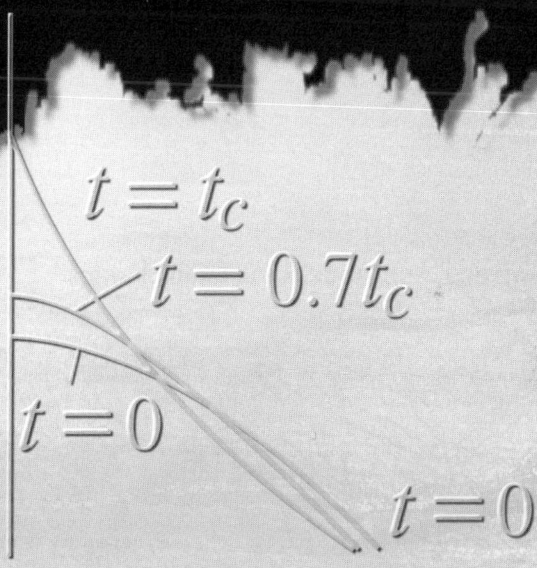

Lokenath Debnath

The University of Texas-Pan American, USA

ICP

Imperial College Press

Published by

Imperial College Press
57 Shelton Street
Covent Garden
London WC2H 9HE

Distributed by

World Scientific Publishing Co. Pte. Ltd.
5 Toh Tuck Link, Singapore 596224
USA office: 27 Warren Street, Suite 401-402, Hackensack, NJ 07601
UK office: 57 Shelton Street, Covent Garden, London WC2H 9HE

British Library Cataloguing-in-Publication Data
A catalogue record for this book is available from the British Library.

SIR JAMES LIGHTHILL AND MODERN FLUID MECHANICS

ISBN-13 978-1-84816-113-9
ISBN-10 1-84816-113-1

Printed in Singapore by B & JO Enterprise

To my grandson **Kirin**,
granddaughter **Princess Maya**,
son **Jayanta** and daughter-in-law **Runi**,
with love and affection.

Preface

Sir James Lighthill (1924-1998) was one of the towering figures in twentieth century applied mathematics and fluid mechanics, internationally recognized as founder of aeroacoustics and biofluid dynamics. He was one of the most brilliant and influential persons who produced some of the most significant applied mathematicians and fluid dynamists in many fields of mathematical sciences. He was notable for the originality, diversity, versatility and independence of his critical thinking. He made outstanding contributions to fluid mechanics, applied mathematics and their applications to physical oceanography, meteorology, geophysical fluid dynamics, biofluid mechanics, and hydrology. He has provided a unique leadership with tremendous interest in diverse uses of mathematics in natural sciences, economics, finance, business, and world modeling and made an attempt to stimulate interest in mathematics among the general public of the world.

Michael James Lighthill was born on January 23, 1924 in Paris. In 1936, he won a prestigious scholarship to Winchester College and then in 1939, was awarded a major scholarship by Trinity College of Cambridge University. After graduating from Trinity College in 1943 with a B.A. degree, Lighthill joined the British National Physical Laboratory as Junior Scientific Officer in the Aerodynamic Division. In 1945, he became a Fellow of the Trinity College. In 1950, at the age of 26, he was appointed to the Beyer Chair of Applied Mathematics at the University of Manchester. In 1953, at the age of 29, he was elected Fellow of Royal Society of London. In 1964, he was appointed Royal Society Research Professor at Imperial College of Science and Technology, London. In 1969, Sir James Lighthill was elected Lucasian Professor of Applied Mathematics at the University of Cambridge to succeed Physics Nobel Laureate, P.A.M. Dirac who held the Lucasian Professorship from 1932 to 1969. The Lucasian Chair once held by Sir Isaac Newton, and the present holder is Stephen Hawking. In 1979, he accepted the position of Provost of the University College London (UCL). After retirement from the position of Provost in 1989, he remained at UCL as Honorary Research Fellow until his death on July 17, 1998.

The devotion of time and energy to mathematics education as well as mathematics instruction was not widely thought to be a worthy pursuit for a serious research mathematician. Sir James characteristically ignored the prevailing view and in 1971-1975 he went further by serving as President of the International Commission on Mathematical Instruction (ICMI) to upgrade mathematics teaching and learning at all levels. Through his presentations and publications of articles on pedagogy, Sir James provided a unique leadership role in promoting the importance of teaching and learning at all levels, and the interaction between mathematics and society. In addition, he helped organize a number of symposia that include 'New topics in applicable mathematics in secondary schools' in Luxembourg in 1973, and 'Mathematics and Language' in Kenya in 1974. As President of ICMI which was then a sub-commission of the International Mathematical Union (IMU), Lighthill had a profound influence on teaching of mathematics and science with newer applications in modern time.

His deep and enduring concern on enormous loss of human lives, economic and marine resources by natural hazards including tropical cyclones, hurricanes, tsunamis, floods, thunderstorms, earthquakes, underwater explosions, landslides and volcanic eruptions reveals the unique character of this man. He spent considerable time and energy to deliver lectures and helped organize topical conferences and symposia on tropical cyclone disasters, monsoon dynamics, weather forecasting, and related areas at national and international levels for the benefit of the developed and developing countries of the world. His leadership role not only provided the research challenge for the multinational and multidisciplinary group of scientists, but also for publication of Proceedings of the topical symposia which brought together current developments as well as possible new directions of advance. Coastal flooding by storm surges and river valley flooding both of tropical cyclones origin are especially disastrous in many countries of Southeast Asia including Bangladesh, China, India and Japan. Sir James responded to this challenge with enthusiasm; organizing international symposia on Monsoon Dynamics in India in 1977, and on Tropical Cyclone Disasters in China in 1992.

Sir James Lighthill stands in great world tradition represented by many giants in continuum mechanics including George Biddell Airy (1801-1892),

J.M. Burgers (1895-1981), Lord Kelvin (1824-1907), Theodore von Kármán (1881-1963), A.N. Kolmogorov (1903-1987), Ludwig Prandtl (1875-1953), L.D. Landau (1908-1968), Lord Rayleigh (1842-1919), John Scott Russell (1808-1882), Osborne Reynolds (1842-1912), G.I. Taylor (1886-1975) and George Gabriel Stokes (1819-1903).

Indeed, Sir James has done more than anyone in the last century to advance applied mathematics and change the stature of fluid mechanics. His stimulating lecture on Fluid Dynamics as a Branch of Physics at the Fluid Dynamics Division of the American Physical Society in 1960 as well as its subsequent publication in volume 15 of *Physics Today* in 1962 is a wonderful example. The concluding statement of this article is a delight to quote: "... by summing up my lecture in a single sentence: It needs categorically to be reaffirmed that the continuum mechanics of a fluid innocent of electric current has as vital and exciting a present and future as any other branch of physical science."

In his famous and authorative chapter published in the Encyclopedia of Twentieth Century Physics in 1995, Sir James described fluid mechanics as a subdiscipline of physics with special emphasis that it is, indeed, another great success story of twentieth century physics. This is one of his most influential articles ever written on fluid mechanics. As global advocate, Lighthill described many major theoretical, experimental and computational progress on the subject during the twentieth century, and cited many unsolved problems and open questions in this rapidly growing field with a number of possible new directions of advance for the 21st century.

In his memorable lecture at the discussion meeting held jointly by the Royal Society and the British Academy in 1986 on Predictability in Science and Society, Sir James Lighthill vigorously criticized the view that dynamical systems governed by Newton's laws of motion do not necessarily exhibit the predictability property. At the same time, he provided the first systematic and persuasive arguments in support of the complete predictability of systems governed by the equations of Newtonian dynamics.

Throughout his research publications and writings, Sir James was perhaps influenced by the emphasis which Lord Rayleigh expressed as follows: "In the mathematical investigation I have usually employed such methods as present themselves naturally to a physicist. The pure mathematician will complain, and (it must be confessed) sometimes with justice, of deficient rigor. But to this question there are two sides. For, however important it may be to maintain a uniformly high standard in pure mathematics, the physicist may occasionally do well to rest content with arguments, which are fairly satisfactory and conclusive from his point of view. To his mind, exercised in a different order of ideas, the more severe procedure of the pure mathematician may appear not more but less demonstrative. And further, in many cases of difficult to insist upon highest standard would mean the exclusion of the subject altogether in view of space that would be required."

Sir James Lighthill was undoubtedly one of the most brilliant and influential fluid dynamists of the twentieth century. He revolutionized applied mathematics with his remarkable contributions to modern fluid dynamics. There is no doubt at all about Lighthill's profound and everlasting impact on mathematical sciences and the scientific community of the world. His lifelong concern for quality mathematics instruction at national and international levels, and for inevitable loss by natural hazards reveals the unique character of this man. He will be remembered forever not only for his great scientific achievements, but also his unique contributions to the welfare of the human race. In many ways, Sir James was the epitome of the applied mathematical community.

This short book is intended as a memorial tribute to this great man and the mathematical scientist of the twentieth century. So my desire as well as interest in writing this book commemorating Sir James Lighthill the man and the mathematical scientist is founded solely on my deep respect and admiration for this great man and renowned applied mathematician whom I had the opportunity of knowing very well for a long period of time.

It is my pleasure to express my grateful thanks to many friends, colleagues, and students around the world who offered their suggestions and help at various stages of the preparation of the book. My special thanks to Ms. Veronica Chavarria who cheerfully typed the manuscript with constant changes and revisions. In spite of the best efforts of everyone involved, some typographical errors doubtless remain. I wish to thank Ms. Anna Tong and Ms. Lai Fun Kwong of Imperial College Press, London for their help and cooperation. Finally, I am deeply indebted to my wife, Sadhana for her understanding and tolerance while the book was being written.

Lokenath Debnath
Edinburg, 2008

Sir James Lighthill: Chronology

January 23, 1924	Michael James Lighthill was born in Paris.
1936	Won a prestigious scholarship to Winchester College.
1939	Awarded a major scholarship by Trinity College of Cambridge University, but he was not allowed to enter until he was 17.
1943	After graduating from Trinity College, Cambridge with a B.A. degree, he joined the British National Physical Laboratory (NPL) as Junior Scientific Officer in the Aerodynamics Division.
1943-1945	Published his first two research papers on the supersonic airfoil theory.
1944	Promoted to a Senior Scientific Officer at NPL.
1945	Married Nancy Dumaresq, a mathematician, when he was 21.
	Selected a Fellow of Trinity College, Cambridge University.
1946	Appointed as Senior Lecturer in Mathematics at the University of Manchester.

1948	Appointed Member of the Fluid Motion subcommittee of the Aeronautical Research Council, and delivered a general lecture on High Speed Flow of Gases at the 7th International Congress of Theoretical and Applied Mechanics, London.
1950-1959	Selected for the Beyer Chair of Applied Mathematics at the University of Manchester at the age of 26 to succeed Professor Sydney Goldstein, a famous fluid dynamist and applied mathematician.
1952	Appointed Chairman of the Fluid Motion Sub-Committee, and Member of the Aeronautical Research Council.
1953	Elected Fellow of the Royal Society of London at the age of 29.
1958	Elected a Foreign Member of the American Academy of Arts and Sciences.
	Published his first classic book on *Introduction to Fourier Analysis and Generalised Functions*, Cambridge University Press, Cambridge, 79 pages.
1958-1979	Served as one of the Founding Associate Editors of the *Journal of Fluid Mechanics.*
1959	Delivered the Ramsden Memorial Lecture on Shock Waves that was published in *Mem. Manchester Lit. Phil. Soc.*, **101** (1959) 1-16.
1959-1964	Served as Director of the Royal Aircraft Establishment (RAE) at Farnborough, England.
1960	Delivered the 48th Wilbur Wright Memorial Lecture on Mathematics and Aeronautics at the Royal Aeronautical Society. Published this lecture in *J. Roy. Aeronaut. Soc.*, **64** (1960) 375-394.

Presented a stimulating general lecture on Fluid Dynamics as a Branch of Physics at an Annual Banquet of the Fluid Dynamics Division of the American Physical Society in November 1960. Published his lecture in *Physics Today*, **15** (1962) 267-275.

1961 Delivered the Bakerian Lecture to the Royal Society of London on Sound Generated Aerodynamically which was published in *Proceedings of the Royal Society*, **A267** (1962) 147-182.

Elected Member of the American Institute of Aeronautics and Astronatics.

Delivered the Fifth Ludwig Prandtl Memorial Lecture on a technique for rendering approximate solutions to physical problems uniformly valid that was published in *Z. Flugwissen*, **9** (1961) 267-275.

Elected Fellow of the Royal Aeronautical Society of England, and an honorary Fellow of AIAA.

1962 Delivered the Wright Brothers Memorial Lecture on Jet Noise in the United States. This lecture was published in *AIAA Journal*, **1** (1963) 1507-1517.

1963 Received the Commander of Order of Léopold.

Awarded the Timoshenko Medal from the American Society of Mechanical Engineers.

1964 Awarded the Royal Medal of the Royal Society of London.

Delivered the Inaugural Lecture on Group Velocity at the Institute of Mathematics and its Applications, in King's College, London on November 25, 1964. This lecture was published in *J. Inst. Maths. Applics.* **1** (1965) 1-28.

1964-1966	Served as the Founding President of the Institute of Mathematics and its Applications for a two-year term.
1964-1969	Served as Royal Society Research Professor at Imperial College of Science and Technology, London.
1965	Received the Gold Medal from the Royal Aeronautical Society.
1965-1969	Served as Physical Secretary and Vice President of the Royal Society of London for a two-year term.
1965-1970	Served as Member of the Natural Environment Research Council and Chairman of its Oceanography and Fisheries Research Committee.
1966	Delivered a lecture on Dynamics of Rotating Fluids: A Survey at the IUTAM Symposium on Rotating Fluid Systems at La Jolla, California, U.S.A., March 28, 1966. This lecture was published in *J. Fluid Mech.*, **26** (1966) 411-431.
	Organized a Discussion Meeting on Nonlinear Theory of Wave Propagation in Dispersive Systems in the Royal Society, May 19-20, 1966.
1967	Published a special issue of A Discussion Meeting on Nonlinear Theory of Wave Propagation in Dispersive Systems, *Proceedings of the Royal Society of London*, Volume **A299**, No. 1456, 145 pages.
1969	Delivered the Symons Memorial Lecture on Unsteady Wing-Driven Ocean Currents. This lecture was published in *Quart. J. Roy Met. Soc.*, **95** (1969) 675-688.

1969-1979	Served as the Lucasian Professor of Applied Mathematics (Newton's Chair) at the University of Cambridge to succeed Physics Nobel Laureate, P.A.M. Dirac.
1970	Elected Member of the American Philosophical Society.
1971	Knighted as Sir James Lighthill by the Queen of England.
	Delivered a lecture on Physiological Fluid Dynamics: A General Survey at the International Centre for Mechanical Sciences, Udine, Italy.
1971-1974	President of the International Commission on Mathematics Instruction (ICMI).
1972	Delivered a general lecture on Aquatic Animal Locomotion: A Survey of Recent Theoretical Developments at the International Congress of Theoretical and Applied Mechanics, Moscow.
	Delivered the Fourth Annual Fairey Lecture on Propagation of Sound through Moving Fluids at the Institute of Sound and Vibration Research, University of Southhamption, July 17, 1972. This lecture was published in *Journal of Sound and Vibration*, **24** (1972) 471-492.
1973	Served as Principal Lecturer at the NSF-CBMS Research Conference on Mathematical Biofluiddynamics in July 16-20, 1973 at Rensselaer Polytechnic Institute, Troy, New York.
1974	Delivered the Fifth Fluid Science Lecture on Aerodynamic Aspects of Animal Flight at the British Hydromechanics Research Association.

1975	Delivered John Von Neumann Lecture on Flagellar Hydrodynamics in the United States of America. This lecture was published in *SIAM Rev.*, **18** (1976) 161-230.
	Published a Research Monograph on *Mathematical Biofluiddynamics*, SIAM, Philadelphia, 281 pages.
1976	Received the American Institute of Aeronautics and Astronautics (AIAA) Award.
	Elected Member of the French Academy of Sciences.
	Elected Member of the United States National Academy of Science.
1977	Elected Member of the United States National Academy of Engineering.
	Organized the joint IUTAM/IUGG International Symposium on Monsoon Dynamics in IIT, Delhi, India, December 5-9, 1977.
1978	Published a comprehensive Research Monograph on *Waves in Fluids*, Cambridge University Press, Cambridge, 504 pages.
	Published a Paperback Edition of *Newer Uses of Mathematics*, Penguin, Harmondsworth, U.K., 431 pages.
	Delivered a general lecture on Acoustic streaming at the Institute of Acoustics Spring Meeting, Cambridge, England on April 6, 1978. This lecture was published in *J. Sound and Vibration*, **61** (1978) 391-418.
1978-1981	Member of the Advisory Council on Research and Development.

1979-1989	Served as Provost of the University College London (UCL) with exceptional academic and administrative skills.
1980	Delivered the Inaugural Frederic Constable Lecture on Academic Innovation.
1981	Won the Harvey Prize for Science and Technology of the Israel Institute of Technology.
	Served as Coeditor of a Proceedings of the joint IUTAM/IUGG International symposium on *Monsoon Dynamics* (with R.P. Pearce), Cambridge University Press, Cambridge, 735 pages.
1982	Received a Gold Medal from the Institute of Mathematics and Its Applications, United Kingdom for his outstanding contributions to applied mathematics.
1984-1988	Served as President of the International Union of Theoretical and Applied Mechanics (IUTAM).
1986	Published a modern book on *An Informal Introduction to Theoretical Fluid Mechanics*(*IMA Monograph Series*), Oxford University Press, Oxford, 260 pages.
	Delivered a profound lecture at a discussion meeting held jointly by the Royal Society and the British Academy in 1986 on 'Predictability in Science and Society'. This lecture was published in *Proc. Roy. Soc. London,* **A407** (1986) 35-50.
1989	Delivered the Rayleigh Lecture on Biomechanics of Hearing Sensitivity at the 1989 ASME Winter Meeting in Chicago, U.S.A. The lecture was published in *ASME J. Vib. Acoust.* **113** (1991) 1-13.

	Retired from the position of Provost and remained at University College, London as Honorary Research Fellow until his death.
	Delivered the Inaugural Sydney Goldstein Memorial Lecture on Some Challenging New Applications for Basic Mathematical Methods in the Mechanics of Fluids that were Originally Pursued with Aeronautical Aims in the United Kingdom. Published this lecture in *Aeron. J.* **94** (1990) 41-52.
1990-1995	Served as Chairman of a special committee on the International Decade for Natural Disaster Reduction of the International Council of Scientific Union.
1992	Delivered the Humphry Davy Lecture on La mécanique dans la biologie moderne which was published in *La vie des Sciences* **9** (1992) 117-136.
	Organized the ICSU/WMO International Symposium on Tropical Cyclone Disasters in Beijing, China; October 12-16, 1992.
1993	Published a research Proceedings of the joint ICSU/WMO International Symposium on *Tropical Cyclone Disasters* (with Zhang Zhimin, Greg Holland and Kerry Emanuel), Peking University Press, Beijing, China, 588 pages.
	Delivered the Inaugural Theodorsen Lecture on Some Aspects of the Aeroacoustic of High-Speeds Jets at NASA Langley Research Center, Hampton, Virginia, U.S.A. This lecture was published in *Theoret. Comput. Fluid Dynamics*, **6** (1994) 261-280.

1995 Published one of the most influential 117-page long survey articles on Fluid Mechanics in the *Twentieth Century Physics, Volume II* (Eds. Laurie M. Brown, Abraham Pais and Sir Brian Pippard) Institute of Physics, and American Institute of Physics Press, New York, Volume II, 795-912.

1996 Delivered an invited lecture at International Congress of Theoretical and Applied Mechanics.

1997 The *Collected Papers of Sir James Lighthill* in four volumes were published by the Oxford University Press.

1998 Delivered his last of several invited lecturers at the British Applied Mathematics Colloquia on A Century of Shock-Wave Dynamics.

Died after completing a nine-hour swim around the Channel Island of Sark on July 17, 1998.

Awarded posthumously the highly prestigious Copley Medal of the Royal Society of London.

Contents

Chapter 1

An Early Life History and Career of Sir James Lighthill

> "... by summing up the argument of my lecture in a single sentence: It needs categorically to be reaffirmed that the continuum mechanics of a fluid innocent of electric current has as vital and exciting a present and future as any other branch of physical science."
>
> "... as Sir Cyril Hinshelwood has observed ... fluid dynamicists were divided into hydraulic engineers who observed things that could not be explained and mathematicians who explained things that could not be observed."
>
> *James Lighthill*

The old city of Paris has many great traditions. It is the capital and largest city of France. It is one of the most beautiful cities in the world. Three of the most famous Parisian landmarks are the twelfth century Notre Dame de Paris on the Île de la Cité, the nineteenth century Eiffel tower and the Napoleonic Arc de Triumphe. The 1063 feet tall Eiffel Tower is the most recognized symbol of Paris and it is located on the bank of the River Seine in north-central France. At night, floodlights shine on Paris' lovely parks, gardens, palaces and monuments. The gleaming beauty of Paris has given it the nickname *City of Light.* It has a reputation as a "romantic city". Paris is widely recognized as one of the world's global cities. It is a leading global cultural, educational, business and political centre and is renowned for its classical architecture as well as its unique role as a major international influence in modern fashion, gastronomy, museums and arts.

More importantly, Michael James Lighthill was born on January 23, 1924 in the Rue Puccini, Paris. His father, Earnest Balzar Lighthill, was mining engineer who worked all over the world but came to England when he retired in 1927 at the age of 59. His mother, Marjorie, was a daughter of Yorkshire engineer, L. W. Holmes and she was 18 years younger than her husband. James had a brother, Olaf, seventeen years older and a sister, Patricia, nine years older than himself. Obviously, James was a new member of a middle class family and grew in his family under the dominant influence of parental love and care.

James was a special and precocious child, showing a clear disposition to originality and independence and early evidence not only of his mathematical talent and of his photographic memory, but also in many other areas including language, music and chess. At the age of ten, James beats his father in a chess game, even though his father was a very good chess player.

James spent early years at private schools and won a Scholarship to Winchester College that was one of the most famous and ancient public schools in England. It was a remarkable coincidence that Freeman Dyson, a twelve-year old scholar at Winchester, was James' classmate at Winchester. Freeman Dyson spent many years and is still at the Institute for Advanced Study in Princeton. Professor Dyson is one of the brilliant mathematical physicists of the twentieth century. James and Freeman became very close friends and spent considerably amount of time together to study mathematics. At the age of fourteen, they spent considerable amount of time to study *Principia Mathematica* (1910-1913) by A. N. Whitehead (1861-1947) and Bertrand Russell (1872-1970) who published three monumental volumes of the *Principia* containing a completely rigorous development of the foundations of mathematics. At the same time, in order to learn mathematical analysis, James read the three-volume *Cours d'analyse de l'Ecole polytechnique* by Camille Jordan (1909). Undoubtedly, this study was a most profound basis for Lighthill's subsequent mastery of mathematical analysis and its applications.

After a rigorous education at Winchester College with an outstanding record, at the age of fifteen, James and Freeman were both awarded prestigious scholarships to Trinity College, Cambridge to pursue a B.A. degree in Mathematics. But the college would not allow them to join until they were seventeen years old. After spending two years of study in more mathematics, they entered the Trinity College in 1941 and began a new life as undergraduate students. Since they had already learned sufficient undergraduate material in mathematics, they chose to attend only the lectures for Part III of the Mathematical Tripos almost entirely in pure mathematics intended for graduate students. James was particularly influenced by the presence of famous pure mathematicians including G. H. Hardy (1877-1947), J. E.

Littlewood (1885-1977) and A. S. Besicovitch (1891-1970) in Cambridge at that time. Lighthill's Cambridge supervisor was Professor Besicovitch who became worried about James' allegiance to pure mathematics that would be completely lost if he moves out from Cambridge. Originally from St. Petersburg, Besicovitch was a renowned pure mathematician for his major contributions to the theory of almost periodic functions and other areas of function theory, and especially, for his pioneering research work in geometric measure theory, where he proved many of the fundamental results. Since there was then little incentive at Cambridge to study applied mathematics, Besicovitch reluctantly recommended James to join Sydney Goldstein's (1903-1989) group at the British National Physical Laboratory (NPL), and wrote to Goldstein begging him not to ruin James. The Mathematical Tripos in those days was a very competitive examination for which many of the candidates were coached personally. Candidates for this examination were put into the first, second, and third class according to the marks obtained and were also ordered within each class. Candidates placed into the top class 1 are called '*Wranglers*'. After two years of study, James and Freeman took both Part II and Part III of the Mathematical Tripos Examination. Not surprisingly, they obtained a first class in the former and a Distinction in the latter.

After successfully completing his B.A. degree from Cambridge in two years, James joined the British National Physical Laboratory as Junior Scientific Officer at Aerodynamics Division in 1943 and soon became a Senior Scientific Officer at NPL. Before he became twenty in 1945, James published his first research papers on two-dimensional supersonic aerofoil theory in the Aeronautical Research Committee Report and Memorandum (ARCRM). He continued his research at NPL and then published several other papers on airfoil theory in ARCRM in 1944 and 1945. His dream of becoming a successful applied mathematician has led him to participate and excel in many other diverse activities at NPL, and to become a well-rounded and knowledgeable research scientist with highest possible standards.

In 1945, at the age of twenty one, James married his undergraduate classmate, Nancy Dumaresq who was studying mathematics at Newnham College in Cambridge. Nancy also got a job at the Royal Aircraft Establishment (RAE) at Farnborough. James was asked to go to the NPL at Teddington and to conduct his research with the distinguished fluid dynamicist, Sydney Goldstein (1903-1989) who convinced James Lighthill to pursue research in fluid mechanics as this subject has a wide variety of challenging research problems. Because of his outstanding research work at NPL, James was awarded major research fellowship at Trinity College, Cambridge in 1945. During his stay for a year at his 'Alma Mater', Lighthill was fully influenced by Sydney Goldstein and G. I. Taylor (1886-1972) to continue his research on challenging problems in fluid mechanics. G. I. Taylor has already become one of the great physical scientists of the twentieth century for his numerous notable theoretical and experimental work in fluid and solid mechanics with applications to oceanography, meteorology, mechanical, civil and chemical engineering, hydraulics and materials science. I was fortunate to see this distinguished scientist and great man many times at the Department of Applied Mathematics and Theoretical Physics of the University of Cambridge during my stay in Cambridge in 1968-1969. The awarding of that prestigious fellowship to James Lighthill combined with the encouragement of Goldstein and Taylor launched a remarkable research career and led directly to lifelong association with fluid mechanics.

In 1946 Sydney Goldstein joined the University of Manchester as the Beyer Chair of Applied Mathematics succeeding Sir Horace Lamb, another renowned applied mathematician and fluid dynamicist, and persuaded James Lighthill to go to there as a Senior Lecturer. Almost simultaneously, Max Newman (1897-1984), a distinguished topologist, was appointed to the Fielden Professorship of Pure Mathematics of the University of Manchester. Sir John Stopford, then Vice Chancellor of Manchester, convinced the Senate and Academic Council that Sydney Goldstein and Max Newman would create at Manchester a dynamic and internationally renowned new Department of Mathematics, interacting admirably with other units of the University. After spending a year in Cambridge, Lighthill joined the University of Manchester as Senior Lecturer in Mathematics at the age of twenty two. Within a short period of four years, he succeeded Goldstein in the Beyer Professor of Applied Mathematics at Manchester at the age of twenty six. This has been almost an extraordinary accomplishment in the United Kingdom in those days. It was, indeed, with a definite commitment in mind, Lighthill arrived at Manchester in 1945. It became the starting point for his forceful participation in moving Manchester forward to still greater levels of excellence in applied mathematics and fluid mechanics. During his thirteen years that were spent helping to the University of Manchester which was outstandingly strong in applied mathematics and fluid mechanics, Lighthill pursued also many other interests. In particular, he was actively involved in maintaining and still further developing those great Manchester traditions that were initiated by Sydney Goldstein and Max Newman. He continued his active interests in interdisciplinary research with engineering scientists and expanded his interaction with other parts of the University. In addition, the Fluid Motion Laboratory became an active and productive Department of Mechanics of Fluids, still working in close cooperation and collaboration with mathematics faculty and post graduate students. His over a decade long stay at Manchester was very productive and rewarding in his life in terms of research, publications and supervision of doctoral students—many of them are now well known in applied mathematics. Highly respected by his applied mathematics peers, he was instrumental in organizing a strong research group in fluid mechanics and applied mathematics in England. Above all, he created an enduring cordial atmosphere of mutual support and encouragement for fluid dynamics research at the highest level. In their biographical memoir, both David Crighton (1942-2000) and T. J. Pedley recognized Lighthill's remarkable success as a teacher-scholar in Manchester as being due to his profound interest in applied mathematics and fluid mechanics and his faculty for inspiring his colleagues, and especially the best research graduate students, with some of his own enthusiasm.

In 1952 Lighthill was elected Fellow of Royal Society of London at the age of twenty nine for his pioneering research on sound generated aerodynamically. He was awarded the Bronze Medal of the Royal Aeronautical Society for his outstanding contributions to aerodynamics in 1955.

During his happy and extremely productive period of 13 years in Manchester, Lighthill was deeply involved in developing a modern pure and applied mathematics research center in collaboration with several eminent pure mathematicians including Max Newman and others. He served as research supervisor of many Ph.D students and other research scholars. Subsequently, almost all of his students and research associates became famous and eminent applied mathematicians for their own original research contributions. Usually, Lighthill did not wish to write joint papers with his doctoral and post doctoral students. Amazingly, Lighthill produced a large number of long research papers and two major books by himself during his tenure at the University of Manchester. In addition, he delivered many research expository and survey lectures as well as famous memorial lectures by invitation from many national and international professional and scientific organizations. Indeed, almost all of his research papers and lectures were loaded with new ideas and results, and represent pioneering contributions to a wide variety of different areas of fluid mechanics and applied mathematics. In 1999, D. G. Crighton described the long range impact of Lighthill's research as: "... worked extensively on gas dynamics, including effects of shock and blast waves. He also launched two major new fields in fluid mechanics."

The first of these new fields was aeroacoustics which proved to be vital importance in the reduction of noise from jet engines. In his research on aeroacoustics, Lighthill discovered his famous *eighth-power law* which states that the acoustic power radiated by a jet aircraft is directly proportional to the eighth power of the jet speed. In fact, Lighthill's theory and his eighth-power law provided a revolutionary impact on the modern mechanics, in general, and on aircraft industry, in particular. The second new field introduced by Lighthill was modern nonlinear acoustics in which he described propagation of shock waves, flood waves in rivers and traffic flow on highways. His 1956 classic 101-page long chapter on "Viscosity Effects in Sound Waves of Finite Amplitude" written in honor of the 70th birthday of great fluid dynamist and physical scientist Sir G. I. Taylor can be considered as the beginning of nonlinear acoustics. On the other hand, in order to extend the classical work of Sir Geoffrey Taylor on shock waves, Lighthill developed a more general and complete theory of plane shock wave formation with conflicting effects of convection (nonlinear) on one hand, and of diffusion and relaxation on the other, where these effects were allowed to first approximation. However, sound waves of finite amplitude with effects of relaxation were investigated by Bethe and Teller (1941). More general theory of the effect of relaxation on general steady flows had been fully treated by Kantrowitz (1946) and Gunn (1946). In his research on this field, Lighthill first initiated a new and modern development of nonlinear acoustic wave propagation and unified the research work of other eminent applied mathematicians including J. M. Burgers (1895-1981), J. D. Cole (1925-1999), H. Hopf (1894-1971), G. I. Taylor, and G. B. Whitham.

According to Professor T. J. Pedley (2001), Lighthill's 13-year stay in Manchester represents his "golden years" as this period was full of intensive intellectual activity during which he has not only published a large volume of research work and two major books, but also produced many future national and international leaders in applied mathematics. In order to promote research interest in a wide range of fluid mechanics, Lighthill initiated a series of annual conferences called the *British Theoretical Mechanics Colloquia (BTMC)* which was first held in Manchester in 1959. These colloquia became the national forums for *British Applied Mathematics Colloquia* in a short period of time. Lighthill delivered many invited lectures at the Annual Conference of BTMC. I myself participated in two BTMC Conferences at Southampton in 1966 and at Oxford in 1968 and enjoyed Lighthill's invited lectures on rotating fluids and dynamics of oceans at these conferences. In appreciation for his remarkable contributions, the Senate and Council of the University of Manchester adopted the valedictory resolution in February 1960 which states:

"Those of us who knew Lighthill in his younger days remember a man of almost frightening brilliance, a prickly mathematician intolerant of any argument which lacked precision and rigor, and devastating in his verbal attack on its author. His standards never changed: but his attitude towards people did. Later he would offer gentle criticism and advice and the most painstaking guidance; and his students, whether the bright or the weakest, received the same patient interest and encouragement."

Lighthill left Manchester in 1959 to take the extremely prestigious position of Director of the Royal Aircraft Establishment (RAE) at Farnborough. His main leadership role included managing the work of the fourteen hundred scientists and engineers, and the eight thousand RAE staff. Despite his large administrative duties and responsibilities as Director, Lighthill was extremely active and productive in his own research, publications and presentations. Indeed, he continued to publish papers and books at an unprecedented rate. For example, he published his lecture on Fluid Dynamics as a Branch of Physics in *Physics Today* in 1962. This lecture was presented at a well attended Banquet Meeting of the Fluid Dynamics Division of the American Physical Society in November 1960. He began his lecture with the following interesting statements: "... why fluid dynamist remained almost completely unabsorbed by physics until considerably less than a century ago, and to note how the process of absorption has taken place only gradually since then, initially as a result of the work of Rayleigh and Prandtl and von Kármán and G. I. Taylor and the rest, and how it has proceeded with accelerated speed over the last thirteen years, greatly as a consequence of the work of members of Fluid Dynamics Division of the American Physical Society.

I'd like to remark again that the great physicists from Newton onwards were notable for unceasingly comparing theory and experiment. ... you may well think unreasonable to criticize Sir Isaac, bearing in mind that he was breaking completely new ground in everything he did, and in particular in his studies of fluid dynamics ..."

"Similar criticisms, couched in the vein of comparing work in this field, can be made of the approaches of Euler, Lagrange, Cauchy and Kelvin to theoretical fluid dynamics. The mathematical reasoning was on a notably high level, but the extent of contact with real fluids was kept to a minimum, and direct attempts at comparison with experiment to a large extend avoided. ..."

"Writers at that time compared what they called the "older" and "newer" theories, namely Newtonian and potential flow theories, and came to the conclusion that the Newtonian must be preferred because it avoided the d'Alembert paradox. This ignored the fact that fluid motion which it predicted different fundamentally from observation....."

"Looking back in the light of our knowledge that viscous forces, though small, play a crucial part in determining the flow about solid bodies, it is worth looking closely at the nineteenth-century work on viscous flow to see how this point was missed. A certain part of this work must be admitted to have been of outstanding value, notably the experiments on resistance in capillary tubes by Poiseuille and Hager, stimulated by physiology, and Stokes's great paper of 1851 on the effect of viscosity on the motion of pendulums, stimulated by the problems of gravity surveying with pendulums in the rather low vacuum conditions which could then be achieved. All these pieces of work obtained excellent agreement with experiment using the condition of zero relative velocity of fluid at a solid surface, which Stokes showed was also to be expected on physical grounds."

"In this field of external aerodynamics, even the appearance of Prandtl's great paper of 1904, which contained all the essentials of the solution to the mystery, was by no means immediately effective in increasing understanding of the physics of the subject among the abstract mathematicians or the empirical engineers. Books do still appear which ascribe flow separation to causes which really are those underlying cavitation! Nevertheless, by the end of the twenties there had been enough university centers of instruction in the physics of fluid motion, notably Göttingen and Caltech, and enough handbooks of the subject like those of Durand, Goldstein, and the German compilations, to have created a recognized and adequately manned discipline which comprehended both theory and experiment."

"I could also mention some other examples to emphasize that it was not only *external* aerodynamics which was slow to be integrated within physics. In the general area of vibrations and waves the intimate binding together of theory and experiment was made rather earlier, largely by Rayleigh, whose *Theory of Sound* is, justly, still one of the best-sellers of science. It is interesting, though, that not till well on into this century was it discovered that the mathematical theory of the attenuation of sound by viscosity and heat conduction gave results orders of magnitude too low, actually owing to the time lag in molecular vibrational energy exchange, and it was about the same time that the physical nature of a shock wave first became clear, while the application of surface wave theory to experimental oceanography had to wait for the nineteen-forties."

"Nevertheless, at the time when this Fluid Dynamics Division was formed, a clearly defined science had grown up on sound physical premises, with a wide range of applications in aeronautics, naval architecture, acoustics, meteorology, oceanography, mechanical engineering and chemical and explosives engineering. The basic physical premises were a continuum fluid and the simple laws of viscosity and heat condition. Ideas derivative from these included boundary layers, vortices, turbulence, sound waves, Mach lines and shock waves, actuator disks, free and forced convection, deep and shallow water waves, phase and group velocity, secondary flow, geostrophic and thermal winds. The consistency of inclusion within physics of fluid dynamics at this period is well shown by the success with which Landau and Lifschitz were able soon after this time to include in their nine-volume course on theoretical physics a volume on fluid mechanics, which recently has been translated into English...."

"Next, we know that liquid helium below its lambda point exhibits a fluid dynamics considerably different from those to which we are accustomed, and a quantum fluid dynamics has been worked out which indicates that in certain circumstances a proportion of the fluid mass is capable of motion but unable to carry entropy or vorticity. This discovery explained a whole range of phenomena, but there are others, beyond, where the low temperature physicists are now postulating a peculiar kind of turbulence in the superfluid...."

"This symposium has reminded us of the continued fascination and inexhaustibility of the study of water waves, and of low-speed aerodynamics. I was interested to find that at the wind tunnels of the National Aeronautics and Space Administration, as well as at those of the Royal Aircraft Establishment, it is the low-speed ones, the good old low-Mach-number wind tunnels, that are absolutely crammed with work!..."

His concluding statement of this stimulating lecture is a delight to read: "I could go on much longer, but I will just conclude by thanking you again for inviting me here, and by summing up the argument of my lecture in a single sentence: It needs categorically to be reaffirmed that the continuum mechanics of a fluid innocent of electric current has as vital and exciting a present and future as any other branch of physical science."

He also published his two major chapters on laminar boundary layer theory in the famous volume in *Laminar Boundary Layers* edited by L.

Rosenhead in 1963. At the same time, he established a new collaboration with the neighboring Institute of Aviation Medicine and created a Department of Space Research at the RAE. Even though he was not less successful obtaining funding for original research in aerodynamics in general, Lighthill gradually realized the major weakness of RAE, especially, the lack of support for modern computers in aerodynamic design of high speed jet aircraft and missiles as well as for research in computational fluid dynamics. Towards the end of his five year stay at RAE, he became dissatisfied due to lack of support for applied mathematics and computational fluid dynamics from the British government and other national agencies. However, even as Director of RAE, he was actively involved in research and development of aerodynamics of the slender delta wing for aircraft, spacecraft, and performance of high speed jet aircraft and missiles.

In 1964, Sir James was selected for the position of Royal Society Research Professor at the Imperial College of Science and Technology at London. In a new and conducive academic atmosphere at Imperial College, he had the usual responsibility of teaching, research and guiding many promising young researchers and other visiting research scholars. As his career progressed he was noted for changing his research from one area to another. At Imperial College, he delivered three courses of lectures on Nonlinear Waves, Geophysical Fluid Dynamics, and Blood Flow in Arteries during 1965-1967. In addition to his many outstanding papers and presentations on these topics, he got involved in the development of strong research in mathematical biofluiddynamics. He was actively involved in understanding of the flow of blood in mammalian cardiovascular systems, of air in the human airways, and of the flying of insects and birds, and developing the standard model of fish swimming using body undulations. His small-amplitude, slender- (or elongated-) body theory describes how thurst is generated from the reactive forces experienced by an undulating body as it exerts sideways acceleration to fluid which is moving backwards relative to the fish at almost steady swimming speed. Lighthill took the opportunity to expand his interdisciplinary collaborative research efforts among engineering scientists, physical scientists, life scientist and medical doctors. His leadership role led to the establishment of a Physiological Flow Studies Unit at Imperial in 1966 under the directorship of Dr. Colin G. Caro. He was delighted to see that this Unit celebrated 25 years of successful work under Professor Caro's leadership in May 1991. From 1966, his work involved a close interdisciplinary collaboration on internal biofluiddynamics by doctors, engineers, physiologists, applied mathematicians within the unit and in closely related departments at Imperial College. However, he again became very unhappy with the level of support given to applied mathematics from government and other external sources for interdisciplinary research. Under his unique leadership, a new *Institute of Mathematics and Its Applications* was established in the United Kingdom in 1965, and he became the first President of this newly created Institute. During 1963-1965, he was awarded Royal Society Medal, Gold Medal from the Royal Aeronautical Society, and Timoshenko Medal from the American Society of Mechanical Engineers for his many outstanding research contributions. He served as Physical Secretary and Vice President of the Royal Society of London during 1965-1969.

In 1967, Lighthill participated in the IAU Symposium with a lecture on "Predictions on the velocity field coming from acoustic noise and a generalized turbulence in a layer overlaying a convectively unstable atmospheric region." He presented the general linear non-dissipative theory of propagation of anisotropic waves through horizontally stratified medium. Included were fundamental wave motions under the influence of compressibility, gravity and a magnetic field and wave generation in and above the convection zone where the turbulence is relatively homogeneous. There are regions above the convection zone where intermittent turbulence is present. Both gravity waves (also known as *internal waves*) and Alfven's magnetohydrodynamic waves can be generated in this region. This is followed by the attenuation mechanisms that exist for linear and nonlinear dissipation of wave energy into heat. When magnetohydrodynamic waves are generated, either directly or by the transformation of sound waves or internal waves into them, magnetic pressures and gas pressures would be comparable so that there exists region for onset of nonlinear effects. In addition to waves generated by linear theory, random nonlinear dissipation mechanisms can generate fast hydromagnetic waves that would in turn produce shock waves. This lecture was published in the *IAU Symposium* No. **28** (1967) 429-453 (*Aerodynamic Phenomena in Stellar Atmosphere*), Academic Press.

From Imperial College, Lighthill returned to Cambridge in 1969 to succeed Paul Dirac (1902-1984), 1932 Nobel Prize winner in Physics as the Lucasian Professor of Applied Mathematics. Paul Dirac was one of the founders of quantum theory and the author of many of its major subsequent developments. He discovered the relativistic equation for the electron which is universally known as the Dirac equation. He is ranked with Isaac Newton (1642-1727), Albert Einstein (1879-1955), James Maxwell (1831-1879), Max Planck (1858-1947), Ernest Rutherford (1871-1937) as one of the greatest theoretical physicists of all time. Dirac also first discovered the existence of a positron — an antielectron having the same mass and opposite charge as the electron which was experimentally confirmed by the 1936 Physics Nobel Prize Winner, Carl Anderson in 1932. The Lucasian Chair is probably the most prestigious Mathematics Chair in the United Kingdom as this Chair was formerly held by many renowned British scientists including Sir Isaac Newton, Sir George Airy (1801-1892), and by Sir G. G. Stokes (1819-1903). Remarkably, Sir G. G. Stokes taught over forty-five future Fellows of the Royal Society, most notably Clerk Maxwell, Lord Rayleigh and J. J. Thompson. It was under Stokes' tenure as the Lucasian Chair that Cambridge became the major center of British applied mathematics and physics, just as the Mathematical Tripos and Natural Science Tripos evolved from a liberal education into specialized training for a scientific research career. Stokes became a popular public figure in Victorian Great Britain since he was a prominent scientist and a loyal Christian. Stokes was President of the Royal Society and served as an editor of the *Philosophical Transactions of the Royal Society* for thirty years and, indeed, was a real statesman of applied mathematics and science. Lighthill was justifiably proud to tell people that his predecessor in the Lucasian Chair was Newton. It is not out of place to point out that Stephen Hawking succeeded Lighthill as the Lucasian Chair in 1979. Stephen Hawking is widely considered as one of the most brilliant scientists since Newton and Einstein. He is a Cambridge based theoretical physicist whose major contributions provided a greater and deeper understanding of the complex nature of the Universe. It is a delight to quote his famous question: "... where did the universe come from? How and why did it begin? Will it come to an end, and if so, how? These are questions that are interest to us all." Through his many publications of bestselling popular science books, he has become the modern world's renowned science story writer in ordinary language.

Within the next two years of his stay in Cambridge, Queen Elizabeth conferred him a Knighthood as Sir James Lighthill in 1971. Among his many other awards and honors were the Bakerian Lecture on Sound generated aerodynamically in 1961, the 48th Wilbur Wright Memorial Lecture on Mathematics and Aeronautics in 1960, the Wright Brothers Lecture on Jet Noise in the United States in 1963, the Symons Memorial Lecture on Unsteady wind-driven ocean currents in 1969, the inaugural Frederic Constable Lecture in 1980, American Institute of Aeronautics and Astronautics (AIAA) Aeroacoustics Award in 1976. During his stay in Cambridge from 1969 to 1979, Sir James vigorously continued his teaching and research on acoustics, more and more wave propagation, geophysical fluid dynamics, biofluid mechanics, ocean and atmospheric dynamics with special reference to prediction of monsoons and tropical cyclones. He served on the British Post Office Board for a period of two years (1972-1974) in order to help promote the long distance telephone direct dialing system, commercial use of television and communications satellites in the United Kingdom.

In addition, Lighthill established two new collaborative research groups including the aquatic animal locomotion group with Sir James Gray of Zoology Department of the University of Cambridge and the animal flight group with T. Weis-Fogh of the Department of Zoology in Cambridge. Lighthill also recognized the brilliant discovery of a new mechanism of lift generation involving formation of trailing vortices at the wing tips, and including the case of a hovering insect like *Encarsia formosa* (a small parasite used in the biological control of greenhouse aphide) moving those tips in circular paths. This work of Weis-Fogh represents a fundamental contribution to knowledge of animal flight, in general and hovering flight, in particular. He described the role of the 'clap' and 'fling' motions in *Encarsia formosa*: that is, the clappings of the wings behind the back of the insect and their subsequent 'fling open'. Indeed, this led to a fundamentally new biofluiddynamic mechanisms by which they operate to generate instantaneously an exceptionally large wing lift. No doubt, Lighthill was an outstanding promoter of interdisciplinary collaboration during his stay in Cambridge as James Gray, G. I. Taylor, T. Weis-Fogh, G. J. Hancock were involved in interdisciplinary research between applied mathematics and other diverse areas including biology, zoology, and biophysics. He himself made fundamental contributions to biofluid mechanics and flagellar hydrodynamics.

Based on his many invited presentations and inaugural addresses, Sir James wrote a large number of articles and texts of addresses that I find informative and interesting additions to his total scientific work. Obviously, the magnitude of the presentations demands some selection of the articles to be included in this book. However, any serious discussion of his all addresses would require more space and time and competence that I have. So I would like to focus on some of his addresses with a little more elaboration.

His 1961 Bakerian lecture to the Royal Society of London on Sound generated aerodynamically deals with Sir James' original work on modern aerodynamics with new practical applications. Included are new developments in the frontiers between acoustics and aerodynamics, reduction of jet aircraft noise and improved knowledge of space-time correlations in turbulent flow that is used to throw new light on the noise radiated by turbulent boundary layers, as well as by jets at high Mach numbers. This lecture was published in the *Proceedings of the Royal Society of London*, volume **A267** (1962) 147-181.

Lighthill was selected to deliver the Fifth Ludwig Prandtl Memorial Lecture on "A Technique for Rendering Approximate Solutions to Physical Problems Uniformly Valid" on April 7, 1961. This Memorial Lecture commemorates Prandtl the great Physical Scientist of the twentieth century who made the revolutionary discovery of the boundary layer theory in 1904, the *annus mirabilis* that totally transformed the nineteenth century fluid mechanics. His discovery conclusively resolved the major controversy of the famous d'Alembert Paradox that an inviscid potential flow in three-dimensions around a rigid body moving at a uniform velocity exerts no resistive force on the body. Mathematically, $\mathbf{F} = -(d\mathbf{P}/dt) = 0$, where $\mathbf{P} = m\mathbf{v}$ is the total momentum of the fluid and $\mathbf{F}$ is the total external force transmitted to the fluid by the body. However, the behavior of the experimentally predicted flow is quite different from that of the potential flow. This fallacy lies not in the direct neglect of viscous forces, but rather in the assumption that there is *no* vorticity in the fluid outside the body. A body moving through a real fluid has behind it a wake containing vorticity. If the flow around a steadily moving body could be made quite close to a potential flow, the resistive force would become very small, but *not* zero.

Prandtl's discovery also resolved another erroneous prediction of the theory that fixed-wing aircraft would be unable to fly, because the air cannot exert any force at all — resistive or lifting force on a steadily moving body. Finally, the boundary layer theory explained how the nature of thin wakes from well designed wing shapes could allow wings to exert large forces (lift forces) normal to the flow, even resistive forces would continue to be small, but not zero, in the general spirit of de Alembert's Paradox. Therefore, major controversies of theory and experiment of the nineteenth century fluid mechanics came to an end. At the same time, Prandtl first introduced a totally new method of treatment of one kind of singular perturbation theory which provided considerable interest far outside his own field of fluid dynamics. Prandtl first developed a new correct approximate expansion method which involved the so-called boundary layer flow near a solid surface of an inviscid fluid flow away from the surface. This matching technique is also called the Prandtl *boundary-layer method* or the singular perturbation expansion method. This method deals with an approximate solution which is valid in a limited region. Another quite different approximate solution is obtained so that it becomes valid in another region that overlaps the first region. After adjusting arbitrary constants, two solutions are matched as closely as possible in the overlapping region.

On the other hand, there is another kind of a new approximate solution valid throughout the region of interest. In this case, an approximate solution valid only outside a certain region is sought, then we seek a method for rendering it uniformly valid. So, there are two different kinds of approximation methods possible for the so-called *singular perturbation problems*. In his lecture, Lighthill gave a modern mathematical treatment of singular perturbation theory which was successfully applied to resolve a long-standing enigma in fluid mechanics of the twentieth century. More generally, the singular perturbation theory is also successfully applied to nonlinear ordinary or partial differential equations with independent contributions of Rayleigh (1910) and Taylor (1910) to the internal structure of shock waves. From a mathematical point of view, Prandtl's original solution, besides representing an extremely early example of a singular perturbation, was successfully applied to fairly general nonlinear field equations. A similar revolution in knowledge of nonlinear effects in the generation and propagation of waves in fluids including sound waves and water waves has been originated from the mathematical analysis of shock waves. Indeed, all these represented outstanding examples of singular perturbation theory that was applied to nonlinear field equations involving partial differential equations. From a physical point of view, Prandtl's new ideas and insights led to the introduction of 'streamlined' shapes deviating from potential flow fields would experience very low resistive forces. Furthermore, Prandtl made the first qualitative as well as quantitative understanding of drag and lift forces on fixed-wing aircraft based on sound physical principles. He progressively revolutionized the major understanding of aerodynamic lift and drag forces and of fluid flows at high Mach numbers. So, the Prandtl's discovery has served as the fundamental basis for all crucially important subsequent developments of modern fluid mechanics. His superb work on fluid mechanics was subsequently published in an extended 1952 English-language edition as *Essentials of Fluid Dynamics*. In order to recognize Prandtl's brilliant contributions to fluid mechanics, Lighthill's comment is worth quoting: "... Indeed, his revolutionary discovery of the boundary layer in 1904 had the same transforming effect on fluid mechanics as Einstein's 1905 discoveries had on other parts of physics."

His next major contribution was the Wright Brothers Memorial Lecture on Jet Noise at the 31st Annual Meeting of the Institute of Aerospace Sciences in January 21, 1963. This lecture deals with the theoretical results with experimental evidence of how the shearing motions in a turbulent jet allow to shed some of their energy as sound radiation that is then propagated away through the atmosphere. Included are both subsonic and supersonic jet noise theories with the reduction jet aircraft noise. This is followed by a recent development, refinement and extension of Lighthill's theory to other areas of fluid mechanics, and to astrophysics where acoustic radiation in stars and in cosmic gas clouds is important. This lecture was subsequently published in *AIAA Journal*, **1** (1963) 1507-1517.

It is noteworthy to mention Sir James' contribution of a review article on Turbulence to the edited volume, *Osborne Reynolds and Engineering Science Today* published by Manchester University Press in 1970. Edited by D. M. McDowell and J. D Jackson, this volume represents a Proceedings of the major symposium held at the University of Manchester to celebrate the 1868 centenary appointment of Osborne Reynolds as the first new University Chair of Engineering. The University of Manchester established a great tradition of engineering science, applied mathematics and fluid mechanics by appointing Osborne Reynolds to this new University Chair of Engineering in 1868, and Sir Horace Lamb to Chair of Applied Mathematics in 1885. Both Reynolds and Lamb made many revolutionary contributions to engineering science, applied mathematics and fluid mechanics. Many original paper of Reynolds dealt with fundamental role of the Reynolds number in instability and transition, and of the Reynolds stresses in fully developed turbulent flows and of the Reynolds analogy in turbulent heat transfer. Among many others, the sixth edition of his book *Hydrodynamics* was completed by Sir Lamb at his home in Selwyn Gardens, Cambridge after retiring from Manchester and it was then published by Cambridge University Press in 1932.

In addition to Reynold's fundamental classical work, Lighthill described some major contributions of Rayleigh, Taylor, Prandtl, von Kármán, Landau and others with special emphasis on instability and transition to turbulence. Taylor first made his pioneering theoretical and experimental studies of stability of viscous fluid flow between two coaxial cylinders rotating at different velocities (see Debnath et al. (2001)). His beautiful experimental work revealed the existence of instability that led to a steady secondary flow in the form of a series of toroidal vorticities (Taylor vortices). These vortices are rotationally symmetric about the axis of the cylinders and spaced periodically with alternating spin along the axis. The basic cellular pattern of Taylor vortices depends on the primary flow before instability being two-dimensional. On the other hand, the stability problem of parallel viscous fluid flows was much more difficult partly because of the serious analytical difficulty of solving the Orr-Sommerfeld equation which governs the problem. In the mean time, Prandtl made exceedingly successful applications of the knowledge about turbulence to problems of external aerodynamics. He also explained the existence of a critical Reynolds number for fluid flow around bluff bodies, above which transition to turbulence in the boundary layers delays separation and significantly reduces drag force. The von Kármán school in Pasadena, California was also very active in finding information on mean velocity profiles in turbulent flow. In the mid thirties, Taylor initiated first to generalize the classical idea of the Reynolds stress which is the mean product, $\overline{u'(\mathbf{x})u'(\mathbf{x}+\mathbf{r})}$ of two disturbance velocities at points separated by a distance vector $\mathbf{r}$. Taylor's most fundamental idea of diffusion by continuous movements was to show how to describe diffusion process in terms of the statistical properties of fluid velocity. To describe turbulent motion of fluid, Taylor represented fluid velocities in the Eulerian manner as continuous random functions of time and position so that it simplifies greatly the expressions for the instantaneous spatial derivatives of velocity which occur in the equation of motion and in the vorticity and the rate of dissipation. This Taylor's new approach was almost subsequently and greatly extended by many others in late thirties and early forties. Taylor also published a series of original papers entitled 'Statistical Theory of Turbulence'. Indeed, all subsequent developments of turbulence have been based on Taylor's original approach and ideas.

The actual difference between 'restrained' and 'abrupt' transition has been recognized and then explained by the great Russian physicist Landau in 1944. Based on his famous nonlinear model equation, he explained that an amplitude A of small disturbances changes *exponentially* like $\exp(\gamma t)$ where

$$\gamma < 0 \quad \text{when} \quad R < R_c \quad \text{and} \quad \gamma > 0 \quad \text{when} \quad R > R_c, \tag{1.1}$$

the Reynolds number R reaches a critical value R_c which is relatively high. The square of the amplitude A proportional to the energy associated with disturbances to the mean flow satisfies the nonlinear Landau equation in the form

$$\frac{dA^2}{dt} = 2\gamma A^2 - \alpha A^4, \tag{1.2}$$

where γ is the rate of exponential increase of amplitude given by the small-amplitude theory and α is constant. For very small A, the right hand side of (1.2) is dominated by the term, $2\gamma A^2$ so that it is totally consistent with the behavior of (1.1) for infinitesimal disturbances. However, the term $(-\alpha A^4)$ in (1.2) provides how the growth of amplitude A may be either become restrained ($\alpha > 0$) or abrupt ($\alpha < 0$).

Substituting $B = (1/A^2) - (\alpha/2\gamma)$ in (1.2) gives the simple form

$$\frac{dB}{dt} + 2\gamma B = 0. \tag{1.3}$$

This admits the general solution

$$B(t) = a\exp(-2\gamma t), \tag{1.4}$$

where a is an arbitrary constant. Consequently, the general solution of (1.2) reduces to the form

$$A^2 = \left[\left(\frac{\alpha}{2\gamma}\right) + a\, e^{-2\gamma t}\right]^{-1}. \tag{1.5}$$

For subcritical Reynolds numbers $R < R_c$, $(\gamma < 0)$ and the energy of disturbances (1.5) tends to zero as $t \to \infty$. For supercritical Reynolds numbers $R > R_c$, $(\gamma > 0)$ and then all solutions (1.5) tend to a finite value of disturbance $A^2 = (2\gamma/\alpha)$. Any very small disturbance grow exponentially, but does not continue indefinitely. In fact, after a certain time, a definite level of disturbance is attained such as Taylor vortices between rotating cylinders for example.

Thus, a restrained transition ($\alpha > 0$) is possible where very small disturbances with ($\gamma > 0$, $R < R_c$) begin by growing exponentially, but, after a certain time, a definite level of disturbance is maintained. On the other hand, an abrupt transition becomes possible, where disturbances with ($\gamma > 0$, $R < R_c$) decays to zero.

The Kármán vortex street in a wake, similar vortex rows in jets, Taylor vortices in flow between rotating cylinders, and various types of cells formed in a layer of fluid heated from below are interesting examples of restrained transition, while transition in pipes, channels and boundary layers is normally abrupt in the sense that an almost fully developed turbulent motion occurs.

Modern theories of turbulence began with the pioneering work on the universal equilibrium theory of the great Russian mathematical and physical scientist Kolmogorov in 1941. He developed a fairly general theory based on the Navier-Stokes equations and the continuity equation, and on the assumption of the existence of the main energy containing 'large eddies' (large scale or small wavenumber) and self-similar 'small eddies' (small scale or large wavenumber) which are primarily responsible for the viscous dissipation of energy. The Fourier spectral analysis of the turbulent energy spectrum $E(k,t)$ representing the contributions to the kinetic energy with respect to the wavenumber (or scale) k is the central problem of the dynamics of turbulence. It is important to point out that the nonlinear (convective) term in the Navier-Stokes equations is the main source of energy transfer from large eddies to small eddies. Denoting the nonlinear transfer term at a wavenumber k by $T(k,t)$, the energy spectrum $E(k,t)$ in the three dimensional isotropic turbulence satisfies the evolution equation (see Debnath (1998))

$$\frac{\partial E}{\partial t} = T(k,t) - 2\nu k^2 E(k), \tag{1.6}$$

where terms of this equation represent contributions of the inertial, nonlinear and viscous term of the Navier-Stokes equation. It follows from the continuity equation that the pressure term in the Navier-Stokes equations has no contributions to (1.6). This means that the net effect of the pressure field is to conserve the total energy in the wavenumber space. Only the nonlinear term in the Navier-Stokes equation is responsible for energy transfer from larger to smaller eddies — a mechanism by which larger eddies decay. The second term on the right hand side of (1.6) represents the dissipation of energy by molecular viscosity. Thus, the action of viscosity leads to a decrease in the kinetic energy of disturbance with the wavenumber which is proportional to the intensity of the disturbance multiplied by $2\nu k^2$.

It follows from the conservation of energy by the nonlinear term in the Navier-Stokes equations that

$$\int_0^\infty T(k,t)dk = 0, \tag{1.7}$$

so that the evolution equation (1.6) gives

$$\frac{d}{dt}\left(\frac{1}{2}\,\overline{u_i^2}\right) = \frac{\partial}{\partial t}\int_0^\infty E(k,t)dk = -\varepsilon(t), \tag{1.8}$$

where $\varepsilon(t)$ represents the rate of energy dissipation and is given by

$$\varepsilon(t) = 2\nu\int_0^\infty k^2 E(k,t)dk. \tag{1.9}$$

This shows that small eddies (or large wavenumber components) are dissipated more rapidly by viscosity than large eddies (or small wavenumber components). The basic assumption of the Kolmogorov theory is that at a very high Reynolds number, the turbulent flow at the very small scales (large wavenumbers) is approximately similar to a state of statistical equilibrium. Further, the motion of the small eddies is assumed to be statistically independent of those of in the energy-containing range. The energy containing scales of the motion may be anisotropic and inhomogeneous, but these features are completely lost in the cascade of energy transfers so that at much smaller scales the motion is locally homogeneous and isotropic. Hence the statistical properties of the turbulent motion in the equilibrium range must be completely determined by the two physical parameters ε and ν only that are relevant to the dynamics of this part of the spectrum only. Using the dimensional analysis, the fundamental length characterizing the energy-dissipating eddies must be $\eta = (\nu^3/\varepsilon)^{\frac{1}{4}}$. This length scale is known as the *Kolmogorov dissipation length* . The same dimensional analysis also reveals that the energy spectrum $E(k)$ depending on k, ε and η takes the form

$$E(k) = C_k\, \varepsilon^{2/3} k^{-5/3} F(k\eta), \tag{1.10}$$

where F is an universal dimensionless function.

According to Kolmogorov's hypothesis, for sufficiently large Reynolds number, there exists a significant range of wavenumbers with $k_\ell \ll k \ll k_d$ where $k_\ell = \ell^{-1}$ (ℓ is the largest scale) and $k_d = \eta^{-1}$ (η is the smallest scale). So, in this range, both energy content and energy dissipation are negligible and the spectral energy $\varepsilon(k) = \varepsilon$ is independent of wavenumber k. The molecular viscosity ν then becomes insignificant, $F(k\eta)$ is asymptotically equal to unity. Consequently, the Kolmogorov energy spectrum in the inertial range reduces to the form

$$E(k) = C_k\, \varepsilon^{2/3} k^{-5/3}. \tag{1.11}$$

This is known as the *Kolmogorov-Oboukhov energy spectrum* for isotropic and homogeneous turbulence, and has received strong experimental support by Grant, Stewart and Moilliet (1962) with a value of $C_k \approx 1.44 \pm 0.06$. Subsequently, Metais and Lesieur (1992) proposed the structure-function model of turbulence with the spectral eddy viscosity based upon a kinetic energy spectrum in space. Their analysis gives the best agreement with the Kolmogorov $k^{-5/3}$ spectrum law and the Kolmogorov constant $C_k \approx 1.40$.

Soon after Kolmogorov's brilliant discovery, considerable progress was made on a detailed study of different physical mechanisms of energy transfer of turbulence. In particular, Onsager (1945) and Heisenberg (1948) elucidated further the viscous dissipation mechanisms of small eddies in the cascade of random energy transfer in turbulence. Of these physical transfer mechanisms, Heisenberg's eddy viscosity transfer was found to be more satisfactory at that time. Based on the assumption that the role of small eddies in the nonlinear transfer process is very much similar to that of molecules in viscous dissipation mechanisms, Heisenberg suggested that small eddies act as an effective viscosity produced by the motions of these small eddies and the mean square vorticity associated with the large eddies. Using this assumption, Heisenberg formulated the energy balance evolution equation in the form

$$\frac{\partial}{\partial t}\int_0^k E(k,t)dk = -2\left(\nu + \frac{\eta_k}{\rho}\right)\int_0^k k^2 E(k,t)dk, \tag{1.12}$$

where η_k is the eddy viscosity defined by Heisenberg in the form

$$\eta_k = (\rho\kappa)\int_k^\infty \left[\frac{E(k,t)}{k^3}\right]^{\frac{1}{2}} dk, \tag{1.13}$$

where κ is a numerical constant.

However, Landau recognized some major difficulties with the Kolmogorov universal equilibrium theory of turbulence, and pointed out that statistics of small scales depend on those of large scales, and hence, it cannot be fully universal in the sense of the Kolmogorov assumptions. His other concern was that intermittency of dissipation increases with Reynolds number. In respond to Landau's concerns, both Kolmogorov (1962) and Oboukhov (1962) suggested some modifications of their original theories which led them to derive the so called *log-normal spectral* law in the form

$$E(k) = C_k \varepsilon^{2/3} k^{-5/3} \ln(k/k_I)^\beta, \tag{1.14}$$

where only a fraction of β of energy is transferred from one scale to another and k_I is the wavenumber at which energy is injected. This is a slight modification of the $(-5/3)$ law in the sense that experimental measurements can hardly detect it. In spite of the above and other criticisms, the Kolmogorov $k^{-5/3}$ energy spectrum law received strong experimental support over a wide range of Reynolds number.

In spite of convincing and accurate arguments of Kolmogorov, the mathematical expression of Onsager's cascade hypothesis has subsequently been modified by Pao (1965) who began with the equilibrium equation

$$\frac{dS(k)}{dk} = -2\nu k^2 E(k), \tag{1.15}$$

where $S(k)$ is the turbulent energy flux from wavenumbers less than k to wavenumbers greater than k, and $2\nu k^2 E(k)dk$ represents the rate of viscous energy dissipation in wavenumbers between k and $k + dk$. Pao's argument was that $S(k)/E(k)$ must be a function of ε and k only because energy transfer from one wavenumber to another is essentially due to inertial effect in which viscosity has no explicit role. Based on dimensional analysis with $S(k)$ representing energy transfer per unit time and $E(k)$ is an energy per unit wavenumber, Pao formulated that

$$S(k)/E(k) = C^{-1}\varepsilon^{\frac{1}{3}}k^{\frac{5}{3}}, \tag{1.16}$$

where the constant must be reciprocal of that in (1.11) so that $S \to \varepsilon$ for small (ηk). Consequently, equations (1.15) and (1.16) can be shown to imply the explicit form of the spectrum for the entire Kolmogorov equilibrium range of wavenumbers in the form

$$E(k) = C\varepsilon^{2/3}k^{-5/3}\exp\left[-\frac{2}{3}\,C(\eta k)^{4/3}\right]. \tag{1.17}$$

This result also received remarkably a good agreement with several experimental observations. In spite of remarkable theoretical and experimental success of the Kolmogorov universal equilibrium theory of turbulence, Lighthill expressed a major concern about isotropic nature of small eddies. Thus, the Kolmogorov theory deals with only isotropic turbulence which is just one of more kinds of turbulence, by no means *generic*. In shear flow, there is a striking contrast between the isotropic small eddies and the big eddies that are highly elongated in the direction of the stream. Indeed, the main energy containing eddies are definitely *anisotropic* in shear flows. So, there are some questions about the validity of the Kolmogorov theory. Despite these criticisms, it must be recognized that the Kolmogorov theory is, after all, an approximate theory based on suitable assumptions and approximations, and has served as the fundamental basis for all subsequent developments in the modern theories of turbulence.

Recognizing all remarkable ideas and methods of Reynolds' and others, Lighthill described all current major developments of chaos and turbulence made by many eminent scientists including G. K. Batchelor, S. Chandrasekhar, W. Heisenberg, Th. von Kármán, A. N. Kolmogorov, L. D. Landau, C. C. Lin, L. Onsager, L. Prandtl, Lord Rayleigh, G. B. Schubauer, G. I. Taylor, and W. Tollmien. He emphasized the major work of Reynolds on turbulence with all subsequent advances that have been made by methods involving extensions of the original ideas in Reynolds' own research and his wide knowledge and expertise in the area of statistical mechanics. Finally, he concluded this article suggesting a number of possible new directions of advance with a statement that turbulence is one of the hardest areas of fluid mechanics and so, future progress will always be slow and "... Only very occasionally are there really new insights. I believe, furthermore, that none which I have described today can be regarded as comparable in magnitude with initial progress made by Reynolds himself. It is satisfying, furthermore, at these centenary celebrations to recognize how many of the later advances have been made by methods involving extensions of the original ideas in Reynolds's own work."

Lighthill delivered his Presidential Address at the Centenary Meeting of the Mathematical Association of Great Britain in London, April 14, 1971. This meeting has been a great event of major significance in an international context. He described many effective activities of the Association during the last 100 years including many continuous improvements in British mathematical education during that period. His lecture also emphasized that the Association exercised a continuing influence on a vast range of activities in mathematics education, while also being able to participate in developing the Art of Teaching Mathematics at all levels. During the last 100 years, the British mathematical scientists provided a major leading role in a wide range of research in mathematics, statistics, physics and engineering science. For example, mathematical analysis was brilliantly developed by G. H. Hardy and J. E. Littlewood, geometry by H. F. Baker and William Hodge, number theory by L. J. Mordell and Harold Davenport, topology by Henry Whitehead and Max Newman, and mathematical philosophy by Alfred North Whitehead and Bertrand Russell. On the other hand, many revolutionary research have been carried out in Great Britain in mathematical physics by Paul Dirac and R. H. Fowler, in engineering science by G. I. Taylor and Sydney Goldstein and in statistics by R. A. Fisher and Frank Yates, and in earth sciences by Harold Jeffreys and Sydney Chapman. According to Sir James, the degree of integration of pure and applied mathematics in British universities has traditionally been very high by world standards. The Mathematics Association has continuously been engaged in improving the current trends in both primary and secondary education, and in expressing concern with how the mathematics being taught can be applied. The emphasis on the idea of building mathematical models representing one aspect of applied mathematics has been widespread all over the world as well as in Britain. One of his major motivations in this lecture was to illustrate diverse and newer applications of mathematics in physical and social sciences, economics, commerce and industry. Sir James also stressed the use of electronic computers in teaching and research that would provide new opportunities for curriculum development. Based on his lecture, he published an article on the Art of Teaching the Art of Applied Mathematics in *Mathematics Gazette* in volume **55** (1971) 471-492. He concluded this lecture by stating his comment as follows: "... to give careful thought and consideration and discussion to all matters concerned with this critical question of how to teach the art of applying mathematics."

His article on the Interaction between mathematics and society was published in the *Proceedings of the Third International Congress of Mathematics Education*, Universität Karlsruhe in 1977. Sir James published another article entitled Teaching how to make and use mathematical descriptions of engineering systems in the *Proceedings of the Second International Congress of Engineering Education* in 1978. Through his presentations and publications of articles on pedagogy, Sir James provided a vital leadership role in promoting not only the importance of teaching and learning at all levels, but also in enhancing the interaction between mathematics and society. His philosophy of mathematics education can be best described by citing his own quotation from his Presidential Address at the Second International Congress on Mathematical Education in Exeter in September, 1972.

"Let's go *beyond mere use* of the concrete example as an aid to understanding or of reference to utility as an aid to widening the circle of those in whom interest is aroused. There is a still more important prize to be won: a prize concerned with a *deeper* integration of mathematics into the total education of the individual."

"I want to suggest that educators may have most benefited their pupils when they have succeeded in giving them a feel for what is involved in the *process of applying mathematics...* Computers may be of great value in problem-solving, but apparently the human brain alone is able to tackle the subtler aspects of creating an effective correspondence between the mathematical world and the world of experiment and observation."

In July 1972, Sir James delivered the Fourth Annual Fairey Lecture on the propagation of sound through moving fluids at the Institute of Sound and Vibration Research in Southampton. Special attention has been given to the interaction of sound waves with turbulence and to further intensification of collaborative efforts between theory and experiment in the subject. This is followed by questions of the relative properties of upstream and downstream propagation of sound both in one-dimensional case and in the more fully three-dimensional cases characteristic of higher frequencies. He concluded his lecture with recent progress on work in the whole field of aero-engine compressor noise reduction as the compressor noise is a major part of the total external noise field of many modern aircraft. This lecture was published in *Journal of Sound and Vibration*, **24** (1972) 471-492.

Sir James Lighthill delivered a special lecture on Ocean Science at a conference held in Greenwich on September 12-14, 1973 to celebrate the hundredth anniversary of the Royal Naval College. In this lecture, he brilliantly described the major role of ocean science in the service of mankind and elaborated the interaction among the many disciplines of mathematics, science and engineering. His major focus includes (i) exploration for oil and gas, (ii) under-sea minerals, (iii) ocean's living resources including protein sources, fish and shell-fish, and other huge untapped resources, (iv) world's fish farming and fish industry, (v) ocean-borne trade and ship design technology, (vi) coastal protection against floods caused by violent winds and high tides, (vii) natural hazards and tsunami waves due to under-water earthquakes, (viii) sand and sediment movements, the effect on coastal erosion and on fisheries of dredging for sand and gravel, (ix) marine pollution and pollution control, world's clear air and world's clean water research and legislation, (x) discharges of oil in oceans and seas, and other industrial wastes, and dangers to marine life, (xi) air-ocean interaction and man's natural environment, (xii) more research on the atmosphere-ocean circulation, weather forecasting, flood prediction and other related topics, (xiii) biogeography dealing with the ranges of species and evolutionary processes, picture of the abundance of different species in different parts of the ocean, and importance of fishery science, (xiv) global geophysics and geology, speeds of ocean-floor spreading, high concentration of minerals in water and commercial exploitation of deep-ocean mineral resources, and (xv) the theory of continental drift and problems of waste disposal. In addition to all of the major topics directly associated with ocean science, he stimulated the participants of this centenary celebration for more advanced study and research on many challenging problems in interdisciplinary ocean science. In this context, his concluding statement is worth quoting: "With that visionary idea, which re-emphasizes my two main themes: ocean science in the service of mankind, and the endless interaction between all the disciplines that make up ocean science, I may perhaps conclude my presentation". Sir James subsequently published this presentation in the *Bulletin of the Institute of Mathematics and Applications* in February 1973 and also in the *Journal of Navigation*, **27** (1974) 91-110. Through publication of these articles in international journals, he presented this material before the multinational and multidisciplinary group of scientists for promotion of cooperative interdisciplinary research on ocean science and marine engineering.

Based on his major work published in the *Annual Reviews of Fluid Mechanics*, **9** (1969) 413-446, Sir James delivered an invited lecture on Aquatic Animal Locomotion at the Thirteenth International Congress of Theoretical and Applied Mechanics held in Moscow in 1973. In this lecture, he presented the interaction between an aqueous medium and the external surfaces of a totally immersed animal swimming through it. He also described importance of hydrodynamic theory relevant to aquatic animal locomotion and of the interdisciplinary research work between zoology and fluid mechanics. Lighthill first discussed his own reactive theory for swimming of animals with elongated bodies which was published in a series of papers based on a small-perturbation expansion about the stationary state. He also discussed the extension of the 'reactive theory' so that it no longer depends on a small-perturbation expansion. Some recent developments in two-dimensional theory that are relevant to lunate tails and their vortex wakes are also presented. Included are other major extensions of theories of motion of individual flagella at very low Reynolds number to the case of ciliary propulsion, where the movements of large number of attached cilia close to synchronism generate relative movement between an animal and the surrounding fluid medium. His lecture also included other topics such as hydrodynamics of ciliary propulsion, and large-amplitude elongated-body theory which deal with his work on motions at high Reynolds number and reactive force theories. Finally, he concluded his lecture by discussing vortex wakes that what aquatic animal locomotion leaves behind, and a diamond-shaped lattice pattern of fishes. Included is also a major contribution by von Kármán and Burgers (1934) which illustrates vortices cast off by the caudal-fin trailing edge as fish moves to the left, and also jet-like streamline pattern produced by those vortices.

Aerodynamic Aspects of Animal Flight was the title of Lighthill's lecture organized by the British Hydromechanics Research Association in 1974. He concentrated in this lecture on the aerodynamic forces, and resulting dynamic interactions, between the movements of animals flying in air and the surrounding air movements. Based on the basic aerodynamic requirements for sustained forward flight to produce sufficient lift to balance body weight and sufficient thrust to balance body drag, Lighthill presented the aerodynamics of sustained forward flight in birds, insects and bats with comments on the many other flight modes of these animals. These modes include manoeuvres, diving, soaring, take off, landing and hovering flight which are most fascinating aerodynamic aspects of animal flight. He concluded this lecture by suggesting a number of new directions of research in this field.

According to Lighthill, "there is immense scope for research in this new field, based on further exploration of the use of the "clap and fling" mechanism and other specialized unsteady aerodynamic effects by flying animals. It would be special interest to determine how widespread is the use of the mechanism by insects so small that they encounter low-Reynolds-number problems. A quite different question is whether the mechanism has been used at all by much larger animals, operating at substantial Reynolds number.... It would be a challenging problem to study by experiment and analysis what are special movements by which these 'thrips' or 'fringe-winged' insects are enabled, at such very low Reynolds numbers, to support their bodies in the air."

In his 1975 John von Neumann Memorial Lecture on Flagellar Hydrodynamics, Sir James presented a general overview of biofluiddynamic aspects of microorganisms with flagella and related organisms. This is followed by eukaryotic flagellar motions and bacterial flagellar motions with the structures of these two fundamentally different kinds of flagella. Special attention is given to application of mathematics to flagellar motions in particular. He concluded his lecture by adding a discussion on flow fields generated by flagella. This lecture was remarkably well attended and a considerable number of participants were not familiar with this new subject. So, many participants took the advantage of the unique opportunity to gain insights into the some of the underlying new ideas and results of the most fascinating and new branch of modern fluid mechanics. In addition to his own research work, however, the material embraced that of other fluid dynamists and zoologists both in the United Kingdom and in the United States. The treatment of the subject was to a large extend new and original contributions of the lecturer. His most remarkable work was subsequently published by *SIAM Review*, **18** (1976) 161-230.

In April 6, 1978, he was invited to deliver lecture on Acoustic Streaming at the spring meeting of the Institute of Acoustics in Cambridge, England. His lecture was published in the *Journal of Sound and Vibration*, **61** (1978) 391-418. Generally speaking, mean motions induced by sound waves are called *acoustic streaming* which is forced by the action of a Reynolds stress that can cause a net force per unit volume, F_j to act on the fluid. This force field can be written as $F_j = -\frac{\partial}{\partial x_i}\left(\overline{\rho u_i u_j}\right)$, where the repeated suffix i is to be summed from 1 to 3. This force field per unit volume is capable of generating a steady streaming motion. In fact, the equation for the

Eulerian mean motion $\overline{u_j}$ produced by this force is

$$\rho_0 \left(\bar{u_i} \frac{\partial \bar{u_j}}{\partial x_i} \right) = -\frac{\partial \bar{p}}{\partial x_j} + \mu \nabla^2 \bar{u_j} + \mu F_j. \qquad (1.18)$$

This equation can be solved with the equation of continuity in the form

$$\rho_0 \frac{\partial \bar{u_j}}{\partial x_j} + \frac{\partial}{\partial x_j} \left(\frac{1}{c^2} I_j \right) = 0 \qquad (1.19)$$

where the first term represents the Eulerian mean motion $\bar{u}_j$, and the second term corresponds to an acoustic energy flux I_j and c is the speed of sound.

This lecture also covered significance of several attenuation mechanisms and the acoustic streaming associated with the attenuation of sound waves. Sir James also elaborated by stating that all really significant streaming motions are studies of the Stuart (1963, 1966) streaming that is calculated from the full equation (1.18). In his pioneering work, Stuart discussed the nature of acoustic streaming based on the Reynolds number R_s on streaming velocity which has many modern ramifications of the classical work of the Rayleigh law of streaming. This is followed by a comprehensive treatment of patterns of turbulent acoustic streaming produced by ultrasonic sources and turbulent jets generated by attenuation of the energy flow in a general acoustic beam. He concluded his lecture with a review of modern developments in the other main type of acoustic streaming associated with the major interactions between sound waves and solid boundaries. These are still dominated by the Rayleigh law of streaming with the so called the *slip velocity* $\bar{u_s} = -\frac{3}{4}\ \omega^{-1} U(x) U'(x)$, where $U(x)$ is involved in any oscillating tangential relative velocity, $U(x) \exp(i\omega t)$ between a fluid and a solid boundary, whether caused by acoustic oscillations in the fluid or by vibrations of the solid. This leads to frictional dissipation within the Stokes boundary layer. The tangential component of the fluid velocity relative to the solid wall is given by

$$u = U(x) \exp(i\omega t) \left[1 - \exp\left\{ -z \left(\frac{i\omega}{\nu} \right)^{\frac{1}{2}} \right\} \right] \qquad (1.20)$$

where $\nu = (\mu/\rho_0)$ is diffusivity of momentum. All these strongly indicate that Rayleigh's classical work has served as the fundamental basis for all modern developments in acoustic streaming.

In 1979, Sir James left Cambridge to serve in the position of Provost of the University College London (UCL). Many people were pleased with his appointment and expressed their feelings by saying that UCL got a Provost who has academic clout, visibility, and tremendous influence in government, industry and society. Due to his enormous wisdom and communication skills, and strength of character, he remained in that administrative position until his formal retirement in 1989. During this period, he was heavily involved in new developments of the College, in promotion of biological and biotechnological research, fund-raising activities, and in significant improvement of women's appointments in senior academic and administrative positions. Even though he was very busy with administrative duties and responsibilities, Lighthill still maintained his scientific research in areas such as extraction of wave energy, dynamical systems, and biomechanics of the human auditory system. In collaboration with a group of scientists in the Institute of Laryngology and Otology at the University College London, Lighthill continued his research on mammalian hearing biomechanics during his stay at UCL. From my personal conversation with him and his wife, Lady Nancy in October 1995, I was delighted to know that Sir James made many significant contributions to the University College London in many different ways.

While he was Provost, he was very active in the presentation of special lectures and in research and national educational affairs. His professional activities included services as President of the International Union of Theoretical and Applied Mechanics (IUTAM), as Member of the Advisory Board for the Research Council, 1980-1986, as Member of Natural Environment Research Council and Chairman of its Oceanography and Fisheries Research Committee, 1965-1970, as Member of Geddes Committee of Inquiry into Shipbuilding Industry, President of the International Commission on Mathematical Instruction (ICMI), 1971-1975, and as Member of the Advisory Council on Research and Development, 1978-1981. Fortunately, the United Nations has declared the period from 1990-1999 as the International Decade for Natural Disaster Reduction (IDNDR) and has adopted tropical cyclones as the major atmospheric hazard. After the International Council of Scientific Unions (ICSU) has strongly encouraged member Unions to support IDNDR, the IUTAM and the International Union of Geodesy and Geophysics (IUGG) have agreed to make a joint plan for a decade long scientific program related to tropical cyclone disasters. Sir James Lighthill provided a leadership role to develop a plan for a better tropical cyclone forecasting and warning system that minimizes the loss of human lives and other socio-economic losses associated with tropical cyclones at the IUTAM/ICSU/IUGG workshop in Vienna in August 1990. As Sir James Lighthill noted in his broad introduction at the ICSU and the World Meteorology Organizations (WMO) the first international symposium in Beijing

China in 1992, scientists from related fluid dynamics disciplines have an important role to play in the research challenge of the International Decade, especially by helping to improve the accuracy of techniques for detailed forecasting of Natural Disasters so that population under threat may learn to rely on the predictions, and consequently, to follow the emergency measures recommended for their protection.

As an active President of ICMI which was then a sub-commission of the International Mathematical Union (IMU), Lighthill has had a great influence on mathematics education and science policy at the international level. The emphasis on new thinking and pedagogical development became his top priority as President of ICMI. In order to promote mathematics teaching at all levels and to upgrade mathematics education all over the world, Lighthill helped organize a number of symposia that include 'New topics in applicable mathematics in Secondary Schools' in Luxembourg in 1973, and 'Mathematics and Language' in Kenya in 1974. He will never be forgotten for his impact on mathematics and its applications, but equally, he made major lasting contributions in raising public awareness of mathematics education, and through conferences and symposia, bringing mathematics education issues to the fore.

In his brilliant lecture on "The recently recognized failure of predictability in Newtonian dynamics" at the Discussion Meeting on Predictability in Science and Society, in 1986, Lighthill, speaking as President of the International Union of Theoretical and Applied Mechanics (IUTAM) presented the first systematic and persuasive arguments in support of the complete predictability of systems governed by the equations of Newtonian dynamics. He also stated that Sir Isaac Newton's discovery of fundamental mathematical and physical laws were published in his first book of *Philosophiae Naturalis Principia Mathematica* (Mathematical Principles of Natural Philosophy) which is considered one of the greatest single contribution ever published in the history of physical sciences. This celebrated volume, usually called *Principia* or *Principia Mathematica* was completed over three hundred years ago and communicated to the Royal Society in the Spring of 1686 and then published in 1687. In it Newton not only put forward a new theory of how bodies move in space and time, but also developed the complicated mathematics needed to analyze these motions. In addition, he also formulated the laws of motion and a law of universal gravitation according to which each body in the universe was attracted toward every other body by a force that was stronger the more massive the bodies and the closer they were to each other. It was exactly the same force that caused objects to fall to the ground. According to his law, gravity causes the Moon to move in an elliptic orbit around the Earth and causes the Earth and the planets to follow elliptical paths around the Sun. It was the first Newton's book to contain a unified system of scientific principles explaining what happens on Earth and in the Universe. Sir James' lecture was well attended by a group of learned audience, but the subject was highly controversial with which a large number of participants were not completely familiar and so many took the advantage of this opportunity to gain insights into the underlying ideas of the most fascinating and fundamental branch of mathematics and physical science. This lecture was published in *Proc. Roy. Soc. London*, **A407** (1986) 35-50.

As a firm believer of Newtonian mechanics, Sir James' statement of public apology is an enlightenment to read:

"Here I have to pause, and to speak once again on behalf of the broad global fraternity of practitioners of mechanics. We are all deeply conscious today that the enthusiasm of our forebears for the marvelous achievements of Newtonian mechanics led them to make generalizations in this area of predictability which, indeed, we may have generally tended to believe before 1960, but which we now recognize were false. We collectively wish to apologize for having misled the general educated public by spreading ideas about the determinism of systems satisfying Newton's laws of motion that, after 1960, were to be proved incorrect. In this lecture, I am trying to make belated amends by explaining both the very different picture that we now discern, and the reasons for it having been uncovered so late."

In his memorable lecture of 1986, Sir James excited the audience by the following statement and quotations:

"Three different new directions of research initiated the discovery of chaotic behaviour in simple systems satisfying Newtons laws during the early part of the 1960s; although, admittedly, it was only 20 years later that the remarkably widespread occurrence of chaotic behaviour had become clear; to such an extent that, for example, at the 1984 Congress of the International Union of Theoretical and Applied Mechanics the specialized topic which was highlighted most strongly in the Congress programme was that of chaotic behaviour.

The first of these three new directions of research arose directly out of Poincaré's work on nonlinear perturbation theory, related to equations in Hamiltonian form for isolated systems of constant energy. Work initiated by the great Russian analyst Kolmogorov (1954), and pursued by his gifted colleague. Arnold (1963), had been aimed first of all towards filling in the gaps in Poincaré's proofs; and, indeed, all this work along with independent studies by Moser (1962) in Germany and America demonstrated that, even in the neighbourhood of resonances, perturbations did assume a distinctly regular pattern in the vast majority of cases. Nevertheless, gaps in that regular pattern did exist; namely, very small ranges of initial conditions for which the motion assumed a form described as chaotic or stochastic (that is, random). It was regarded as interesting that equations of motion that included absolutely no random element should possess solutions which behaved in such a random way. Initially, however, the matter was seen as something of a curiosity because the highly complex proofs used in the perturbation theory were valid only when the perturbations were restricted to being sufficiently small and, in this case, the ranges of initial conditions for which chaotic or stochastic motions occurred were very limited indeed.

The second new direction of research utilized the powerful computers that were by then available to compute solutions not just in cases when the perturbations were small, but also for much larger perturbations. This work, carried out, for example, by Greene (1979) in the U.S.A. and by Chirikov (1979) in the U.S.S.R., demonstrated that as the strength of the perturbations continued to grow there was a sharp increase in the range of initial conditions for which solutions behaved stochastically. Finally, at a certain level of the perturbation strength, the authors observed what they called a transition to global stochasticity, with all solutions behaving chaotically. I shall describe later what this amounts to in detail, but in the meantime will note the obvious fact that the new data made a big increase in the importance to be attached to chaotic or stochastic solutions.

These results on isolated systems of constant energy were of interest not only to astronomy but also to thermodynamics. The second law of thermodynamics envisages, of course, an increase in the randomness of motions experienced by an isolated system of molecules; that is, an increase in its entropy; but physicists had long supposed that large numbers of collisions between molecules were necessary to allow such randomization to occur. Now, with the wider understanding of how chaotic motions can develop, it is possible to see that collisions may not be essential. For example, the ionized gas between the Sun and the Earth with its extremely low density, producing an astronomically large mean free path between molecular collisions, may nevertheless in the presence of magnetic fields experience phenomena that are possible only with an increase in entropy. One of these, which spacecraft have observed, is the so-called 'bow shock wave' where the solar wind of charged particles emanating from the sun is abruptly slowed down where it first encounters the Earth's magnetosphere.

But that is a digression, which may on the other hand have provided a valuable reminder that Newtonian dynamics is applicable not only to systems of solid bodies but also to fluid systems, including ordinary gases and liquids. Ordinary gases and liquids, of course, are subject to the phenomenon called viscosity, which causes the mechanical energy in their shearing motions to be gradually dissipated into heat; precisely as a result of an entropy increase associated with normal molecular collisions on a submicroscopic scale. Yet even the damping of fluid motions by viscosity does not prevent perfectly regular fluid motions from becoming chaotic and this fact was first made precise over a century ago, in 1883, by Osborne Reynolds. He showed how the regular flow of fluid through a pipe suddenly becomes chaotic or turbulent when the force producing the motion becomes sufficiently large relative to the damping forces due to viscosity. He showed, furthermore, that this randomization has nothing to do with the random molecular movements occurring at submicroscopic scales; they, indeed, have a damping effect tending to reduce the trend towards turbulent motions; motions which themselves involve, rather, a chaotic pattern of fluid movement on a strictly macroscopic scale.

Thus, the specialists in dynamics of fluids, such as myself, have long been most fully conscious of the common tendency for regular or laminar motions of fluids to become chaotic or turbulent even though the motions in questions are subject to energy dissipation by the action of viscosity. On the other hand, fluids represent very complicated dynamical systems with an essentially unlimited number of degrees of freedom (each separate particle of fluid is separately free to be arbitrarily positioned relatively to all the other particles) and it had never been clear whether or not this was an essential pre-requisite for chaotic behaviour to develop.

Against that background it may be interesting to note that the third new direction of research which began in the early nineteen-sixties was concerned with some quite simple systems analogous to turbulence. These were dynamical systems with energy dissipation and just two or three degrees of freedom which, although forced in a perfectly regular way, responded in a completely chaotic way when a ratio of forcing effects: damping effects (a ratio similar to the Reynolds number introduced by Reynolds) was sufficiently large. Initially, they were devised by some noted experts in dynamics of fluids, including the great atmospheric scientist E. N. Lorenz, in order to mimic as closely as possible the development of turbulence in fluid systems. Lorenz (1963) introduced the term 'strange attractor' to describe the type of randomized motion which inexorably tends to develop.

More recently, a very general theory of these strange attractors has been produced for such systems, which unlike the isolated energy-preserving systems studied by Poincaré and others, are subjected both to forcing and to damping (see chapter 7 of Lichtenberg & Lieberman 1983). This theory suggests the steps by which regular motions develop into chaotic motions as some forcing:damping ratio changes. Often that takes place via an infinite sequence of so-called 'period-doubling bifurcations' which terminate, after just a finite change in that ratio, in a completely chaotic motion. Numerical computations have excellently confirmed these theories and demonstrated the strong tendency for systems of this type also to develop chaotic motions.

Now at this point I might easily feel tempted to enlarge upon all the immense variety of different types of chaotic systems and of transitions to chaos; and yet none of that would be relevant to the subject of this meeting. My objective in the time that remains to me is, rather, to focus upon certain properties which are common to all chaotic systems and which are relevant to the issue of predictability."

Lighthill used his lecture at this discussion meeting to criticize the view of the recently recognized failure of predictability in Newtonian dynamics. A thorough appreciation of Newton's great scientific achievement was shown by Sir James in his following concluding comments:

"I feel fully justified, therefore, in repeating that systems subject to the laws of Newtonian dynamics include a substantial proportion of systems that are chaotic; and that, for these latter systems, there is no predictability beyond a finite predictability horizon. We are able to come to this conclusion without ever having to mention quantum mechanics or Heisenberg's uncertainty principle. A fundamental uncertainty about the future is there, indeed, even on the supposedly solid basis of the good old laws of motion of Newton, which effectively *are* the laws of motion satisfied by all macroscopic systems. I have ventured to feel that this conclusion would be of interest to a Discussion Meeting on Predictability in Science and Society. For example, there might be some other discipline where practitioners could be inclined to blame failures of prediction on not having formulated the right differential equations *or* on not employing a big enough computer to solve them precisely *or* on not using accurate initial conditions; yet we in mechanics know that, in many cases where the equations governing a system are known exactly and are solved precisely, nevertheless however accurately the initial conditions may be observed prediction is *still* impossible beyond a certain predictability horizon."

In spite of controversial nature of the subject, Lighthill's explanations have been strongly supported by one of the great mathematical scientists of the twentieth century, A. N. Kolmogorov (1903-1987) who published an article on Newton and Contemporary Mathematical thought in the journal *Matematika v Shkole* 1982, No. 6, 58-64 and the same article was reprinted in the book of *Kolmogorov in Perspective* (2000). It is appropriate to quote Kolmogorov's thought:

"Newton not only made fundamental discoveries in the mathematical natural sciences that we need not go into here, since they are widely known, but he actually was the first to create a mathematics for the natural sciences in the sense of a system for the mathematical investigation of all mechanical, physical, and astronomical phenomena. Before Newton one could say only that this or that individual area of the natural sciences could be studied by mathematical methods. Of course, the ideas of Leibniz about the possibility of a mathematization of the whole of human knowledge were even more universal. But they bore no fruit precisely because of their absolute generality and lack of concreteness. On the subject of universal applicability and at the same time restrictedness see my article "Mathematics" in the *Great Soviet Encyclopedia.*"

At the invitation of the American Society of Mechanical Engineers (ASME), Sir James delivered the 1989 Rayleigh Lecture on Biomechanics of Hearing Sensitivity at the ASME Winter Annual Meeting in Chicago, Illinois. Speaking as the immediate Past President of IUTAM, he presented an elaborate survey on several topics including cochlear anatomy, cochlear sensitivity, cochlear macromechanics, cochlear micromechanics, evoked otacoustic emissions, and micromechanical interpretation of hearing sensitivity. He concluded his lecture by stating: "Finally, I emphasize that this lecture on the biomechanics of hearing sensitivity has been concerned, not with how the brain in man and other mammals analyze the data coming to it along auditory nerve fibers, but with the initial capture of that data in the cochlea. The brain, needless to say, can produce all its miracles of interpretation only where it works on good initial data.... I believe that Lord Rayleigh would have been excited and heartened by both aspects of our improved appreciation of the biomechanics of hearing sensitivity."

Lord Rayleigh was undoubtedly a wide ranging physical scientist who mapped out all of the classical known fields of acoustics including the science of hearing in his 1896 monumental treatise '*The Theory of Sound*' in two volumes. These two volumes provided a comprehensive treatment of an enormous body of knowledge obtained from linear wave theory as applied to generation and propagation of sound waves. Indeed, linear theories of wave motion in general became well developed during the nineteenth century. These volumes, a delight to read, are loaded with perennially significant information and ideas including the general derivation of the energy propagation velocity and acoustic streaming. Both Rayleigh (1910) and Taylor (1910) made an independent discovery of the internal structure of shock waves. This discovery subsequently led to the modern study of nonlinear effects in the generation and propagation of waves in fluids including sound waves and water waves. On the other hand, this discovery also led to the singular perturbation approach to nonlinear field equations in fluid mechanics. Indeed, the revolution in fluid mechanics during the second half of the twentieth century originated from the applications of new singular perturbation theory to nonlinear field equations where progress would have been impossible without the use of such a radically new approach. According to Lighthill, Lord Rayleigh was undoubtedly one of the most brilliant and influential physical scientists of the nineteenth century.

Sir James Lighthill was the best choice for the Inaugural Sydney Goldstein Memorial Lecture at the University of Manchester in October 11, 1989. His first association with Goldstein was in 1943 in Cambridge, and then in 1945 when Lighthill joined Goldstein's fluid dynamics group at NPL to do research in aerodynamics and applied mathematics. After his appointment to the Beyer Chair of Applied Mathematics at the University of Manchester in 1945, Goldstein first recruited a strong group of applied mathematicians including Lighthill as Senior Lecturer in Mathematics. Among his many outstanding contributions in teaching, research and service to the University of Manchester, Sydney Goldstein in cooperation with Max Newman accepted the challenge of creating a dynamic and internationally renowned mathematics department at Manchester. At the same time, Goldstein spent his time and energy to build a strong group of applied mathematics and fluid mechanics in Manchester's great tradition established by Osborne Reynolds and Sir Horace Lamb. Many of his later contributions were made as an outstanding leader of a fine group of applied mathematicians including M. J. Lighthill, J. W. Craggs, F. G. Friedlander, C. R. Illingworth, D. S. Jones, R. E. Meyer, G. N. Ward and E. Wild.

In his memorial lecture on 'Some challenging new applications for basic mathematical methods in the mechanics of fluids that were originally pursued with aeronautical aims', Sir James described Sydney Goldstein's major research accomplishments in Manchester including his key role as editor of the big collective work of the two volumes of *Modern Developments in Fluid*

Dynamics: An account of Theory and Experiment Relating to Boundary Layers, Turbulent Motion and Wakes first published by the Oxford University Press in 1938. This two-volume paperbound is a unique classic and an indispensible introduction to laminar and turbulent fluid flows for graduate students and professionals in applied mathematics and engineering science. This basic work is not only offers an enormous essential fundamental information in fluid mechanics, but also provides a solid foundation for other specialized studies in the field. Sir Horace Lamb was originally selected to serve as the Managing Editor of this work, but when he relinquished this task, it was assigned to Sydney Goldstein. On the sudden death of Sir Horace in 1934, Goldstein became his successor in two capacities, one private and one public. He acquired the late Sir Horace's beautiful house in Selwyn Gardens and succeeded Lamb as the Managing Editor of *Modern Developments in Fluid Dynamics* that was originally planned by the Fluid Motion Panel of the Aeronautical Research Council. Most appropriately, this book was dedicated to the memory of Horace Lamb, and it does, indeed, represent an enduring memorial.

Sir James particularly mentioned Goldstein's famous work on the vortex theory of screw propellers and on laminar boundary layer flow near a position of separation which deals with the complete mathematical nature of solutions of the boundary layer equations near separation. In order to stimulate new theoretical and experimental research on turbulence, Goldstein had spent a year in Göttingen with the renowned fluid dynamist, Ludwig Prandtl to learn the enormous advances made by Prandtl's group in experimental work on fluid flow structures of engineering interest and their physical interpretation in terms of vorticity, and then offered a year long advanced course on the theory of turbulence in Manchester. Based on his own research, Goldstein published an original paper 'On the law of decay of homogeneous isotropic turbulence and theories of the equilibrium and similarly spectra' in *Proc. Camb. Phil. Soc.*, **47** (1950) 554-574. This is a highly original research work which provided some new implications of Kolmogorov's equilibrium theory for the decay of homogeneous and isotropic turbulence. The last two topics of Lighthill's lecture dealt with problems of biofluiddynamics including (i) vortex wakes of birds in flapping flight and (ii) some challenging problems in the field of animal locomotion with special reference to the nature of the velocity and pressure fields in near field as well as in far field of swimming fishes. He examplified the nature of the quadrupole for two cases of swimming clupeoid fishes: regular swimming with lateral oscillations of the caudal fin, and sudden movements by which a steadily swimming fish makes a turn. In both cases, Lighthill identified the expected nature of the quadrupole far fields in terms of a dipole acting at the tail or at the head, respectively, and an equal and opposite dipole acting located at the centroid. Sir James concluded the Goldstein Memorial Lecture with an interesting statement: "... in two biological sections of my lecture, I have indeed indicated 'some challenging new applications for basic mathematical methods in the mechanics of fluids, that were originally pursued with aeronautical aims' by great scientists including Sydney Goldstein." This lecture was subsequently published in the *Journal of Aeronautical Society*, **94** (1990) 41-52.

Among his many degrees, honors and awards, he received a Gold Medal from the Institute of Mathematics and Its Applications in 1982, and the Harvey Prize of the Israel Institute of Technology in 1981. In recognition of his notable research contributions to fluid mechanics and applied mathematics, he was elected to twelve learned societies including the US National Academy of Sciences, Russian Academy of Sciences, and the Indian National Science Academy.

Lighthill never received a Ph.D. degree from any university. However, he was awarded honorary doctorate degrees from at least 24 universities including Liverpool (1961), Leicester (1965), Strathclyde (1966), Princeton (1966), Essex (1967), East Anglia (1968), Manchester (1968), Bath (1969), St. Andrews (1969), Surrey (1969), Cranfield (1974), Paris (1975), Aachen (1975), Rensselaer (1980), Leeds (1983), Brown (1984), Southern California (1984), Lisbon (1986), Rehovot (1987), London (1993), Compiègne (1994), Kiev (1994), St. Petersburg (1996) and Florida State (1996).

Many students of the University of Manchester, Imperial College and the University of Cambridge including M. R. Abbott, J. R. Blake, R. W. Blake, H. Cameron, G. D. Crapper, E. Cumberbatch, S. N. Curle, J. M. Fitz-Gerald, N. C. Freeman, J. H. Gerrard, F. A. Goldsworthy, I. M. Hall, G. J. Hancock, T. K. Herring, M. S. Howe, A. I. Mees, J.J.L. Higdon, P. E. Rapp, J.M.V. Rayner, N. Riley, M. J. Simon, F. B. Smith, M. A. Swinbanks, E. J. Varley, and G. B. Whitham have completed their research or Ph.D. thesis under Lighthill's supervision. Subsequently, almost all of them did much more original work in applied mathematics, fluid dynamics and biofluiddynamics and became famous for their outstanding contributions.

After his formal retirement from the University College London in 1989, he was appointed Emeritus Scientist by the College. Sir Eric Ash (1989) describes Sir James Lighthill's contributions to the University College London as Provost:

"James Lighthill was indeed a brilliant scientist; but he was also a polymath, with knowledge, insight and enthusiasm for the arts and humanities. He would invariably take the chair at inaugural lectures and, in thanking the speaker, provide an erudite coda — for any discipline — be it Egyptology, literature (illuminated by his ability to read in most modern European languages), medicine, or our own field of engineering. He was able to inspire his colleagues over the whole range of academic disciplines. Without the slightest doubt, during his watch, Lighthill succeeded in raising academic standards and in enhancing the international recognition accorded to University College."

Even after his retirement, Lighthill continued his research, publications, presentations, and professional service to national and international scientific community. He accepted the Royal Society Lectureship to deliver the Humphry Davy Lecture in 1991 and the Inaugural Perkins Memorial Lecture in 1995. In October of the same year, Sir James and Lady Nancy visited the University of Central Florida (UCF) for a week to deliver three lectures on fluid mechanics as part of the Distinguished Lecture Series in Mathematics at UCF. In 1993, he received the ICASE/LARC Theodorsen Lectureship Award.

In November 1996, Sir James and Lady Nancy came to Florida State University at Tallahassee to attend the International Symposium on Theoretical and Computational Fluid Mechanics which was organized in honor of Sir James Lighthill to celebrate his monumental contributions to fluid mechanics, applied mathematics, and the scientific community of the world. In 1997, the *Collected Papers of Sir James Lighthill* was published by the Oxford University Press. As editor of Sir James' collected papers M. Hussaini states the following in his General Introduction: "For more than half a century, his contributions spanned the fields of aeronautics, astrophysics, atmospheric and oceanographic sciences, and biofluid dynamics. But Lighthill is distinguished not just by the diversity and excellence of his technical contributions; his unique vision and perspicacity are evident in the groundbreaking, original work in every field he touched. In certain areas of aerodynamics and biofluid dynamics — such as wave propagation, aeroacoustics, and animal motion — his first articles were seminal and remain virtually the last word on the subject."

After the symposium at Tallahassee, Sir James and his wife returned to England. Sir James remained at the University College London until his death on July 17, 1998 after completing a nine-hour swim around the Channel Island of Sark against high tides and strong winds. The 1998

Copley Medal of the Royal Society (the highest award of the Royal Society) has been awarded posthumously to Sir James Lighthill.

Sir James Lighthill was undoubtedly one of the most brilliant and influential fluid dynamists of the twentieth century. He revolutionized applied mathematics with his remarkable contributions to modern fluid dynamics. There is no doubt at all about Lighthill's profound and everlasting impact on mathematical sciences and the scientific community of the world. His lifelong concern for quality mathematics instruction at national and international levels, and for inevitable loss by natural hazards reveals the unique character of this man. He will be remembered forever not only for his great scientific achievements, but also his unique contribution to the welfare of the human race. In many ways, Sir James was the epitome of the applied mathematical community.

Chapter 2

Music and Swimming

"The grand piano was open wide,
And strings in it were trembling ...."

"Oh my sea wave,
You are willful.
Both in peace and joyful game.
You are full of lovely life."

"The sea waves have a melody.
And there is harmony in Nature."

Fyodor Tyutcher

"The light of thy music illumines the world. The life breath of thy music runs from sky to sky. The holy stream of thy music breaks through all stony obstacles and rushes on."

Rabindranath Tagore (Gitanjali)

No biographical sketch of James Lighthill's life would be complete without mention of his life-long interest in music and swimming. While he was an undergraduate in Cambridge, James became involved in music for fun and recreation. He organized a small group for a musical performance of the Mozart's G minor piano quartets produced by one of the World's great composers. Even today Mozart's music is well known and admired throughout the world. This group consists of only two members with James on the piano and Nancy Dumaresq on the Cello. He met Nancy, a mathematics student in this musical performance group. It was a pleasant surprise for both of them that their families lived next door to each other in Highgate, North London.

Both James and Nancy were interested in solo music performances. It is not out of place to mention James' life-long interest in music. While Sir James was Provost of UCL, he used to participate in solo music performance in the annual concerts of the UCL Chamber Music Society. He continued his performance activity at UCL even after his official retirement in 1989. In addition to fun and recreation, Lighthill used to play the music on a piano because to perform its functions, the piano makes use of scientific principles, such as those governing the vibration of stretched metal strings and those involving the nature of sound. It can produce both melody and harmony at the same time. And it can produce an extraordinary variety of soft and loud notes with great speed and beautiful tone-color effects. Perhaps, Lighthill was attracted by these qualities of piano and he loved to play the music on a piano.

His love of nature and his fascination with swimming, which evolved into a lifelong hobby, began when he was a very young man. Swimming was the most favorite hobby of James Lighthill not only for fun and physical recreation, but also for appreciation of natural beauty. He loved and enjoyed swimming. His delight in swimming was a source of solace and comfort to his life. Clear evidence of his early love of swimming, tides, waves and currents in an open sea can be described by the above delightful quotations of Fyodor Tyutcher. In addition to music and swimming, his other leisure interests were poetry, literature and extraordinary foreign language skills. In addition to his extraordinary knowledge in English, Lighthill was very proficient in several other languages including French, German, Portuguese, and Russian. His lectures and presentations were well known not only for their literary quality, but also for sufficient new information and ideas in applied mathematics and engineering science.

Lighthill's relatives and friends remember him for his unlimited interest in swimming even in a very high tide, in violent currents and waves. They characterized him as an adventurous swimmer as he often took dangerous chances and risks for his life. James often violated safety rules and laws of swimming in restricted areas of Puget Sound, Lundy Island in the Bristol Channel, or in the Ramsey Island off the Southwest corner of Wales. At the age of 49, he was the first person ever to swim around Sark in the Channel Islands where the tides were extremely violent. In fact, he successfully completed the swim around the Sark five times. He used to swim using the called 'Old English backstroke' a kind of back-butterfly with breast-stroke leg motions. His love of swimming remained with him throughout his life. However, he often took a big risk to swim in channels or islands even when waves and currents were too dangerous as he had anticipated. After giving a main lecture at the British Theoretical Mechanics Colloquium in St. Andrews in 1970, he went swimming in St. Andrews Bay where he was found too far away from the coastline. So, a rescue helicopter was called out to save him from this dangerous situation, but he refused to be rescued saying that he was only out for a few miles swim and not in any big trouble. There were many other instances which he was involved either in violation of safety rules or arrested for swimming in a restricted areas of channels or islands. Several times he was involved in traffic violations including speeding or jumping off from a moving train which led him to pay the fine or under arrest by the police. Surprisingly, he was unwilling to plead guilty of such offences, rather he used to contest in the public court. On several occasions he successfully defined himself on charges of speedy driving, and not applying the brake on any downhill section of road because of his duty to society not to waste energy. All such deliberate or careless mistakes have been major concerns for his relatives and friends, especially for his wife, Nancy. Sometimes, Nancy used to chide him for his careless mistakes. She was, however, very loyal to her husband and would not tolerate any criticism of him from others. On July 17, 1998, he died after completing a nine-hour swim around the Channel Island of Sark against very high tides and violent winds. Before his death, Sir James was staying in a nearby hotel on Sark with his wife of 53 years, Nancy and their son. In conclusion of this short account of James Lighthill's swimming activities, we may note his remarkable appreciative view of the beauty of nature especially waves, tides, and currents generated in seas and oceans by winds and the solar heating. Indeed, the ocean as well as the atmosphere was his great research laboratory.

Chapter 3

Personal Reminiscence

"The important thing in science is not so much to obtain new facts as to discover new ways of thinking about them."

William Lawrence Bragg

"A teacher can never truly teach unless he is still learning himself. A lamp can never light another lamp unless it continues to burn its own flame. The teacher who has come to the end of his subject, who has no living traffic with his knowledge but merely repeats his lessons to his students, can only load their minds; he cannot quicken them."

Rabindranath Tagore (Gitanjali)

"Science is nothing without generalisations. Detached and ill-assorted facts are only raw material, and in the absence of a theoretical solvent, have but little nutritive value."

Lord Rayleigh

Perhaps it is appropriate to record here my personal relationship and association with Sir James Lighthill. Before I came to England for higher study and research in applied mathematics, I heard his name as one of the most prominent applied mathematicians, and read his famous book *Introduction to Fourier Analysis and Generalized Functions* and some research papers in order to prepare myself for pursuing research in fluid dynamics at Imperial College, London. After joining the Imperial College as a postgraduate student, I first came into contact with him in the Fall of 1965. During the next two years of my stay at Imperial College, I had the rare opportunity of attending his three courses of lectures on nonlinear waves, geophysical fluid dynamics, and blood flow in arteries. All of his three courses of lectures were primarily based on his own research notes and research papers either already published or under preparation at that time in Imperial College. His courses of lectures at Imperial and other conference presentations were well attended by students and faculty. The participants of his lectures and presentations remember him in his characteristic poses: standing before a blackboard or microphone or discussing problems in office. His gesture, smile, loud voice, and the expression of the eyes always reflected the kind of thought or the nature of the problem under discussion. He was always trying to cover more material within the time available for him in an organized, forceful, and enthusiastic manner. He will be always remembered gratefully for his effective teaching and presentations by his students, fellow faculty and professional colleagues who knew and worked with him.

All of these lectures dealt with remarkably new mathematical and physical ideas, results and predictions. I thoroughly enjoyed and learned a great deal from his lectures, and immediately decided to work on linear and nonlinear water waves for my Ph.D. dissertation. Subsequently, I wrote my Ph.D. dissertation on Transient Wave Motions in Fluids and successfully finished my degree. Sir James' unlimited enthusiasm, great challenge and interest in teaching and research will remain a lifelong memory with me and my other classmates at Imperial College. As I had stated earlier, his influence on me can, in fact, be traced back to my postgraduate days at the University of Calcutta well before I first met with him at Imperial College.

Probably, it is not out of place to mention my experience in Imperial College and then at the University of Cambridge. During my stay at Imperial from 1965 to 1967, I attended many other courses of lectures in both mathematics and mathematical physics given by many professors. They included Professors J. W. Dungey, W. K. Hayman, W. A. Hepner, H. Jones (then Head of Mathematics), F. G. Leppington, D. W. Moore, S. Rosenblat, Abdus Salam, R. S. Scorer, R. F. Streater, J. T. Stuart, F.W.G. Warren and Ffowcs Williams. In addition, I went to the University College to attend two courses of seminars given by Professor K. Stewartson. During my one year stay in Cambridge as a postdoctoral research fellow, I attended courses of lectures given by Professors P. M. Dirac, T. Brooke Benjamin, and still I vividly remember Lighthill's most interesting and stimulating lectures and his lecture notes prepared for postgraduate students and faculty members. Also what I do remember is the excitement they generated, but most importantly his numerous examples of applications intended to deepen our understanding of the mathematical theory with physical insight.

Sir James was an extremely dynamic and inspiring teacher in the classroom. Indeed, he was my most favorite teacher in my student days in London and Cambridge.

In addition to his course of lectures on nonlinear waves in Imperial College, M. J. Lighthill organized a Royal Society Discussion Meeting on Nonlinear Theory of Waves Propagation in Dispersive Systems in May, 1966. He selected a group of distinguished fluid dynamists and applied mathematicians including, G. B. Whitham, T. B. Benjamin, J. E. Feir, O. M. Phillips, M. S. Longuet-Higgins, A. E. Gill, and K. Hasselmann, P. G. Saffman, and K. Kenyon to deliver lecturers on different aspects of nonlinear dispersive waves. They provided a digest of recent developments, open questions and unsolved problems in this rapidly growing and important field. In his Introductory remarks, Lighthill states: "This Discussion Meeting is devoted to the task of summarizing, consolidating and bringing into a self-consistent state the knowledge which exists in a relatively new area of study of the mechanics of waves, an area in which most of the work has been done during the past six years. This work on dispersive systems, represents the second phase of development of the nonlinear mechanics of waves."

It is interesting to note that Lighthill was very proud to present his first Ph.D student of Manchester University, G. B. Whitham as the first speaker of this Discussion Meeting due to Whitham's pioneering work on nonlinear dispersive waves. In his opening remarks, Lighthill states: "The student in question was, as you know, successful in unraveling the problem (Whitham 1956) and it is a pleasure to us all to have him as our first speaker today on the still more difficult questions that arise in dispersive media. I believe it is still somewhat uncertain at this moment just what all the practical applications of these new theories will be (just as in 1947 it was uncertain in the acoustic case): and this is partly because the results of the new theories have not yet been worked out in a wide range of examples." Indeed, Lighthill was somewhat less optimistic in spite of many difficulties in developing new theories of nonlinear dispersive waves.

It is my pleasure to record here, his influence in my own research on this new subject and my direct involvement in organizing three NSF-CBMS research conferences on nonlinear waves and Integrable systems , with Alan Newell as the Principal Lecturer in 1982, with T. Brooke Benjamin as the Principal Lecturer in 1992, and with A.S. Fokas, as the Principle Lecturer in 2005, in the United States. In addition to a large number of participants, there have been a large number of one-hour invited speakers in these conferences included Jerry Bona, A. S. Fokas, Roger Grimshaw, Wallace D. Hayes, E. Infeld, J. C. Luke, R. S. Johnson, C. C. Mei, Martin Kruskal, J. H. Toland, David Kaup, Peter Olver, Harvey Segur, Michael Weinstein.

In his closing remarks, Lighthill said: "This has been a most interesting Discussion to attend, and will, I believe, in its written version, be a work of marked permanent value, especially because the contributors, as well as giving penetrating and comprehensive reviews of their own parts of the subject, have carefully elucidated also the relations between the parts. Methods apparently of very diverse character for analyzing the nonlinear development of dispersive waves have been described.

Not only those now working in the field, but also those still to be attracted into it by all its interesting potentialities, will find the present compendium of knowledge on the subject uniquely valuable. I am exceedingly grateful to everyone who has helped in bringing it into being." Influenced by Lighthill's own closing remark: "This has been a most interesting discussion to attend, and will, I believe, in its written version, be a work of marked permanent value....", I have had the opportunity to publish as a follow-up to the NSF-CBMS Conferences', four edited volumes (1983-84, 1992). There have been a large number of distinguished contributors to these volumes including: N. E. Andreev, R. Bingham, M.I.G. Bloor, John Brindley, Peter Bryant, R. K. Bullough, J.G.B. Byatt-Smith, P. J. Caudrey, H. H. Chen, P. C. Clemmow, Henri Cornille, Alex D.D. Craik, Antonio Degasperis, A. E. Dixon, R. K. Dodd, V. M. Eleonsky, A. S. Fokas, Mitsuaki Funakoshi, Jean-Michael Ghidaglia, R. J. Gribben, Roger Grimshaw, Mark D. Groves, Richard Haberman, John R.C. Hsu, E. Infeld, Susumu Inoue, Alan Jeffrey, R. S. Johnson, D. J. Kaup, T. Kawata, K. Konno, N. E. Kulagin, E. W. Laedke, C. N. Lashmore-Davies, Y. C. Lee, S. Lichter, A. K. Liu, Philip L. F. Liu, M. S. Longuet-Higgins, Yan-Chow Ma, Boris A. Malomed, C. R. Menyuk, T. Miloh, Irene M. Moroz, H. C. Morris, P. I. Naumkin, N. S. Novozhilova, H. Okuda, Z. Peradzynski, D. J. Pilling, Phoolan Prasad, Michael C. Reed, D. N. Riahi, P. Ripa, A. J. Roberts, Colin Rogers, G. Rowlands, N. Rudraiah, P. G. Saffman, N. Saitoh, Jean-Claude Saut, Peter Schuur, W. F. Shadwick, M. C. Shen, P. K. Shukla, P. V. Silin, V. P. Silin, Peter Smereka, K. H. Spatschek, M. D. Spector, C. H. Su, A. Thyagaraja, J. Timonen, Morikazu Toda, O. Troshkin, Y. Tschiya, Y. Tsuchiya, Jean-Mark Vanden-Broeck, Rene Van Dooren, M. Venkatachalappa, Miki Wadati, H. Wang, S. Watanabe, T. Yasuda and H. C. Yuen.

Lighthill recognized that considerable research was reflected in many contributions scattered throughout a vast literature on the mathematical theory of nonlinear dispersive wave phenomena and their applications. Yet it is increasingly difficult for scientists to do research since important articles often appear in journals of widely different character. This difficulty might be alleviated by publishing a single volume containing original papers together with research-expository and survey articles selected from key areas. Hence, the special volume of the Discussion Meeting that was published in the *Proceedings of the Royal Society of London, Series A, Mathematical Sciences*, Vol. **299**, No. 1456 (1967). This brought together new and current developments in nonlinear dispersive wave phenomena and their applications, and hence, it represents a marked permanent value to a wider group of audience. This volume also includes a Lighthill's own paper entitled "Some Special Cases Treated by the Whitham Theory." This paper was concerned with applications of the Whitham Theory to stationary waves on a steady flow, form of Lagrangian for gravity waves on deep water, comparison of Lighthill's Theory with experimental results of Benjamin and Feir, results from the Whitham Theory in particular examples and comparison with reliable experimental observations. Although, Lighthill was somewhat skeptical about both outstanding Whitham's Theory and Benjamin and Feir's experiments and Theory, he could not provide any better theoretical results or experimental observations.

Based on my research work, I first gave a seminar on Transient Development of Water Waves at Imperial College in the Spring of 1966. In my talk, I pointed out an error in a paper by J. W. Miles (1962) on the asymptotic solution of an initial value problem in water waves, and tried to obtain correct asymptotic solution with right physical interpretation. The audience at my seminar consisted of many postgraduate students and faculty members including Professors Lighthill and R. S. Scorer. At the end of my presentation, Professor Scorer raised a doubt in my asymptotic analysis, and then also in asymptotic methods that I had used to solve the problem. He was almost sure that the transient component of the asymptotic solution should decay to zero in the limit as time $t \to \infty$ so that the steady state solution can be obtained. In an immediate response to Scorer's comments, Professor Lighthill said that Debnath's claim seemed to be correct. The transient term obtained by Debnath does not decay to zero in the limit as $t \to \infty$, possibly due to the Dirac delta function disturbance used as a wave generating mechanism. Obviously. I was delighted to hear the comments of Professor Lighthill in support of my claim. At the request of Lighthill, one of his postgraduate students, M. S. Howe borrowed my research notes and repeated the calculation in order to detect any error in my asymptotic

Fig. 3.1 At the University of Central Florida, October 1995. Left to right: Sir James Lighthill and Lokenath Debnath.

analysis. Fortunately, no error was found. Mr. Howe and I discussed my research problem with Professor Lighthill. Interestingly enough, Lighthill was impressed with my work and then gave a nice physical interpretation of my solution.

During my stay at Imperial College, I had many opportunities to discuss my research problems with Sir James Lighthill and other members of his group which included several British and Indian scholars including Professor P. L. Bhatnagar, Drs. C. Devanathan and A. C. Srivastava. I also attended two annual meetings of the British Theoretical and Applied Mechanics Association at Southampton in 1966 and at Oxford in 1967. In both of these meetings, Sir James Lighthill was a keynote speaker. I enjoyed his outstanding lectures on rotating fluids and on nonlinear waves.

It is appropriate to mention one of his unusual personal characteristics. There are many examples which show that Lighthill himself was hardly interested in putting his name as a coauthor on papers of his research students, even though he supervised research and helped his students how to solve problems. Many friends of mine and I myself discussed our research problems with him on numerous occasions during our stay in London, Oxford and Cambridge, but he was unwilling to publish any joint papers with us.

It is not out of place to mention Sir James Lighthill's positive influence on other Indian researchers including Professors S. D. Nigam, M. P. Singh, M. G. Chopra and V. P. Saxena. Professor M. P. Singh and his research group at Indian Institute of Technology (ITT), Delhi conducted collaborative research with the Physiological Flow Studies Unit at Imperial College for many years. Dr. M. G. Chopra and V.P. Saxena spent some time in Cambridge to conduct research under the direction of Lighthill. In particular, Dr. Chopra published several papers based on the ideas and results of Lighthill on fish swimming. With Lighthill's support, Professor M. P. Singh also established a new Center for Atmospheric Sciences in ITT, Delhi. Sir James also made special contact with Indian Scientific organizations including the Indian National Science Academy (INSA) which elected him a Foreign Fellow. In collaboration with INSA and Professor M. P. Singh, Lighthill organized the joint IUTAM (International Union of Theoretical and Applied Mechanics) and IUGG (International Union of Geodesy and Geophysics) International symposium on Monsoon Dynamics with Applications to Weather Forecasting and Flood Prediction in New Delhi in December 5 to 9, 1977. This was the most successful international symposium with interdisciplinary character that deals with very strong interactions between the atmosphere and ocean resulting in occurrence of hurricanes and typhoons which involve physical processes at various spatial space and time scales. His 1977 visit to India with an International symposium gave an impetus of inestimable value to the development of applied mathematics research in India, in particular and the whole Asia, in general, and brought lasting benefit to the subcontinent. The subject of Monsoon Dynamics is a great value to scientists in India and in other Asian countries as it combines profound scientific and applied mathematical interest with great significance to India, and dependence on observational data available in India. In his Keynote address to this symposium with a special emphasis to multidisciplinary and multinational research, Lighthill states that: "So the keynote of this symposium, which I have been asked to give you with my tuning fork as it were, is its interdisciplinary character, merging the disciplines of fluid dynamics, atmospheric physics, oceanography, estuary dynamics and hydrology; merging the basic sciences of global geophysical processes with applied sciences of local weather forecasting and flood prediction; merging the professional activities of operational departments in the field and those of university research workers."

As I have indicated before, Sir James was Lucasian Professor of Applied

Mathematics in Cambridge from 1969 to 1979, and then became Provost of the University College London from 1979 to 1989. During this period of twenty years, I visited Cambridge for a year and then Oxford for six months to conduct research in applied mathematics. When I was at the Department of Applied Mathematics and Theoretical Physics in Cambridge, I again had the opportunity to attend some of his lectures and discuss research problems with him. During my stay in Oxford in 1980, I went to the University College London to discuss my research problems on nonlinear diffraction of water waves based on his famous 1979 paper on "Waves and Hydrodynamic Loading." Subsequently, I wrote a paper on "A Theory of Nonlinear Waveloading on Offshore Structures" in 1981, and then a chapter on "Nonlinear Diffraction of Water Waves" for my book entitled *Nonlinear Water Waves* published in 1994. Meetings with him, which I would look forward to with pleasure, were relaxed—even when dealing with difficult mathematical problems. Despite his very busy schedule, Sir James still was willing to discuss research problems with his students and colleagues with interest and enthusiasm. His kind, warm and dynamic personality was appreciated by students, colleagues, and friends. He was also noted for his leisure interests in swimming, languages, and music throughout his life. Despite his great achievement in mathematical sciences and in public life, he never lost his sense of humor, and his devotion to his family and friends, both personal and scientific.

Sir James traveled extensively throughout the world and whenever possible, Lady Nancy accompanied him. The year 1995 was their golden anniversary of their successful marriage. They are an elegant and admired couple who are exemplary ambassadors for Great Britain, for British applied mathematics and science, and indeed for mathematical sciences in general. Based on my limited personal experience, I may say that Nancy is simple, modest and gentle in nature, whereas James is a proud and assertive man with a razor-sharp mind. She knew a lot of Sir James' academic and scientific world, but was devoted to him and proud to be his wife. She happily shared in his scientific achievements, entertaining activities such as music and his adventurous activities such as swimming. Undoubtedly, she was a suitable partner for James and their marriage was a happy one. They have one son and four daughters. She contributed a great deal to his well-being and thereby enabled him to use mathematical and scientific talent fully for teaching, research and professional service. According to T. J. Pedley, "Nancy remained "The light of his life" until his death."

During my tenure as Chair of Mathematics at the University of Central

Florida, I established a Distinguished Lecture Series in 1983. Many Nobel Prize winners, Field Prize winners, Abel Prize winners, members of the National Academy of Sciences and Fellows of the Royal Society of London participated in this Lecture Series for a period of fourteen years. At my personal invitation, Sir James and Lady Nancy Lighthill visited our department in the Fall of 1995. During this week long visit at Central Florida, he delivered three lectures on hurricane dynamics, bacterial biofluid dynamics, and some aspects of aeroacoustic of extreme-speed jets. As usual, these lectures were loaded with many new, important and interesting ideas and results, and have stimulated interest in applied mathematical research. During this visit to Orlando, we had the opportunity to discuss many issues of mutual interest including research topics, the future of applied mathematics research, education and training in the UK and in the USA. His analytical mind is subtle and penetrating and he articulates his thoughts on applied mathematics or music, Beethoven or butterflies with such clarity and with such well-balanced reasoning that are almost unbeatable by any standard. One day, I escorted both James and Nancy on a sightseeing tour at the Walt Disney World in Orlando and it seemed to me they enjoyed the tour during their Golden Wedding Anniversary year. When I told them at the entrance of the Walt Disney World that this is a great place to celebrate this wonderful occasion - they both smiled at me and James said that they were in a right place and in the right time. However, I enjoyed their company for a week and was impressed with their high levels of thinking with grace, good humor and friendliness. We had an ample but great time and this time we became good friends and were free to discuss many topics of diverse interests including past stories of our stay at London, Cambridge, and Oxford. In response to one of my questions about publication of the second edition of his famous and delightful book on *Fourier Analysis and Generalized Functions* that was first published in 1958, Lighthill said that forty thousand copies of this book were sold so far. He had neither plan nor intention of making its second edition. This short book is a classic in which he developed the theory of Fourier transforms, generalized functions and asymptotic analysis of Fourier transforms accessible to everyone.

After their return to London, I have had regular correspondences with him. In each of his many letters written to me, he signed his first name in the end. It is my pleasure to display one such letter here to record his friendliness and kind invitation to participate in the International Symposium organized in honor of Sir James at Tallahassee, Florida in November 1996. Unfortunately, I missed this unique opportunity of attending this

Fig. 3.2 At the University of Central Florida, October 1995. Left to right: Lokenath Debnath, Sir James Lighthill and Lady Nancy Lighthill.

symposium to pay a special tribute to my favorite professor and friend due to my prior commitment to visit India as Fulbright Professor from July to December of 1996. After my return from India, I resumed correspondences with him and was delighted to see his *Collected Works* in four volumes edited by Professor M. Yousuff Hussaini and published by Oxford University Press. My last letter to him was in late June or early July in 1998 with a special proposal for publishing a book entitled *Sir James Lighthill and Modern Fluid Mechanics*. Unfortunately, I did not receive any response to this letter, possibly due to his sudden death in July. I was deeply disturbed by this shocking news of his death. There is absolutely no doubt that this is a great loss for the entire mathematical as well as scientific community of the world. Personally, I lost an outstanding teacher, and a great friend to whom I would remain grateful for his help and encouragement in my life. Indeed, I was fortunate to come in contact with such a great applied mathematician in both personal and professional levels. So my desire as well as interest in writing this book commemorating Sir James Lighthill the man and the mathematical scientist is founded solely on my deep respect and admiration for this great man and renowned applied mathematician whom I had the opportunity of knowing very well for a long period of time.

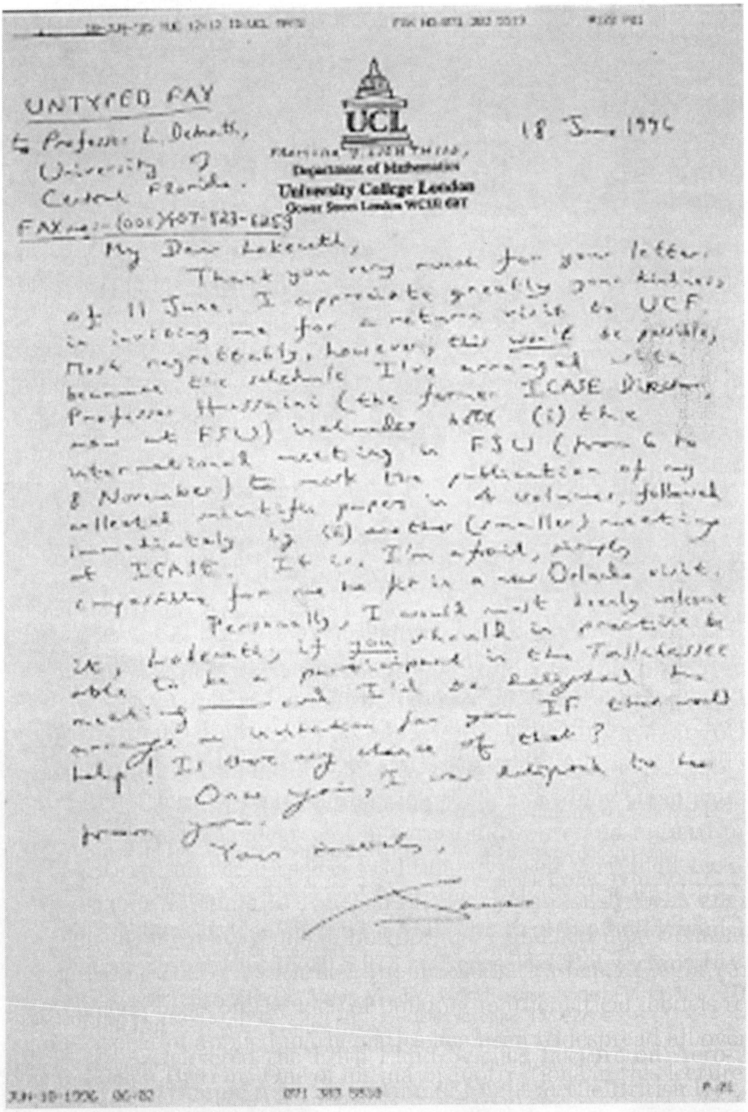

UNTYPED FAX

to Professor L. Debnath,
University of
Central Florida.
FAX no:- (001) 407-823-6253

Professor Sir James Lighthill,
Department of Mathematics
University College London
Gower Street London WC1E 6BT

18 June 1996

My Dear Lokenath,

Thank you very much for your letter of 11 June. I appreciate greatly your kindness in inviting me for a return visit to UCF. Most regrettably, however, this won't be possible, because the schedule I've arranged with Professor Hussaini (the former ICASE Director, now at FSU) includes both (i) the international meeting in FSU (from 6 to 8 November) to mark the publication of my collected scientific papers in 4 volumes, followed immediately by (ii) another (smaller) meeting at ICASE. It is, I'm afraid, simply impossible for me to fit in a new Orlando visit.

Personally, I would most dearly welcome it, Lokenath, if you should in practice be able to be a participant in the Tallahassee meeting — and I'd be delighted to arrange an invitation for you IF that would help! Is there any chance of that?

Once again, I am delighted to hear from you.

Yours sincerely,

JUN-18-1996 06:02 071 383 5519 P.01

University College London

Chapter 4

Sir James Lighthill's Books

> "To Paul Dirac, who saw it must be true, Laurent Schwartz, who proved it, and George Temple, who showed how simple it could be made."
>
> *M. J. Lighthill*

4.1 *Introduction to Fourier Analysis and Generalised Functions*, Cambridge University Press, Cambridge, 1958

Based on his lecture notes for undergraduate courses on the generalised functions, Fourier series and integrals at the University of Manchester, Lighthill published his first 79-page classic book entitled *Introduction to Fourier Analysis and Generalised Functions* in 1958. His dedication page reads: "To Paul Dirac, who saw it must be true, Laurent Schwartz, who proved it and George Temple, who showed how simple it could be made." This remarkable student's edition published by Cambridge University Press contains considerable modern development of new mathematical techniques in the area of Fourier analysis and generalised functions. Motivated by the fact that Fourier series, Fourier integral and generalised functions play a fundamental role in solving differential, integral and functional equations, Lighthill took the initiative to teach an undergraduate course at the University of Manchester, and subsequently wrote this book. Fascinated by the mystery of the Heaviside function (1890), renamed Dirac's delta function (1926), Laurent Schwartz (1915-2002) opened the way to the theory of Fourier transforms of (tempered) distributions that has proved to be remarkably useful in the study of partial differential equations. His new and beautiful ideas have all the simplicity and inevitableness which characterize his work as the highest class. This also reminded us of the words of Subrahmanyan Chandrasekhar: "The simple is the seal of the true. And beauty is the splendor of truth." Schwartz's discovery of the theory of distributions is a unique example of his extraordinary skill in the presentation of mathematics. His two volumes of brilliant work entitled, "*Théorie des distributions*" were published in 1950-1951. It also represents a model of fairmindedness in its coverage of the work of his predecessors, contemporaries, and students, notably Lars Hörmander (1962 Fields Prize Winner). Schwartz's mathematical achievement attracted international attention and crowned him with the award of a Fields Medal at the 1950 ICM Meeting in Harvard. In receiving the Fields Medal at the ICM meeting, he was cited especially for his great contribution to the development of the theory of distributions. According to the citation for his Fields Medal, Schwartz has "clearly seen, and been able to shape, the new ideas in their purity and generality," adapted to the needs of further research.

On one hand, Lighthill was fascinated by the brilliant work of Laurent Schwartz on rigorous theory and applications of distribution, on the other hand, he recognized that George Temple's (1901-1992) elegant and simplified version of the theory of generalised functions based on good functions is more easily intelligible to students of mathematics, science and engineering. Indeed, the good functions are important in Fourier analysis because the inversion, convolution and differentiation theorems, as well as many others, apply in particular simple forms with no problems of convergence. Based on Temple's approach to generalised functions, Lighthill wrote this book which is of great value in that it brings together many simple and major results of Fourier analysis and generalised functions. This short book has five short chapters and the summary of these chapters are given below:

Chapter 1. Introduction is concerned with an introductory remarks on Fourier series which are used for analysing oscillations periodic in time, or waveforms periodic in space, and also for representing functions of plane or cylindrical polar coordinates. This is followed by definition of the Fourier transform and the inverse Fourier transform as

$$g(y) = \mathfrak{F}\{f(x)\} = \int_{-\infty}^{\infty} e^{-2i\pi xy} f(x)dx, \tag{4.1.1}$$

$$f(x) = \mathfrak{F}^{-1}\{g(y)\} = \int_{-\infty}^{\infty} e^{2\pi ixy} g(y)dy. \tag{4.1.2}$$

The difficulty with this definition is that many simple and and useful functions including a constant function c, $\sin ax$ and $\cos ax$ have *no* Fourier transform in the ordinary sense. This difficulty can be resolved in the theory of generalised functions. For example, the Fourier transform of 1 is the Dirac delta function, $\delta(x)$ which is defined by

$$\int_{-\infty}^{\infty} \delta(x) f(x) dx = f(0) \tag{4.1.3}$$

for any continuous function $f(x)$. No ordinary function has the property (4.1.3), but a sequence of functions, $\delta_n(x)$ defined by

$$\delta_n(x) = \sqrt{\frac{n}{\pi}}\, e^{-nx^2}, \qquad n = 1, 2, 3, \cdots, \tag{4.1.4}$$

has progressively taller and thinner peaks at $x = 0$ as shown in Figure 4.1 with the area under the curve equal to 1, where $\delta_n(x) \to 0$ as $n \to \infty$ at any $x \neq 0$ and $\delta_n(0) \to \infty$ as $n \to \infty$. However, in the limit as $n \to \infty$,

$$\lim_{n\to\infty} \delta_n(x) = \delta(x). \tag{4.1.5}$$

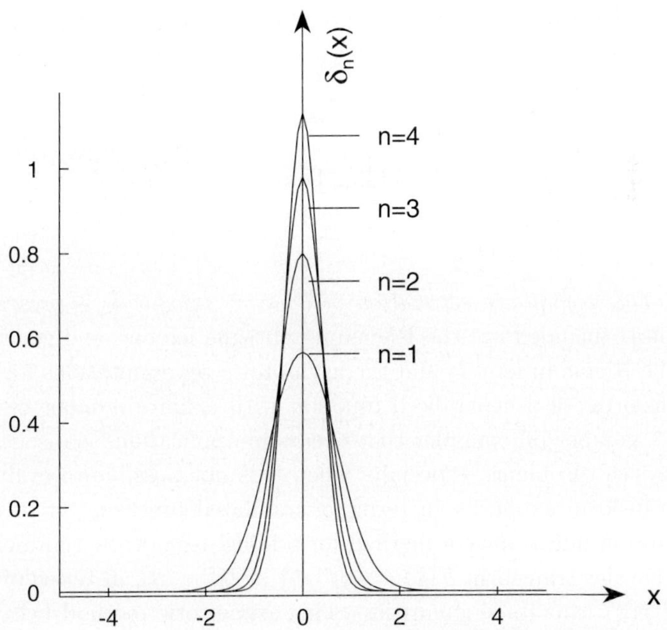

Fig. 4.1 The sequence of delta functions, $\delta_n(x)$.

So, the delta function can be considered as the limit of a sequence of ordinary functions. In the limit, this sequence has the property (4.1.3). Thus, the Dirac delta function is not a function in the classical sense. However, $\delta(x)$ can be treated as a function in the generalised sense, and in fact, $\delta(x)$ is called a *generalised function*. Physically, $\delta(x)$ can be considered as the distribution of charge (or mass) along the x-axis, and represents a *unit point charge* (or *unit point mass*). Similarly, $\delta'(x)$ represents a dipole of unit electric moment, since as a particular case of (4.1.3), we have

$$\int_{-\infty}^{\infty} x\,\delta'(x)dx = -1. \tag{4.1.6}$$

Chapter 2. The theory of generalised functions and their Fourier transforms deal with good functions and fairly good functions, generalised functions, the Dirac delta function and its derivatives, and ordinary functions as generalised functions. Many properties of generalised functions including their limits, derivatives, Fourier transforms, and Parseval's theorem are discussed with examples.

Chapter 3. Definitions, properties and Fourier transforms of particular generalised functions represent topics of this chapter. Included are particular generalised functions

$$|x|^{\alpha},\ |x|^{\alpha} sgn\, x, \quad \text{and} \quad x^{\alpha}H(x) = \frac{1}{2}\left[|x|^{\alpha} + |x|^{\alpha} sgn x\right], \tag{4.1.7}$$

$\alpha(> -1)$ is a non-integral power and $H(x)$ is the Heaviside unit step function. This chapter is completed by adding Fourier transforms of the elementary functions possessing algebraic or algebraic logarithmic singularities at $x = 0$, and rational functions.

Chapter 4. The asymptotic estimation of Fourier transforms is presented with examples. Included are the Riemann-Lebesgue lemma, and generalisations of the Riemann lemma and the asymptotic representations for the Fourier transforms of a generalised function with a finite number of singularities. A number of examples that occurs in applications is discussed. In many physical problems, especially when it is not possible to evaluate a Fourier transform explicitly in terms of tabulated functions, it is very important and useful to have a method for finding asymptotic representation of the Fourier transform $F(k) = \mathfrak{F}\{f(x)\}$ as $|k| \to \infty$, in terms of the behavior of $f(x)$ near its singularities. This asymptotic method becomes very effective, and particularly simple when generalised functions are used. Thus, the asymptotic representation is very useful for most functions occurring in many applications.

Chapter 5. Fourier series is the main subject of this final chapter. Convergence and uniqueness of trigonometric series as series of generalised functions, and the existence of Fourier-series representation for any periodic generalised function are discussed in some detail. Included are Poisson's summation formula and the asymptotic behavior of the Fourier coefficients c_n as $|n| \to \infty$ of a given periodic generalised function $f(x)$ with examples.

Without loss of mathematical clarity and rigor, Lighthill clearly described the theory of generalised functions so that mathematicians could easily understand and use the method of asymptotic expansion of Fourier integrals that seems superior to previous techniques. He also gave a special attention to convergence and uniqueness of trigonometric series as a series of generalised functions. This edition became very successful as a very large number of mathematicians and scientists around the world made regular use of this simple and elegant book on modern Fourier analysis and its applications. However, the book does not include one fundamental concept of convolution and the convolution theorem which are widely used in the study of nonhomogeneous differential and functional equations.

Lighthill's undergraduate Cambridge classmate and an eminent mathematical physicists, Professor Freeman Dyson, now in the Princeton Institute for the Advanced Study, described the importance of Lighthill's invaluable book as:

"This book presents the ideas of distribution theory in so clear and self-contained a form that they can be offered directly to undergraduates.... It is written in a language that [they] can understand, with discussion of examples that physicists can use, and yet there is not the slightest departure from strict mathematical rigour.... It is to be hoped that this book will quickly be adopted as the appropriate introduction to Fourier analysis; it is likely also to have a long life as a work of reference for research students."

Freeman J. Dyson in *Physics Today*

4.2 *Higher Approximations in Aerodynamic Theory*, Princeton University Press, Princeton, 1960

At the invitation of W. R. Sears, the editor of Volume VI of the Princeton Series on High Speed Aerodynamics and Jet Propulsion (entitled *General Theory of High Speed Aerodynamics*), Lighthill wrote a 147-page article on *Higher Approximations* in 1954. This article was later reprinted as a separate paperback edition entitled *Higher Approximations in Aerodynamic Theory* in 1960. This monograph is an example of his extraordinary skill in the presentation of the improved approximate theories of high speed aerodynamics. It is also a model of fair mindedness in its coverage of the work of his predecessors, contemporaries, and research associates including J. B. Broderick, K. O. Friedrichs, M. Van Dyke, S. Goldstein, M. J. Legras, H. S. Tan, G. N. Ward and G.B. Whitham. This edition deals with a detailed description of all moderate departures from the linearized theory in which shocks are weak and planar (or conical), apart from those associated with dissipative effects. The major objective was to develop new improved perturbation expansions and higher approximations in subsonic and supersonic aerodynamic flows with nonlinear effects included.

Based on a quantitative study of nonlinear effects, Lighthill described the theories of high speed aerodynamics in this paperback. It has essentially nine articles including the last article 9. Cited References. Before we describe each article in some detail, it seems important to give a short summary of these articles. The material in the first three articles deals with classical linearized theories and results of aerodynamics with the major objective to develop the improved approximate theories in the sense that errors due to the nonlinearity of the equations of adiabatic flow can be reduced. The topics presented in Articles 4 to 8 are partly published in late 1940s and early 1950s including the major work of Lighthill and his associates on supersonic aerodynamics. Indeed, Lighthill made major contributions to high-speed aerodynamics, slender-body theory and higher approximations that had tremendous influence on the performance of modern jet aircraft and missiles. This is followed by a brief description of each article of this book.

Article 1. Introduction deals with serious comments on approximate solutions of the equations and boundary conditions of the adiabatic theory, weaknesses of the linearized theories and the improved approximate theories with nonlinear effects.

Article 2. Subsonic Flows are concerned with study of flow patterns with Mach number until before shock waves develop. More precisely, this article deals with the steady adiabatic flow of a uniform subsonic stream past an obstacle not thin enough so that effects of compressibility can be included from the Prandtl-Glauert perturbation expansion. The method of successive approximations is employed to investigate the problems of subsonic flows. Although the method is mathematically valid and natural, but not the best for many physical problems of interest. However, the method can hardly be useful for certain values of the Mach numbers when shock waves appear as convergence of solution becomes a problem.

Article 3. Supersonic Two-Dimensional Airfoil Theory: Surface Pressures are devoted to the theory of higher approximations than the acoustic small perturbation expansion theory used in supersonic flows. In such flows, Busemann's second order approximation is fairly accurate for most cases as it gives better results for lift, drag and moment coefficients. Under suitable conditions, the simple wave theory of the surface pressures is also adequate for all practical cases, up to a Mach number 3. At very high Mach numbers the simple wave theory produces unsatisfactory results, but shock-expansion theory, even for shapes for which it it not exact, seems to be considerably better.

Article 4. Supersonic Two-Dimensional Airfoil Theory: Complete Flow Pattern are concerned with the study of complete flow pattern when a uniform, steady, two-dimensional supersonic stream flows past an airfoil. A systematic theory for general supersonic two-dimensional flow past an airfoil due to K. O. Friedrichs (1948) is discussed in some detail. The Friedrichs theory is a good approximation to simple wave theory. The pressure distribution in an N wave and the analysis of drag in terms of wave energy are also presented. All energy involved in the N wave as a reflected and transmitted pressure wave are reabsorbed in the N wave by overtaking of the rear shocks by acoustic waves moving behind them. After the trail pressure wave has caught up with the rear shocks, the asymptotic position and strength of the rear shock can be determined from the ultimate total energy flux in the N wave.

Article 5. Supersonic Two-Dimensional Airfoil Theory: Influence of a Blunt Leading Edge deals with an extension of the theory of Articles 3 and 4 to the supersonic flow past airfoils with blunt leading edges. However, the mathematical results of Articles 3 and 4 do not apply directly to blunt-nosed airfoils. So, appropriate semiempirical modifications of these results can, in principle, be obtained to give some information about every aspect of the flow for simpler problems of sharp-nosed airfoils.

Article 6. Supersonic Projectile Theory: Complete Flow Pattern is concerned with problems of axisymmetric flow past projectiles similar to those of Articles 3-5. Special attention is given to the Whitham theory for axisymmetric flow pattern of a supersonic projectile. With regard to the mass flow behind the projectile, to a second approximation, a net mass outflow far from the axis must balance the reduced mass outflow behind the projectile due to increased entropy. So, it is expected that pressure waves generated by reflection from the front shock and transmission through the near shock play a key part in this balance.

Article 7. Supersonic Projectile Theory: Surface Pressures deal with the errors in the linearized theory of surface pressures on a supersonic projectile. This is followed by a study of a second approximation for the pressure on projectiles of revolution in axisymmetric flow. Then the effect of yaw is included. A second approximation to nonaxisymmetric flows, and the yawed flow past projectiles of revolution are also presented in this article. It is worth noting that Van Dyke's (1951) remarkable contribution to the first- and second-order theory of supersonic flow past bodies of revolution are smooth because of a significant extension of his theory to second order approximation. Therefore, his work not only supersedes Broderick's (1949) second order theory in term of accuracy, but also it provides a satisfactory agreement with experimental observations except for boundary layer effects. Indeed, Van Dyke's work is particularly satisfactory because his theory continues to provide excellent agreement with exact calculations by the method of characteristics even at a Mach number greater than 3. For nonaxisymmetric flows past bodies (of any cross section) which both smooth and slender in all directions normal to the stream, Ward's (1949) theory is fairly general in the sense that all features of other theories of flows past smooth bodies emerge clearly from his general approach. However, there was a major question how the extension of the Ward theory to general slender bodies which are not smooth can be made. On the other hand, Lighthill's theory of flow past slender bodies of revolution with discontinuous slope represents a new contribution. In his second paper, Broderick (1949) extended further his solution for the supersonic flow past a cone to include the flow near the shock. However, Broderick's conclusion concerning flow near the shock follow from Whitham's subsequent work (1952) which is more satisfactory because it includes the second approximation and can be applied to the general theory of conical fields. Furthermore, Whitham's theory states the fact that the square root singularity in the first and second approximate solutions correspond to a real singularity in the potential flow, that is, a limit line. In fact, the velocity fields near the limit line vary like the square root of distance from it. But this singularity is actually slightly ahead of where it is predicted to be on these approximate solutions, and hence, the flow field must be shifted forward so that the predicted singularity occurs in the right position.

Article 8. Supersonic Three-Dimensional Wing Theory is developed with an extension of the ideas of the foregoing articles to fully three-dimensional flow problems. The surface pressure distribution in the three-dimensional wing theory has not been obtained to a better approximation than the linearized theory. Based on the solution of the linear theory problem, Lighthill developed a general method to determine the strength of the shock (of conical shape, nearly but not quite right circular) in an arbitrary conical field. In general, an adiabatic flow field can, even behind shocks, be calculated to good approximation by solving the equation of potential motion. However, the complete solution of these equations has in general singular regions, at each point of which it gives three different values of each flow field. These regions must in practice be traversed by a shock, ahead of which the flow fields assume one of the three values, and behind which they assume of the others. In case of the two-dimensional flow past a sharp-nosed airfoil, the singular region associated with the front shock is of wedge shape which is bounded by a Mach line associated with the flow deflection at the nose on the upper surface, and by another associated with the undisturbed stream. For axisymmetric flow past a cone, the region is bounded by the undisturbed Mach cone and by a cone of larger angle, which may be called the *limiting cone* (that is, the upstream limit of disturbances in the potential flow) and which is an envelope of Mach lines. Such an envelope, not necessarily cone, also exists for any supersonic flow past slender projectiles.

4.3 *Mathematical Biofluiddynamics*, SIAM, Philadelphia, 1975

Lighthill was greatly influenced by the 1929 outstanding paper of Sydney Goldstein (1903-1989) on the vortex wake of a screw propeller, and the brilliant work of Sir Geoffrey Ingram Taylor (1886-1975) on aquatic animal locomotion and its hydrodynamics during 1950s to initiate new research on aquatic animal locomotion. Later on, at the request of Sir James Gray, Professor of Zoology at the University of Cambridge, Lighthill became interested in hydrodynamical study of animal swimming, aquatic animal propulsion and animal flight as well as low Reynolds number flagellar hydrodynamics. Based on a highly idealized model of microorganism swimming with appropriate assumptions and approximations, his first paper "On the Squirming Motion of Nearly Spherical Deformable Bodies Through Liquids at Very Small Reynolds Number" was published in 1952. Subsequently, in collaboration with his research students, J. R. Blake, G. J. Hancock and J. J. L. Higdon, Lighthill introduced another new field known as mathematical biofluiddynamics and made some major contributions to this field during his stay at Imperial College and Cambridge. In 1969, Lighthill wrote a major expository and survey article on Hydromechanics of Aquatic Animal Propulsion: A comprehensive survey of this field published in the first volume of the *Annual Review of Fluid Mechanics* in which he states: "This survey notes several such cases (see especially Section 11), but concentrates principally on those extensive portions of the animal kingdom where capabilities of motion in water at speeds comparable with those of other animals of similar mass has successfully been achieved and retained."

In his first section on the scope of this survey, Lighthill describes:

"The mechanical discussion concentrates similarly on the aqueous medium, and on its interactions with the movements of the animal's external (and, where relevant, internal) surfaces. By contrast, little is said either about the internal mechanics of the animal, or about means of coordination and stimulation of those movements (which include the sense organs and neural channels of communication as well as the muscles and other motile organs). Within the hydromechanical field, furthermore, the survey concentrates on the problems of propulsion at the expense of related problems of stability and control.

The subject is delimited is, however, of particular hydromechanical interest, because about 10^9 years of animal evolution in an aqueous environment, by preferential retention of specific variations that increase ability to survive and produce fertile offspring, have inevitably produced rather refined means of generating fast movement at low energy cost, which merit study."

His research and major research reviews of aquatic animal propulsion, animal flight and low Reynolds number of flagellar hydrodynamics provided a major impetus for further study and research in this new field. On one hand, Lighthill made major contributions to biological fluid dynamics, on the other hand, he totally transformed the study and research of this field.

In 1973 Lighthill served as the Principal Lecturer of an NSF-CBMS research conference organized at Rensselaer Polytechnic Institute on Biofluiddynamics. Based on his expanded version of ten lectures delivered at this conference, Lighthill wrote a major research monograph on *Mathematical Biofluiddynamics* that was published by SIAM in 1975. This book covers a fuller account of both external (animal locomotion) and internal (physiological) biofluid dynamics and their recent developments. In addition to *Chapter 1. Introduction to Biofluiddynamics*, this volume has two parts with a total of thirteen chapters.

Part I. External Biofluiddynamics

Chapter 2. Hydromechanics of aquatic animal propulsion: A Survey,

Chapter 3. Mathematics of aquatic animal locomotion at low Reynolds number,

Chapter 4. Aquatic animal propulsion of high hydromechanical efficiency,

Chapter 5. Large-Amplitude elongated-body theory of fish locomotion,

Chapter 6. Aquatic animal locomotion: A Survey of recent theoretical developments,

Chapter 7. Some current investigations of aquatic animal motions,

Chapter 8. Animal flight,

Chapter 9. On the Weis-Fogh mechanism of lift generation.

With regard to recent developments in aquatic animal motions, Lighthill pointed out that the study of general feature of aquatic animal locomotion has received much less attention at high Reynolds numbers. So, he described new opportunities for investigations of external biofluiddynamics in general, and of aquatic animal locomotion at high Reynolds numbers, in particular. In his 1972 lecture at the International Congress of Theoretical and Applied Mechanics in Moscow, Lighthill presented another remarkable expository research and survey article on Aquatic animal locomotion and its recent developments with future potential for advanced study and research. On one hand, he elaborated his own research and interest, on the other hand, he described the potential interactions between an aqueous medium and the external surfaces of a fully immersed animal swimming through it. His following quotation explains vividly his thoughts and views on interdisciplinary nature of fluid mechanics and zoology:

"Generally speaking, progress in Zoologically Applied Mechanics can be made only through close collaboration between a zoologist and a mechanician. The animal kingdom is vast, and the problem of discerning and describing what types of animal motion have in different parts of it been significant, above all in the evolutionary sense, is one for the zoologist alone. At the same time, study of the complicated interactions between those motions and a surrounding aqueous medium is one for the hydrodynamicist alone. It is therefore when a zoologist and a hydrodynamicist have got to know each other well enough to be able to talk together about the problems, and gradually to learn enough of each other's language so as to be able to communicate effectively, that collaborative progress involving hydrodynamically sound analysis of zoologically significant motions becomes possible.

Especially significant for aquatic animal locomotion was such a collaboration which started around 1950 between Sir Geoffrey Taylor with his enormous mechanics experience and Sir James Gray whose zoology school at Cambridge had for over 20 years been a prime source of knowledge, and of research workers, in almost every aspect of animal locomotion (Gray 1968). It was in fact on aquatic animal locomotion and its hyrdodynamics that Taylor and Gray particularly talked, and this led to important papers by Taylor (1951, 1952a, 1952b) on cases of such motion both at very low Reynolds numbers and at high Reynolds numbers."

All of the above reveal that Lighthill strongly advocated for the tremendous needs for interdisciplinary and crossdisciplinary research among fluid dynamicists, zoologists, biologists and engineering scientists in order to gain understanding of many enormously complex problems of great interest.

Animal flight as well as its modern developments is another major area of external biofluiddynamics. This area has also received considerable attention by a diverse group of theoretical and applied scientists. Lighthill also expressed his strong support in favor of interdisciplinary research by fluid dynamics, zoologists and biologists.

In Chapter 8 on Animal Flight of his book on *Mathematical Biofluiddynamics* (1975), Lighthill began his introduction as:

"1. The conquest of the air

Within the field of external biofluiddynamics, the aquatic aspects with which Chapters 1 to 7 have been concerned comprise the earliest developments in the evolution of animal locomotion, as well as many interesting later developments:at least 6×10^8 years of animal evolution in the marine environment can be traced in the extensive fossil record from around the beginning of the Cambrian period to the present time. By contrast, the *aerial* aspects of external biofluiddynamics, to which Chapters 8 and 9 are devoted, began their process of evolution much later: the 'conquest of the air' by flying animals is a considerably more recent event than the origin of swimming.

Nevertheless, the chronology of animal flight is longer, and the epoch of the conquest of the air has to be placed earlier, than many people would imagine. Already 3×10^8 years ago, in the middle of the Carboniferous period, the problems of sustained powered flight had been solved by the insects. An excellent account of some of the processes involved in the evolution of animal flight has been given by Smart & Hughes (1972) ..."

Based on the major role of the 'clap' and 'fling' motions in *Encarsia formosa* (an important parasite widely used in the biological control of greenhouse aphids), Weis-Fogh (1973) proposed a fundamentally new biofluiddynamics that describes a totally new mechanism of lift generation in inviscid two-dimensional flow. The Weis-Fogh mechanism of lift generation is of fundamental importance in all other animal flight problems (for example, the biofluiddynamics of butterflies). His mathematical analysis of the hovering motions of the chalcid wasp *Encarsia formosa* revealed that its performance is remarkably superior to that of most hovering animals as a result of the Weis-Fogh mechanism of lift generation - which was not studied earlier in aerodynamics. Lighthill was very impressed with this brilliant work of Weis-Fogh, zoologist at Cambridge and maintained a collaborative research effort with Professor T. Weis-Fogh. Based on his collaboration with T. Weis-Fogh, Lighthill learned many new ideas for normal flight of birds, bats and large insects. In collaboration with his Ph.D. student, J. M. V. Rayner, Lighthill developed these ideas further. At the same time, Rayner published a vortex theory of animal flight in the *Journal of Fluid Mechanics*, **91** (1979) 397-730. This theory described the stack of vortex rings shed by a hovering animal. However, Rayner's vortex-ring model needs extension to take other physical effects into account.

Lighthill's impact on external biofluiddynamics has been most influential. On one hand, he provided his own extensive research and his work in collaboration with others, on the other hand, he provided his leadership role for further collaborative research and developments in mathematical biofluiddynamics. At the same time, he took a major initiative to disseminate the current research, information and knowledge about the subject. For example, his book on *Mathematical Biofluiddynamics* is an excellent collection of not only his own research contributions, and research reviews, but also of research reprints of articles of other experts in biofluiddynamics.

Part II of his book on *Mathematical Biofluiddynamics* has five chapters:

Chapter 10. Physiological fluid dynamics: A general survey,

Chapter 11. Respiratory flow patterns,

Chapter 12. Pulse propagation theory,

Chapter 13. Blood flow and arterial disease,

Chapter 14. The microcirculation.

In these chapters, Lighthill presented not only his own original research and research expository survey, but also described his collaborative research with others and some major work of other experts in the field of physiological fluid dynamics. It it evident from his research, publications and presentations that Lighthill was a strong supporter of interdisciplinary research by fluid dynamics, biologists, physicians and physiologists. Indeed, during his stay at Imperial College he was deeply involved in collaborative research with several colleagues including a physiologist, Dr. Colin Caro and a fluid dynamicist, T. J. Pedley. With his leadership, a new Physiological Flow Studies Unit was established at the Aeronautics Department of the Imperial College in 1966. This unit has now become a part of the Department of Biological and Medical Systems of the Imperial College and has a strong national and international reputation for study and research on blood flow in arteries with special reference to cardiovascular diseases. During his stay in Imperial College, Lighthill delivered also a course of lectures on Blood Flow in Arteries in 1967. This was followed by publication of his two research papers on physiological fluid dynamics in the *Journal of Fluid Mechanics*. One of these was a research expository and survey article, and the other was a mathematical model for the motion of red blood cells in elastic tubes. In closing of Chapter 13 on Blood Flow and Arterial Disease, Lighthill (1975) recorded his deep admiration and great respect for interdisciplinary research as follows:

"One aim in this chapter has been to give readers some idea, however incomplete, of the vast of non-biofluiddynamic factors that enormously complicate any interactions between blood flow and arterial disease. It will be evident that numerous different biological skills, including those of the physician, the physiologist and the biochemist, have to be represented in an effective team conducting research into such problems. A further complication derives from the quite sophisticated biofluiddynamic knowledge and skills that also have been indicated as relevant to such problems: the team needs a biofluiddynamicist too! It may be quite hard to weld together effective interdisciplinary teams to attack problems of arterial disease in relation to blood flow, but the great human importance of mounting such attacks should remind us that the effort needed is indeed worth making."

In 1971, Lighthill delivered an invited lecture on the microcirculation at the International Centre for Mechanical Sciences, Udine, Italy. Chapter 14 of his book (1975) deals with his lecture at Udine. The microcirculation deals with blood flow in vessels of diameter less than 0.1mm (100 microns), and especially with flow in the arterioles, capillaries, and venules with diameter between 5 microns and 50 microns. According to Lighthill, the microcirculation is concerned with quantitative aspects of peripheral resistance, that is, of the dependence of the pressure drop on flow rate. This dependence turns out to be nonlinear. Another essential feature of the microcirculation involves the function of the capillaries in various exchange processes. His lecture was essentially dealt with those special characteristics of the microcirculation originating from the fact that blood is a highly concentrated suspension of red cells in plasma.

After his retirement from the University College as Provost, Lighthill himself returned to further study and research on biofluiddynamics and published a series of papers on balistiform and gymnotiform locomotion. He published one paper on Biofluiddynamics of balistiform and gymnotiform locomotion jointly with his former zoology student, R. W. Blake in 1990. Subsequently, Lighthill (1993) published a paper on Estimates of pressure differences across the head of a swimming clupeid fish in the *Philosophical Transactions of the Royal Society, London*, **341B** (1993) 129-140. His last paper dealing with fish swimming on the role of the lateral line in active control of drag reduction by clupeoid fishes was published in a volume *Biological Fluid Dynamics* edited by C. P. Ellington and T. J. Pedley, (*Soc. Exp. Biol. Symp.* **49** (1995) 35-48).

4.4 *Waves in Fluids*, Cambridge University Press, Cambridge, 1978

Throughout his whole life, Sir James spent considerable amount of time and effort for the modern development of an important body of fundamental ideas of general application of wave motions in fluids with their diverse applications. Much of the modern work has been made possible by his remarkable advances in theoretical, experimental and computational methods. The 1978 year saw another milestone with his publication of his modern comprehensive treatise entitled, *Waves in Fluids* published by Cambridge University Press. The study of wave motions in fluids (liquids or gases) is of major importance in almost all branches of continuum mechanics, science and engineering. It is not only of direct interest to fluid dynamics, but is indispensable for all modern advances in aerodynamics, meteorology, oceanography and biofluiddynamics. Sir James wrote this book with emphasis on mathematical theory, but his presentation is clearly tied to physical observations and data — giving a wealth of experimental results and carefully describing methods and solutions. It has become a major source of reference for many subsequent investigations of wave motions in fluids, continuing for several decades afterwards. It apparently pleased Sir James when a question was asked by Professor T. J. Pedley (2001) of Cambridge University about his stimulating book: "When asked if he thought *Waves in Fluids* would be found useful in the future, Lighthill replied that Lord Rayleigh's *Theory of Sound* was still a mine of information and understanding one hundred years after its publication, and he did rather see himself as a modern Lord Rayleigh. Why not?"

In this major graduate level textbook and research reference he states in his Prologue:

"This book is designed as a comprehensive introduction to the science of wave motions in fluids (that is, in liquids and gases); an area of knowledge which forms an essential part of the dynamics of fluids, as well as a significant part of general wave science; and, also, has important applications to the sciences of the environment and of engineering. The subject's extent and variety are enormous: the different types of waves in fluids, the different fundamental ideas that have been developed to interpret their properties, and the different applications of those properties are so extensive that a comprehensive introduction in one volume demands very careful selection.

The design adopted for the book, in four chapters and an epilogue, has two principal aims. First, as the chapter titles suggest, it allows an analysis in depth of four important and representative types of waves in fluids (sound waves; one-dimensional waves in fluids; water waves; internal waves) to precede brief descriptions of some other important types in the epilogue. At the same time, the subject matter of the four chapters is chosen so that, as far as possible, all the most generally useful fundamental ideas of the science of waves in fluids can be developed at length, one after another. The main exceptions are certain very difficult, advanced ideas which could not, even in a comprehensive introduction, be treated so fully; they are merely sketched, with references to more extensive treatments, in the epilogue."

Included are an interesting Prologue, an Epilogue, and a large Bibliography with commentary. His Prologue gives the reader a detailed description of four chapters, while his Epilogue contains a substantial amount of additional material on other types of linear and nonlinear waves in fluids, combined with a short sketch of some more recent developments. An extensive annotated bibliography with a large selection of major references seems to be very useful to the reader. In his comprehensive book, Lighthill provided the reader with a thorough mathematical analysis of the most important and representative types of waves including shock waves, sound waves and water waves of all kinds, and the so-called inertial and internal waves due to density stratification in oceans and the atmosphere. In the cover page of his book, Lighthill writes: "... book will be invaluable to engineers, physicists, geophysicists, applied mathematicians or any research worker concerned with wave motions or fluid flows."

Indeed, almost half of this book deals with water waves which represent one of the most striking observable phenomena in Nature. The subject of water waves was the most favorite subject of Sir James Lighthill as he described in his course of fascinating lectures in Imperial College in 1966 (I was fortunate to attend this course of lectures). According to Lighthill, throughout the history of the study of dispersive wave phenomena, the theory of water waves has provided a real challenge and a source of inspiration for many new theoretical, experimental and computational developments.

This treatise has four large chapters:

Chapter 1. Sound Waves. Taking into account compressibility and inertia of a fluid, an extensive linear theory of sound waves is developed in this chapter. The three-dimensional linear wave equation for the velocity potential is derived. This is followed by a wide variety of plane wave solutions and a clear understanding of highly complex sound-generation mechanisms or sources of sound that can often be built up from the study of a few fundamental solutions of the wave equation. Included is the study of propagation of sound generated by a simple source (or monopole source), a dipole source and a compact source regions in general. Some discussion of sound waves generated by a radially pulsating sphere is made. He also considered another radiation problem amenable to fairly simple analysis at any frequency. This problem deals with fluid with part of a single plane boundary making prescribed small displacements. The analysis of this problem plays an important role from both theoretical and practical view points. The final section of this chapter is concerned with mechanisms how acoustic energy is dissipated into heat and how these modify the linear theory of sound waves.

Chapter 2. One-Dimensional Waves in Fluids. This chapter is devoted to the general theory of longitudinal wave propagation in systems involving uniform lengths of fluid-filled tube or channel separated by compact elements: junctions, cavities, constrictions. Included are branching systems, such as cardiovascular systems or rivers with their tributaries. This is followed by a study of effects of gradual variation of fluid density and compressibility, or of cross-sectional area and distensibility, along a non-compact length of tube or channel. An analysis of dissipation of wave energy and its effects on the above problems is presented in some detail. The nonlinear theory of plane waves, simple waves, shock waves (as effectively a discontinuity) and their structure and thickness, and hydraulic jumps are presented in several sections of this chapter. The nonlinear wave propagation with gradually varying cross-section and composition is treated here by the transformation method. The final section of this chapter deals with the theory of nonlinear geometrical acoustics and its application to estimating nonlinear effects on three-dimensional propagation of weak disturbances.

Chapter 3. Water Waves. This long chapter deals with a general and profound view of the linear theory of surface gravity waves on water. A special attention was given to the dispersive wave propagation in fluids where the wave speed (or phase velocity of waves) depends on the wavelength (or wavenumber), and possibly on the direction of propagation. Based on the classical equations of water waves with the bottom and free surface boundary conditions, the linear theory of sinusoidal waves on deep water and on water of arbitrary, but constant, depth, h are presented in great detail. The velocity potential and the free surface elevation that represent a sinusoidal wave propagation in the positive x-direction with wave speed $c = (\omega/k)$ are given by

$$\phi(x,z,t) = \operatorname{Re}\left(\frac{iag}{\omega}\right)\frac{\cosh k(z+h)}{\cosh kh}\exp[i(\omega t - kx)], \tag{4.4.1}$$

$$\zeta(x,t) = \operatorname{Re} a \exp\left[i(\omega t - kx)\right], \tag{4.4.2}$$

where $a = \left(\frac{\omega A}{ig}\right)\cosh kh$, is the wave amplitude and A is an arbitrary constant and the dispersion relation between the frequency ω and wavenumber k for surface gravity waves on water of depth h is given by

$$\omega^2 = gk\tanh kh. \tag{4.4.3}$$

The wave speed $c(k)$ is given by

$$c(k) = \left(\frac{\omega}{k}\right) = (gk^{-1}\tanh kh)^{\frac{1}{2}} = \left[\left(\frac{g\lambda}{2\pi}\right)\tanh\left(\frac{2\pi h}{\lambda}\right)\right]^{\frac{1}{2}}. \tag{4.4.4}$$

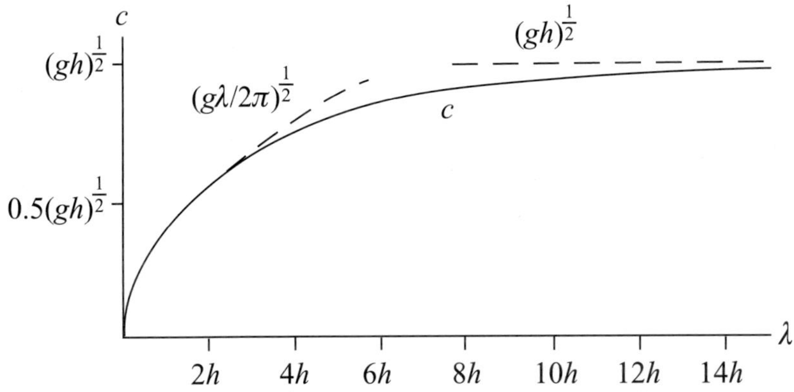

Fig. 4.2 Plots of wave speed $c(\lambda)$ against the variable wavelengths λ showing the transition between deep water value $(g\lambda/2\pi)^{\frac{1}{2}}$ and the shallow water value $(gh)^{\frac{1}{2}}$.

This result is in perfect agreement with both the limiting forms, $\sqrt{g/k}$ in the deep-water limit ($kh \gg 1$) and $\sqrt{gh}$ in the shallow-water limit ($kh \ll 1$) because $\tanh kh \to 1$ as $kh \gg 1$ and $\tanh kh \to kh$ as $kh \ll 1$. Figure 4.2 shows the plot of $c(\lambda)$ against λ according to (4.4.4).

Indeed, the wave speed c for water waves of frequency ω on gradually varying depth h exhibits the transition between the shallow-water value $\sqrt{gh}$ and the deep-water value (g/ω). The dispersion relation (4.4.4) with k replaced by $\frac{\omega}{c}$ can be written in terms of c

$$c(k) = \left(\frac{g}{\omega}\right) \tanh\left(\frac{\omega h}{c}\right). \tag{4.4.5}$$

This determines how the wave speed of $c(\lambda)$ varies for fixed frequency ω as h changes gradually as shown in Figure 4.3.

Lighthill gave a detailed mathematical and physical introduction to group velocity which is different from the wave speed c in any dispersive system, where $c(k)$ varies with wavelength (or wavenumber). Three sections of the book deal with the general theory of group velocity for *isotropic* (c does not depend on direction) dispersive systems. In one-dimension, the group velocity $U(k)$ is defined by $(d\omega/dk)$ and represents the velocity of energy propagation of wavenumber k. This interpretation depends on the idea that, in a system satisfying linear partial differential equations with constant coefficients, there can be no transfer of energy from one part of the spectrum to another. In case of surface gravity waves on water of uniform

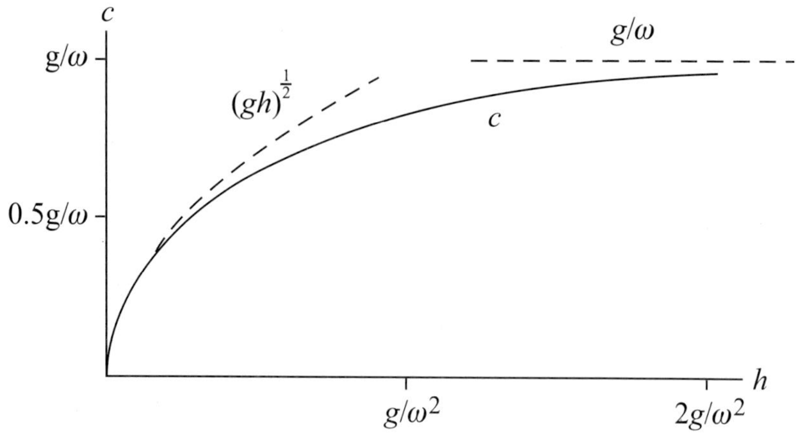

Fig. 4.3 Plot of the wave speed c for waves of frequency ω in water of variable depth h showing the transition between the shallow water speed $(gh)^{\frac{1}{2}}$ and the deep water speed (g/ω).

deep h, the group velocity is given by

$$U(k) = \frac{d\omega}{dk} = \left(\frac{g}{2\omega}\right)(\tanh kh + kh\,\mathrm{sech}^2\, kh), \tag{4.4.6}$$

which can be expressed in terms of c as

$$U(k) = \frac{1}{2}c\,[1 + 2kh\ \mathrm{cosech}(2kh)]\,. \tag{4.4.7}$$

Evidently, the group velocity, $U(k)$ is, in general, different from the wave velocity.

In the special case of deep water waves,

$$U = \frac{d\omega}{dk} = \frac{d}{dk}\left(\sqrt{gk}\right) = \frac{1}{2}\left(\frac{g}{k}\right)^{\frac{1}{2}} = \frac{1}{2}\left(\frac{\omega}{k}\right) = \frac{1}{2}\,c. \tag{4.4.8}$$

Thus, the group velocity is half of the wave velocity. This is a remarkable simple result in the theory of deep water waves.

In shallow water of depth h, $\omega^2 = (gh)k^2$ and the wave and group velocities are

$$c = \frac{\omega}{k} = \sqrt{gh} \qquad \text{and} \qquad U = \frac{d\omega}{dk} = \sqrt{gh}. \tag{4.4.9}$$

In this shallow water case, the wave velocity does not depend on wavenumber k and so, the shallow water waves are *nondispersive*. Both the group velocity and the wave velocity are the same.

In general, for any nondispersive waves, the wave velocity c is independent of wavenumber k. Thus, $\omega = ck$ is a linear function of k and so

$$U = \frac{d\omega}{dk} = \frac{d}{dk}(ck) = c. \tag{4.4.10}$$

Thus, nondispersive waves include not only shallow water (long) waves on water of uniform depth h, but also sound waves.

Another formula of the group velocity can be obtained by rewriting the general expression for U in terms of the linear relation between ω $(= ck)$ and k $\left(= \frac{2\pi}{\lambda}\right)$ as

$$U(k) = \frac{d\omega}{dk} = \frac{d}{dk}(ck) = c + k\frac{dc}{dk} = c - \lambda\frac{dc}{d\lambda}. \tag{4.4.11}$$

This shows that $U > c$ for any wave for which $c(k)$ increases as k increases.

Using the dispersion relation of water waves on water of uniform depth h_0, $\omega = (gk \tanh kh_0)^{\frac{1}{2}}$, we can write the approximate values of the wave velocity and the group velocity in the form

$$c(k) = \frac{\omega}{k} = \sqrt{\frac{g}{k}}(\tanh kh_0)^{\frac{1}{2}} = c_0\left(1 - \frac{1}{3}k^2h_0^2\right)^{\frac{1}{2}} \simeq c_0\left(1 - \frac{1}{6}k^2h_0^2\right) \tag{4.4.12}$$

$$U(k) = c + k\frac{dc}{dk} \simeq c_0\left(1 - \frac{1}{2}k^2h_0^2\right), \qquad c_0 = \sqrt{gh_0}. \tag{4.4.13}$$

This shows that the energy of water waves propagates just little slower than the wave crests in these slightly dispersive waves. However, observations reveal that the real situation is more complicated that this. Lighthill indicated that a traveling discontinuous increase in depth of water (or *hydraulic jump*) can drag along behind it a regular wavetrain with crests that propagate at the same velocity, where its energy moves more slowly and it is gradually carried backward relative to the jump and hence, there is a complete loss of energy. This fact is supported by observations of a hydraulic jump (see Figure 48 of Lighthill's book, page 181) on the river Severn in England. Turbulent bores with violently foaming front (see Figure 49 of Lighthill's book, page 182) are also observed in another part of the same river Severn. In spite of all complexities associated with undulation and turbulence as methods of energy removal from a hydraulic jump, it remains valuable for many purposes of approximate analysis to treat the jump as effectively as a discontinuity.

In general, for a large class of wave motions observed in nature, the wave velocity at which crests and troughs are propagated is quite different from the group velocity at which wave energy is propagated. In many dispersive wave systems, the sinusoidal waves of each wavenumber have a definite wave velocity (though not the same for each), the *Fourier analysis* can effectively be used to predict the development of disturbances of arbitrary shape. Such disturbances can be analysed, in fact, into a linear combination of sinusoidal waves. We consider a more general combination of waves of different wave lengths in a homogeneous system, where $\omega = \omega(k)$, such a linear combination is

$$\zeta(x,t) = \int_{-\infty}^{\infty} F(k) \exp\left[i\{kx - \omega(k)t\}\right] dk, \tag{4.4.14}$$

where the initial form of the wave at $t = 0$ is

$$f(x) = \zeta(x,0) = \int_{-\infty}^{\infty} F(k) e^{ikx} dk, \tag{4.4.15}$$

which represents a disturbance $f(x)$ with compact support.

The asymptotic evaluation of the integral for $\zeta(x,t)$ when t is large and x may also be large establishes rigorously the properties of the group velocity. In fact, the asymptotic form of $\zeta(x,t)$ for large t is given by

$$\zeta(x,t) \approx F(k_0) \left[2\pi/t|\omega''(k_0)|\right]^{\frac{1}{2}} \exp\left[i\{t\omega(k_0) - k_0 x + \frac{\pi}{4}\, sgn\, \omega''(k_0)\}\right], \tag{4.4.16}$$

where the wavenumber k_0 satisfies the equation

$$\omega'(k_0) = \frac{x}{t}. \tag{4.4.17}$$

Thus, waves of wavenumber k_0 are found at positions moving forward at the group velocity $\omega'(k_0)$. The amplitude of the waves generated is given by

$$A(x,t) = |F(k_0)| \left[2\pi/t|\omega''(k_0)|\right]^{\frac{1}{2}}, \tag{4.4.18}$$

where their phases θ becomes

$$\theta(x,t) = t\omega(k_0) - k_0 x + arg\, F(k_0) + \frac{\pi}{4}\, sgn\, \omega''(k_0). \tag{4.4.19}$$

The phase function $\theta(x,t)$ satisfies the following relations

$$\frac{\partial \theta}{\partial x} = -k_0 \quad \text{and} \quad \frac{\partial \theta}{\partial t} = \omega(k_0). \tag{4.4.20}$$

Based on the asymptotic representation of the isotropic dispersive wave solution, Lighthill expressed the asymptotic solution of the water-surface displacement function $\zeta(x,t)$ in the form of the elementary plane wave solution

$$\zeta(x,t) = A(x,t) \exp\left[i\theta(x,t)\right], \tag{4.4.21}$$

where $A(x,t)$ is the slowly varying amplitude function and $\theta(x,t)$ is a smoothly varying *phase function* that satisfies the equation

$$\frac{\partial\theta}{\partial x} = -k \quad \text{and} \quad \frac{\partial\theta}{\partial t} = \omega. \tag{4.4.22}$$

This phase function is approximately linear at a particular position x_0 and at particular time t_0 so that

$$\theta(x,t) \approx \theta(x_0,t_0) - k_0(x-x_0) + \omega_0(t-t_0), \tag{4.4.23}$$

where $-k_0$ and ω_0 are the values of $\frac{\partial\theta}{\partial x}$ and $\frac{\partial\theta}{\partial t}$ respectively at (x_0,t_0).

Equation (4.4.21) with $A(x,t)$ slowly varying function makes $\zeta(x,t)$ locally a sinusoidal wave with wavenumber k_0 and frequency ω_0. Evidently, eliminating θ from (4.4.22) yields a partial differential equation

$$\frac{\partial k}{\partial t} + \frac{\partial\omega}{\partial x} = 0 \tag{4.4.24}$$

satisfied by the wavenumber and the frequency. This can be thought of as an equation of continuity for the phase function where k represents a phase density (phase per unit length) and ω is a phase flux (phase passing a fixed point in unit time).

For isotropic dispersive waves, $\omega(= kc(k))$ is a nonlinear function of k so that

$$\omega = \omega(k) \tag{4.4.25}$$

which is assumed to hold for nearly sinusoidal waves.

Substituting (4.4.25) into (4.4.24) gives

$$\frac{\partial k}{\partial t} + U(k)\frac{\partial k}{\partial x} = 0, \tag{4.4.26}$$

where $U = U(k) = (\frac{d\omega}{dk})$ is the *group velocity*. Equation (4.4.26) implies that k is constant along paths in the (x,t) plane satisfying $\frac{d\omega}{dk} = U$, but if k is constant, then U is constant so that these paths are straight lines $x - Ut = \text{constant}$.

Substituting $\omega = kc$ (c is the wave speed) transforms (4.4.24) into the standard hydrodynamical continuity equation for longitudinal motions with $k = \rho A$ as the mass per unit length and $\omega = ku = k\rho A$ as the mass flow rate.

The dispersion relation (4.4.3) is obtained for water of constant depth. However, if $h = h(x)$ so that the depth of water changes only in the direction normal to wave crests. Consequently, this case corresponds to a *nonhomogeneous dispersion relation* in the form

$$\omega = \omega(k,x). \tag{4.4.27}$$

This allows us to define an associated group velocity for fixed x by

$$U(k,x) = \frac{\partial}{\partial k}\,\omega(k,x). \tag{4.4.28}$$

It is evident that equation (4.4.26) cannot be deduced from (4.4.24). However, multiplying (4.4.24) by U and using the fact that

$$U\,\frac{\partial k}{\partial t} = \frac{\partial \omega}{\partial k}\cdot\frac{\partial k}{\partial t} = \frac{\partial \omega}{\partial t} \tag{4.4.29}$$

and keeping x fixed, the resulting equation (4.4.24) reduces to the form

$$\frac{\partial \omega}{\partial t} + U\,\frac{\partial \omega}{\partial x} = 0. \tag{4.4.30}$$

This equation implies that ω is constant along paths in the (x,t) plane satisfying

$$\frac{dx}{dt} = U(k,x). \tag{4.4.31}$$

So, in this nonhomogeneous case, these paths are in general, *not straight lines* because neither U nor k is constant along a path.

Another class of water waves problems deal with the propagation of dispersive waves generated by unsteady motions of solid objects immersed in water or by various unsteady disturbances at the water surface including storms and entry of solid body. However, we next consider the generation of stationary waves in a stream of velocity V by a nonzero boundary value η independent of time so that

$$\eta = Vf(x) = V\int_{-\infty}^{\infty} F(k)e^{ikx}dk. \tag{4.4.32}$$

This more general boundary value (4.4.32) generates a wave motion that can be represented in the form

$$\zeta = \int_{-\infty}^{\infty} \frac{V\,F(k)\,e^{ikx}dk}{D(kV,k)}, \tag{4.4.33}$$

where the dispersion relation with $\omega = kV$ is

$$D(\omega,k) = 0. \tag{4.4.34}$$

This kind of problem is generally known as *steady-state* or *stationary* wave problems. In general, it is not possible to determine a unique solution of physical interest because there is an inherent difficulty associated with the problem, that is, the general solution (4.4.34) has real polar singularities on the path of integration with respect to k (the real axis). In order to overcome this difficulty, several mathematical methods have been employed for the treatment of the steady-state problem, all of which are aimed at deriving a unique solution of physical interest by introducing some effective artificial device.

In order to derive a unique solution of physical interest, namely to find the wave solution truely determined by the applied disturbance itself, the problem is solved subject to what usually called the *Sommerfeld Radiation Condition* which simply states that no waves are generated at infinity. More precisely, this condition strictly excludes the possibility of any waves crossing the field from one side to the other and of any incoming waves generated at infinity. In other words, the waves exclusively behave like outgoing progressive waves propagating away from the seat of disturbance.

As an alternative to the radiation condition, Lighthill successfully developed a fairly general method that requires the substitution of $\omega = kV - i\varepsilon$ for the given frequency ω in the Fourier integral solution, where ε is a small positive quantity that is later allowed to tend to zero. This substitution moves the poles off the real k-axis of integration, and the resulting k-integral can then be evaluated unambigously with the aid of the Cauchy residue theory. Mathematically, the residue - contribution to the integral is made up of the poles provided the group velocity is positive. Ultimately, the limit of the resulting solution as $\varepsilon \to 0$ would lead to the unique solution of physical interest. Physically, Lighthill's procedure ensures the unique solution satisfying the condition for outward propagation of energy. Thus, it excludes the possibility of any incoming waves generated at infinity. It is worth noting that Lighthill's method, an alternative way of applying the radiation condition, is fairly general and also very useful for higher-dimensional wave problems.

Using the Lighthill method, solution (4.4.33) can be expressed as

$$\zeta = e^{\varepsilon t} \int_{-\infty}^{\infty} \frac{V\, F(k)\, e^{ikx} dk}{D(kV - i\varepsilon, k)}. \tag{4.4.35}$$

The integral can be evaluated unambiguously because the poles are off the path of integration and the final answer follows from the limit as $\varepsilon \to 0$.

An alternative to Lighthill's method, there are three other methods available in the literature of wave propagation problems. One was due to Stoker (1957) who employed a complicated complex variable method combined with the radiation condition at infinity. Another equivalent formulation of the steady-state problem was introduced in 1904 by Lamb (1849-1934) who had assumed a fictitious dissipative force proportional to velocity ($\mu\phi_z$) due to Lord Rayleigh (1842-1919). Such a small dissipation is always present in a real physical system and has the real ability to move the poles off the real axis of integration so that the wavenumber for free waves of fixed real frequency becomes complex. So the Fourier integral can then be evaluated with the Cauchy theory of residues. Thus, the ultimate steady-state wave motions can be obtained in the limit as $\mu \to 0$. The use of the Fourier analysis of generalized functions developed by Debnath (1994) and Debnath and Rosenblat (1969) turns out to be useful for the study of surface gravity wave problems in water of arbitrary, but uniform depth.

Finally, Lighthill concluded this chapter by adding a section on ship waves which deals with wave pattern generated by a ship moving at a constant velocity V over some large expanse of still water. The nature of a ship-wave pattern depending on the Froude number $F = (V^2/gh)$ is of great technological importance because P_w, the power required to generate the wave pattern carried along by the ship, can be a significant fraction of the total output of the ship's engines. Some general features of P_w including methods for determining it by ship model experiments are discussed in some detail. Lighthill then closed this section by drawing the crest shapes in the Kelvin ship-wave pattern and the boundary of Kelvin wedge as shown in Figure 4.4. All the crest shapes have cusps on the wedge boundary. He discussed methods of analysis how the Kelvin ship-wave pattern is modified by shallow-water effects that allow the Kelvin wedge to become wider.

Chapter 4. Internal Waves deal with gravity waves in continuously stratified fluids, that is, fluids with continuously varying density $\rho_0(z)$. Based on the Boussinesq approximation, the equation for $q = \rho_0(z)w$ as

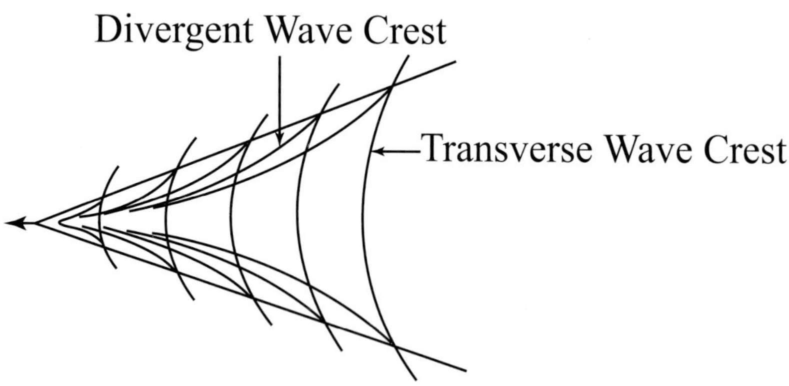

Fig. 4.4 The Kelvin ship-wave pattern, and the boundary of Kelvin wedge.

the upward mass flux (z-component of $\rho_0\mathbf{u}$) is given by

$$\frac{\partial^2}{\partial t^2}(\nabla^2 q) + N^2(z)\nabla_H^2 q = 0, \tag{4.4.36}$$

where ∇^2 is the three-dimensional Laplacian, ∇_H^2 is the two-dimensional (horizontal) Laplacian and $N(z)$ is the Brunt-Väisälä frequency defined by

$$N(z) = \left(-\frac{g}{\rho_0}\frac{d\rho_0}{dz}\right)^{\frac{1}{2}}. \tag{4.4.37}$$

For constant $N(z)$, the equation for q has a simple plane wave solution

$$q = Q(z)\exp\left[i(\boldsymbol{\kappa}\cdot\mathbf{x} - \omega t)\right], \tag{4.4.38}$$

provided the following dispersion relation is satisfied:

$$\omega^2 = N^2(k^2 + \ell^2)/(k^2 + \ell^2 + m^2) = N^2\cos^2\theta, \tag{4.4.39}$$

where $\boldsymbol{\kappa} = (k, \ell, m)$ is the wavenumber vector, $\mathbf{x} = (x, y, z)$, and ω is the frequency. The dispersion relation shows that the internal waves can have any frequency between zero and a maximum value N. The plane waves involves unidirectional oscillations of fluid at an angle θ to the vertical so that (4.4.39) leads to

$$\omega = N\cos\theta. \tag{4.4.40}$$

There is a striking contrast between the surface gravity waves and internal waves. The frequency of the former depends only on the magnitude κ of the wavenumber vector, whereas the frequency of the latter is independent of the magnitude of the wavenumber vector and depends on the angle θ that the wavenumber vector makes with the vertical.

Lighthill then derived the equations for the combined propagation of sound and internal waves in a stratified fluid. All oceanic internal waves are completely decoupled from sound waves, and $N(z)$ can also be viewed as a maximum possible frequency of internal waves. Equation (4.4.36) admits plane wave solutions in the x-direction as

$$q = Q(z)\exp[i(\omega t - kx)] \tag{4.4.41}$$

provided that $Q(z)$ satisfies the ordinary differential equation

$$\omega^2 Q''(z) + k^2[N^2(z) - \omega^2]Q(z) = 0. \tag{4.4.42}$$

This equation where $N(z) > \omega$ in just one interval of values of z, possesses trapped wave solutions for values of k belonging to certain increasing wavenumber sequence k_0, k_1, k_2, $\cdots$.

The maximum horizontal wave speed (ω/k) is found with the smallest wavenumber $k_0 = k_0(\omega)$. The corresponding vertical mass flow, $Q_0(z)$, is positive for all z. This implies that the entire thermocline region rises and falls in phase, it is said to make *sinuous oscillations* since k_0 varies with ω, these horizontally travelling waves, therefore, show isotropic dispersion with a group velocity $U = [k_0'(\omega)]^{-1}$. All slower oscillations with k_1, k_2, k_3, $\cdots$ are more complicated. For each of these, the group velocity is $[k_n'(\omega)]^{-1}$.

An analysis similar to that of one-dimensional case can be generalized for three-dimensional situation by writing a physical quantity q in the waves in the form

$$q = Q(\mathbf{x}, t) \exp[i\theta(\mathbf{x}, t)], \tag{4.4.43}$$

where $Q(\mathbf{x}, t)$ is a slowly varying amplitude and $\theta(\mathbf{x}, t)$ is the phase function. At a fixed point $\mathbf{x} = (x, y, z)$, equation (4.4.43) requires that the phase θ increases by 2π in one wave period so that $\left(\frac{\partial \theta}{\partial t}\right)$ can be considered as the local frequency ω. Similarly, at a given time, components of the local wavenumber vectors $\boldsymbol{\kappa} = (k, \ell, m)$ are defined by

$$\frac{\partial \theta}{\partial x} = -\,k, \qquad \frac{\partial \theta}{\partial y} = -\,\ell, \qquad \frac{\partial \theta}{\partial z} = -\,m. \tag{4.4.44}$$

These equations for the phase $\theta(\mathbf{x}, t)$ with $\frac{\partial \theta}{\partial t} = \omega$ ensure that locally the wave is nearly of the sinusoidal form (4.4.38). Indeed, at a particular position $\mathbf{x}_0$ and a particular time t_0, the phase function $\theta(\mathbf{x}, t)$ is approximately linear so that the quantity q is proportional to $\exp[i(\omega t_0 - \boldsymbol{\kappa}_0 \cdot \mathbf{x}_0)]$ as in sinusoidal plane wave solution.

For internal waves, the group velocity vector is defined by

$$\mathbf{U}(\boldsymbol{\kappa}) = \left(\frac{\partial \omega}{\partial k}, \frac{\partial \omega}{\partial \ell}, \frac{\partial \omega}{\partial m}\right). \tag{4.4.45}$$

In fact, $\mathbf{U}$ is the gradient $(\nabla_\kappa\, \omega)$ of the frequency in the wavenumber space $\boldsymbol{\kappa} = (k, \ell, m)$. For example, for the dispersion relation (4.4.39), the group velocity vector is

$$\mathbf{U}(\boldsymbol{\kappa}) = \left(\frac{Nm}{\kappa^2}\right)\left[\frac{km, \ell m, -(k^2 + \ell^2)}{\kappa(k^2 + \ell^2)^{\frac{1}{2}}}\right], \quad \kappa^2 = k^2 + \ell^2 + m^2. \tag{4.4.46}$$

Obviously, the vector inside the square bracket is a *unit vector* normal to $\boldsymbol{\kappa} = (k, \ell, m)$ and $\mathbf{U}(\boldsymbol{\kappa})$ represents the velocity of energy propagation for sinusoidal internal waves.

Following the above ideas, it is possible to describe the general theory of ray tracing in nonhomogeneous anisotropic wave systems. In a homogeneous system, we can use (4.4.43), in the suffix notation, in the form

$$q = Q(x_1, x_2, x_3, t) \exp\left(i\theta(x_1, x_2, x_3, t)\right], \tag{4.4.47}$$

which is locally sinusoidal to good approximation. The phase function θ satisfies

$$\frac{\partial\theta}{\partial x_i} = -k_i, \qquad \frac{\partial\theta}{\partial t} = \omega, \tag{4.4.48}$$

the local frequency, ω and the local wavenumbers, k_i are connected by a dispersion relation of the form

$$\omega = \omega(k_1, k_2, k_3). \tag{4.4.49}$$

For a nonhomogeneous wave system, the dispersion relation can be written in the form

$$\omega = \omega(k_1, k_2, k_3, x_1, x_2, x_3), \tag{4.4.50}$$

and the energy propagation velocity has the approximate form, similar to sinusoidal waves,

$$U_j(k_j) = \frac{\partial\omega}{\partial k_j}. \tag{4.4.51}$$

In a nonhomogeneous system, it is expected that wave energy propagated at the group velocity (4.4.51) suffers from *refraction*, that is, changes of wavenumber due to the nonhomogenity. In view of (4.4.48), the dispersion relation (4.4.50) takes the form

$$\frac{\partial\theta}{\partial t} = \omega\left(-\frac{\partial\theta}{\partial x_1}, -\frac{\partial\theta}{\partial x_2}, -\frac{\partial\theta}{\partial x_3}, x_1, x_2, x_3\right). \tag{4.4.52}$$

Differentiating (4.4.52) with respect to x_i gives

$$\frac{\partial^2\theta}{\partial x_i \partial t} = \left(\frac{\partial\omega}{\partial k_j}\right)\left(-\frac{\partial^2\theta}{\partial x_i \partial x_j}\right) + \frac{\partial\omega}{\partial x_i}, \tag{4.4.53}$$

where the term $\frac{\partial\omega}{\partial x_i}$ arises due to fact that ω depends on x_i for constant k_i.

In view of (4.4.48) and (4.4.51), equation (4.4.53) assumes the form

$$\frac{\partial k_i}{\partial t} + U_j\frac{\partial k_i}{\partial x_j} = -\frac{\partial\omega}{\partial x_i}. \tag{4.4.54}$$

Obviously, the left hand side of (4.4.54) represents the rate of change of k_i with time t at a position propagating with the group velocity U_j. Thus, changes that satisfy

$$\frac{dx_j}{dt} = U_j \tag{4.4.55}$$

lead to the equation

$$\frac{dk_i}{dt} = -\frac{\partial\omega}{\partial x_i}. \tag{4.4.56}$$

This represents the rate of change of wavenumber along paths (4.4.55) traversed at the energy propagation velocity U_j, that is, along *rays*. Equation (4.4.56) describes the *refraction of wave energy*.

Lighthill also developed three dimensional Fourier analysis with the stationary phase approximation to determine the asymptotic development of waves generated by an initial disturbance in an anistropic system subject to linear partial differential equations. He considered the general problem due to an oscillating source of frequency ω whose spatial distribution is given by

$$f(x_1, x_2, x_3) \exp(i\omega_0 t), \tag{4.4.57}$$

where the Fourier representation of $f(x_1, x_2, x_3)$ over the entire wavenumber space is

$$f(x_1, x_2, x_3) = \int_{-\infty}^{\infty}\int_{-\infty}^{\infty}\int_{-\infty}^{\infty} F(\boldsymbol{\kappa}) \exp(-ik_j x_j) d\boldsymbol{\kappa}, \tag{4.4.58}$$

where $f(x_1, x_2, x_3)$ vanishes outside a limited forcing region around the origin.

Thus, the function $q(\mathbf{x}, t)$ representing the wave motion is given by the Fourier integral

$$q(\mathbf{x}, t) = \exp(i\omega_0 t) \int_{-\infty}^{\infty}\int_{-\infty}^{\infty}\int_{-\infty}^{\infty} \frac{F(\boldsymbol{\kappa}) \exp(-ik_j x_j)}{B(\omega_0,\ \boldsymbol{\kappa})} d\boldsymbol{\kappa}, \tag{4.4.59}$$

where the equation $B(\omega_0,\ \boldsymbol{\kappa}) = 0$ represents the related dispersion relation for the waves.

In particular, for the two-dimensional water waves with oscillations of the form

$$q(\mathbf{x}, t) = a\ \exp\left[i(\omega t - \boldsymbol{\kappa} \cdot \mathbf{x})\right], \tag{4.4.60}$$

where $\mathbf{x} = (x, y)$ and $\boldsymbol{\kappa} = (k, \ell)$, a boundary value $\eta(\mathbf{x}, t)$ on the boundary plane z =constant can be calculated so that

$$\eta(\mathbf{x}, t) = a\ B(\omega,\ \boldsymbol{\kappa}) \exp\left[i(\omega t - \boldsymbol{\kappa} \cdot \mathbf{x})\right]. \tag{4.4.61}$$

If the undisturbed dispersive system is defined by the boundary condition $\eta = 0$, the dispersion relation is

$$B(\omega, k, \ell) = 0. \tag{4.4.62}$$

On the other hand, an oscillatory boundary disturbance in the form

$$\eta = f(x, y) \exp(i\omega_0 t) = \exp(i\omega_0 t) \int_{-\infty}^{\infty}\int_{-\infty}^{\infty} F(\boldsymbol{\kappa}) \exp(-i\boldsymbol{\kappa} \cdot \mathbf{x}) d\boldsymbol{\kappa}, \tag{4.4.63}$$

would generate the wave solution in the form

$$q(\mathbf{x}, t) = \exp(i\omega_0 t) \int_{-\infty}^{\infty} \int_{-\infty}^{\infty} \frac{F(\boldsymbol{\kappa})}{B(\omega_0,\ \boldsymbol{\kappa})} \exp(-i\boldsymbol{\kappa} \cdot \mathbf{x}) d\boldsymbol{\kappa}. \tag{4.4.64}$$

The propagation of three-dimensional dispersive wave generated by forcing effects, that may be steady, oscillatory or transient in character, is usually described by a partial differential equation with constant coefficients

$$B\left(-i\,\frac{\partial}{\partial t},\ i\,\frac{\partial}{\partial x},\ i\,\frac{\partial}{\partial y},\ i\,\frac{\partial}{\partial z}\right) q = f(x, y, z) \exp(i\omega_0 t), \tag{4.4.65}$$

where B is a polynomial in four variables and q represents a small disturbance to the stable undisturbed state. The free oscillations satisfy the partial differential equations (4.4.65) with $f(x, y, z) \equiv 0$, and the corresponding plane wave solution is

$$q(\mathbf{x}, t) = a\ \exp\left[i(\omega t - \boldsymbol{\kappa} \cdot \mathbf{x})\right], \tag{4.4.66}$$

provided the dispersion relation

$$B(\omega,\ \boldsymbol{\kappa}) = 0 \tag{4.4.67}$$

is satisfied for $\omega = \omega_0$. Since the undisturbed state is stable, no solution of (4.4.67) exists with $\boldsymbol{\kappa}$ real and the imaginary part of ω positive.

Therefore, the mathematical problem described by (4.4.64) or (4.4.65) represents the unique solution of physical interest provided the radiation condition at infinity is satisfied. This means that the problem can be solved with ω_0 replaced by $\omega_0 - i\varepsilon$ where ε is a small positive quantity which is allowed to tend to zero. So, the limit of the resulting answer as $\varepsilon \to 0$ leads to the wave solution of physical interest which satisfies the *radiation condition* at infinity, that is, waves exclusively behave like outgoing waves propagating from the source.

Similarly, the general theory of internal waves generated by an oscillating source can be solved without any further difficulty. In fact, the equation

$$\nabla^2(q_{tt}) + N^2(z)\left(\nabla_H^2\, q\right) = \frac{\partial^3 Q}{\partial^2 t\, \partial z}, \tag{4.4.68}$$

describes internal waves generated by the source distribution $\left(\frac{\partial Q}{\partial t}\right)$ which represents the rate of change of mass outflow rate per unit volume.

In particular, when $N(z)$ is constant and the source distribution independent of y in the form

$$\frac{\partial Q}{\partial t} = f(x, z) \exp(i\omega_0 t), \tag{4.4.69}$$

equation (4.4.68) becomes

$$\left(N^2 - \omega_0^2\right) q_{xx} - \omega_0^2 q_{zz} = i\omega_0 \left(\frac{\partial f}{\partial z}\right) \exp(i\omega_0 t). \tag{4.4.70}$$

This is a special case of (4.4.65) with the dispersion relation ($\ell = 0$)

$$0 = B(\omega, k, m) = \left[\omega^2 m^2 - (N^2 - \omega^2)k^2\right] (\omega m)^{-1}. \tag{4.4.71}$$

Or,

$$\omega^2 = \left[N^2 k^2 / \left(k^2 + m^2\right)\right]. \tag{4.4.72}$$

This means that the wavenumber curve $\omega = \omega_0$ represents a pair of straight lines in the (k, m) plane, at an angle $\theta = \cos^{-1}(\omega_0/N)$ to the k-axis.

On the other hand, three-dimensional internal waves in stratified fluids generated by an oscillating source

$$\frac{\partial Q}{\partial t} = f(x, y, z) \exp(i\omega_0 t), \tag{4.4.73}$$

are governed by equation (4.4.68). When $N(z)$ is constant, equation (4.4.68) becomes

$$\left(N^2 - \omega_0^2\right) \left(\nabla_H^2\, q\right) - \omega_0^2\, q_{zz} = i\, \omega_0 \left(\frac{\partial f}{\partial z}\right) \exp(i\omega_0 t). \tag{4.4.74}$$

The associated dispersion relation is

$$0 = B(\omega, k, \ell, m) = (\omega m)^{-1} \left[\omega^2 m^2 - (N^2 - \omega^2)(k^2 + \ell^2)\right]. \tag{4.4.75}$$

Thus, the wavenumber surface $\omega = \omega_0$ represents a cone with semi-vertical angle $\theta = \cos^{-1}(\omega_0/N)$.

This is followed by a brief discussion of the local behavior of waves near caustics which is a boundary between a region with a wave pattern (due to interference between two groups of waves) and a neighboring region including no waves. It is a relatively easy matter to describe the local behavior of one-dimensional wave near a caustic. It is convenient to recall the asymptotic representation (4.4.16) where the wavenumber $k = k_c$ satisfies $U(k_c) = \omega'(k_c) = \frac{x}{t}$. When $U'(k_c) = \omega''(k_c) = 0$, the asymptotic solution (4.4.16) is no longer valid. This means that the group velocity U is stationary at $k = k_c$. This is what happens with water waves at their minimum group velocity. The path

$$U(k_c) = \frac{x}{t} \tag{4.4.76}$$

carrying waves of wavenumber k_c is a *caustic*, because values of the group velocity U on one side of such a stationary value $U(k_c)$ are found for *two wavenumbers* k, and values on the other side for none.

For asymptotic evaluation for large t, the Fourier wave integral (4.4.14) can be expressed in the form

$$\zeta = \int_0^\infty F(k)\exp[i\,t\,\psi(k)]dk, \tag{4.4.77}$$

where

$$\psi(k) = \omega(k) - \left(\frac{x}{t}\right)k \tag{4.4.78}$$

and $t\,\psi(k)$ has zero first derivative that is $\frac{x}{t} = \omega'(k) = U(k)$.

Since $U'(k_c) = \omega''(k_c) = 0$, $\psi'(k_c) = 0$, and $\psi''(k_c) = 0$. We assume that $\psi'''(k_c) \neq 0$. Consequently, the integral (4.4.77) can be evaluated using the method of steepest descent around $k = k_c$ so that the asymptotic representation of ζ can be expressed in terms of the Airy function, $Ai(z)$. The final asymptotic expansion of (4.4.77) takes the form, as $t \to \infty$,

$$\zeta \sim 2\pi F(k_c)\left\{\exp[i\,t\,\psi(k_c)]\right\}\left[\frac{1}{2}\,t\,\psi'''(k_c)\right]^{-\frac{1}{3}} Ai(X), \tag{4.4.79}$$

where

$$X = \left[\frac{1}{2}\,t\,\psi'''(k_c)\right]^{-\frac{1}{3}} t\,\psi'(k_c),$$

and the integral representation of the Airy function is

$$Ai(X) = \frac{1}{2\pi}\int_{\infty\,\exp(\frac{5\pi i}{6})}^{\infty\,\exp(\frac{\pi i}{6})} \exp\left[i\left(sX + \frac{1}{3}s^3\right)\right] ds, \tag{4.4.80}$$

where $(k - k_c) = s\left[\frac{1}{2}\,t\,\psi'''\,(k_c)\right]^{-\frac{1}{3}}$

For negative X, the asymptotic value of $Ai(X)$ is

$$Ai(X) \sim \frac{1}{\sqrt{\pi}}|X|^{-\frac{1}{4}}\cos\left(\frac{2}{3}|X|^{\frac{3}{2}} - \frac{\pi}{4}\right). \tag{4.4.81}$$

and for positive X, the asymptotic representation of $Ai(X)$ is

$$Ai(X) \sim \frac{1}{2}\,\frac{1}{\sqrt{\pi}}\,X^{-\frac{1}{4}}\exp\left(-\frac{2}{3}\,X^{\frac{3}{2}}\right), \tag{4.4.82}$$

It follows from graphical representation of $Ai(X)$ that it makes a smooth transition from its wavy asymptotic form (4.4.82) for $X < 0$ to its exponentially vanishing value (4.4.81) for $X > 0$ as shown in Figure 4.5.

Finally, we conclude this chapter by adding briefly dispersive waves generated by forcing effects, which may be steady, oscillatory or transient in nature, traveling at a constant velocity **V**. A sinusoidal source of waves

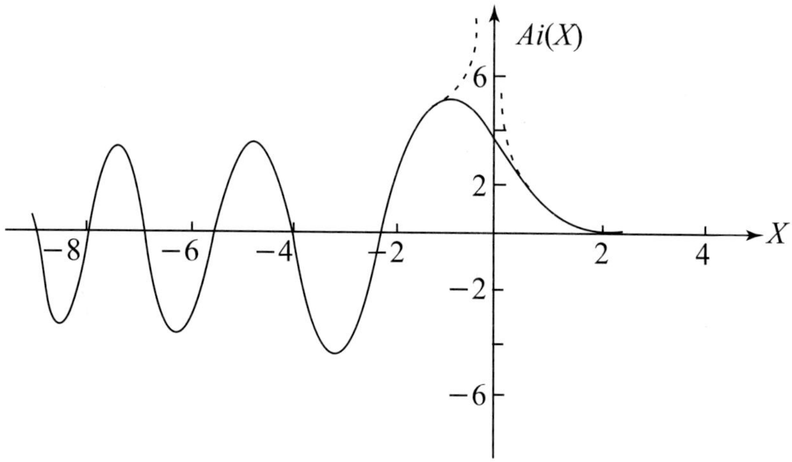

Fig. 4.5 The Airy integral $Ai(X)$ shown solid line and its asymptotic forms (4.4.81) for $X < 0$ and (4.4.82) for $X > 0$ by broken lines.

oscillating with frequency ω_0 and traveling at velocity $(-\mathbf{V})$ can be represented in the form

$$f(\mathbf{x} + \mathbf{V}t)\exp(i\omega_0 t) \tag{4.4.83}$$

which is a generalization of forcing term (4.4.57), where $F(\boldsymbol{\kappa})$ is the Fourier term of $f(\mathbf{x})$ defined by (4.4.58). The partial differential equation

$$B\left(-i\frac{\partial}{\partial t},\ i\frac{\partial}{\partial x},\ i\frac{\partial}{\partial y},\ i\frac{\partial}{\partial z}\right)q = c^{i\omega_0 t} f(\mathbf{x} + \mathbf{V}t), \tag{4.4.84}$$

where (4.4.58) is used to rewrite the right hand side of (4.4.84), admits the formal solution in the form

$$q = \int_{-\infty}^{\infty}\int_{-\infty}^{\infty}\int_{-\infty}^{\infty} \frac{F(\boldsymbol{\kappa})\exp\left[i\left\{\omega_0 t - \boldsymbol{\kappa}\cdot(\mathbf{x} + \mathbf{V}t)\right\}\right]}{B(\omega_0 - \boldsymbol{\kappa}\cdot\mathbf{V},\ \boldsymbol{\kappa})}\, d\boldsymbol{\kappa}, \tag{4.4.85}$$

where the wavenumber surface S (or, in two dimensions, wavenumber curve) is defined as the locus $S(\omega_0)$:

$$B(\omega_0 - \boldsymbol{\kappa}\cdot\mathbf{V},\ \boldsymbol{\kappa}) = 0. \tag{4.4.86}$$

In view of reasons stated before, solution (4.4.85) is not unique. However, the unique solution can be obtained by replacing ω_0 by $\omega_0 + i\varepsilon$ in (4.4.85) and then integral (4.4.85) gives a definite value. In the limit as $\varepsilon \to 0$ of the resulting value leads to the unique solution of physical interest.

In two-dimensional surface gravity waves in deep water with the forced oscillations of the free surface, the elevation function ζ takes the form

$$\zeta(x, y, t) = a \ \exp\left[i(\omega t - kx - \ell y)\right]. \tag{4.4.87}$$

In deep water, the corresponding velocity potential $\phi(x, y, z, t)$ satisfying the Laplace equation and the surface value of $\phi_z = \zeta_t$ is given by

$$\phi(\mathbf{x}, t) = i\,\omega a\,(k^2 + \ell^2)^{-\frac{1}{2}} \ \exp\left[i(\omega t - kx - \ell y) + (k^2 + \ell^2)^{\frac{1}{2}} z\right]. \tag{4.4.88}$$

The boundary value of the water pressure relative to atmospheric pressure, p_a is

$$\eta = (p - p_a) = -\rho_0(\phi_t)_{z=0} - \rho_0 g\zeta \tag{4.4.89}$$

which vanishes for free gravity waves.

On the other hand, for the forced motions (4.4.88), η takes the value

$$\eta = B(\omega, k, \ell)\ a \ \exp\left[i(\omega t - kx - \ell y)\right], \tag{4.4.90}$$

where the dispersion relation is given by

$$0 = B(\omega, k, \ell) \equiv \rho_0 \omega^2 (k^2 + \ell^2)^{-\frac{1}{2}} - \rho_0\, g. \tag{4.4.91}$$

This is identical with the one-dimensional ($\ell = 0$) dispersion relation for surface gravity waves in deep water.

An oscillating source traveling with a velocity $(-\mathbf{V}) = -(V, 0)$ generates a non-zero boundary value

$$\eta(x, y, t) = f(x + Vt,\ y) \exp(i\omega_0 t). \tag{4.4.92}$$

The corresponding wavenumber curve $S(\omega_0)$ is given by

$$B(\omega_0 - \boldsymbol{\kappa}\cdot\mathbf{V},\ \boldsymbol{\kappa}) = \rho_0(\omega_0 - Vk)^2(k^2 + \ell^2)^{-\frac{1}{2}} - \rho_0 g = 0, \tag{4.4.93}$$

so that the wavenumber curve $S(\omega_0)$ can be written as

$$(\omega_0 - Vk)^4 = g^2(k^2 + \ell^2). \tag{4.4.94}$$

These wavenumber curves $S(\omega_0)$ representing surface gravity waves produced by a source oscillating with frequency ω_0 and traveling to the left with velocity V are drawn in Figure 101 (see Lighthill (1978) page 405) for various values of the nondimensional frequency $\left(\frac{V\omega_0}{g}\right)$. The wavenumber curve $S(0)$, with $\left(\frac{V\omega_0}{g}\right) = 0$, has two symmetrical branches where normals to them fill a Kelvin wedge of semi-angle 19.5° representing *Kelvin ship waves* as shown in Figure 70 of Lighthill (1978) on page 277. The points of inflexion on $S(0)$ leads to the caustics forming the boundary of that wedge.

For three-dimensional internal waves generated by a source distribution

$$\frac{\partial Q}{\partial t} = f(\mathbf{x}) \exp(i\omega_0 t), \tag{4.4.95}$$

equation (4.4.68) for internal waves leads to the dispersion relation

$$0 = B(\omega_0, \boldsymbol{\kappa}) = (\omega_0 m)^{-1} \left[\omega_0^2 m^2 - (N^2 - \omega_0^2)(k^2 + \ell^2)\right]. \tag{4.4.96}$$

For an obstacle moving vertically upward with velocity $\mathbf{V} = (0, 0, -V)$, the corresponding dispersion relation follows from (4.4.96) in the form

$$B(\omega_0 - \boldsymbol{\kappa} \cdot \mathbf{V}, \boldsymbol{\kappa}) = 0, \tag{4.4.97}$$

where the expression for B follows from the right hand side of (4.4.96) as

$$B(\omega_0 - \boldsymbol{\kappa} \cdot \mathbf{V}, \boldsymbol{\kappa}) = [(\omega_0 + mV)m]^{-1} \left[(\omega_0 + Vm)^2 \kappa^2 - N^2(k^2 + \ell^2)\right]. \tag{4.4.98}$$

Thus, the wavenumber surface $S(\omega_0)$ is given by $B(\omega_0 - \boldsymbol{\kappa} \cdot \mathbf{V}, \boldsymbol{\kappa}) = 0$, that is,

$$(\omega_0 + Vm)^2(k^2 + \ell^2 + m^2) - N^2(k^2 + \ell^2) = 0. \tag{4.4.99}$$

Obviously, $S(0)$ represents the wavenumber surface of revolution for steady disturbance ($\omega_0 = 0$) in the form

$$V^2 m^2 (k^2 + \ell^2 + m^2) = N^2(k^2 + \ell^2). \tag{4.4.100}$$

On the other hand, when an obstacle moves horizontally in a stratified fluid with velocity $-\mathbf{V} = (-V, 0, 0)$, the wavenumber surface $S(\omega_0)$ is given by

$$(\omega_0 - Vk)^2(k^2 + \ell^2 + m^2) - N^2(k^2 + \ell^2) = 0 \tag{4.4.101}$$

so that the corresponding wavenumber surface $S(0)$ is given by

$$V^2 k^2 (k^2 + \ell^2 + m^2) = N^2(k^2 + \ell^2). \tag{4.4.102}$$

Thus, the associated waves generated are more complicated because $S(0)$ is *no longer* a surface of revolution.

Chapter on internal waves is then concluded by describing the acoustic waveguides, waveguides for waves in fluids, in general, and a marine *one-sided waveguide*, in particular, which is associated with propagation along a shoreline of *long water waves*. Thus, such long waves in water of variable depth $h(x, y)$ satisfy the equation of continuity

$$\frac{\partial \zeta}{\partial t} + \frac{\partial}{\partial x}(hu) + \frac{\partial}{\partial y}(hv) = 0. \tag{4.4.103}$$

The equations of motion are

$$\frac{\partial u}{\partial t} = -g \, \frac{\partial \zeta}{\partial x}, \qquad \frac{\partial v}{\partial t} = -g \, \frac{\partial \zeta}{\partial y}, \tag{4.4.104}$$

where $\zeta(x, y, t)$ is the free surface elevation.

Using (4.4.104), the continuity equation reduces to the form

$$\frac{\partial^2 \zeta}{\partial t^2} = \frac{\partial}{\partial x}\left(gh \, \frac{\partial \zeta}{\partial x}\right) + \frac{\partial}{\partial y}\left(gh \, \frac{\partial \zeta}{\partial y}\right). \tag{4.4.105}$$

If $h(x, y)$ is constant, this equation represents the two-dimensional wave equation with shallow water wave speed, $c_0 = \sqrt{gh}$.

When $h = h(y)$, where y represents distance from the shoreline, equation (4.4.105) describes quasi one-dimensional wave equation along the shoreline in the form

$$\frac{\partial^2 \zeta}{\partial t^2} = gh(y)(\zeta_{xx} + \zeta_{yy}) + g \, h'(y) \, \zeta_y. \tag{4.4.106}$$

The plane wave solution

$$\zeta(x, y, t) = A(y) \exp\left[i(\omega t - kx)\right], \tag{4.4.107}$$

exists provided the following equation

$$gh(y)A''(y) + gh'(y)A'(y) + \left[\omega^2 - g \, h(y)k^2\right] A(y) = 0, \tag{4.4.108}$$

is satisfied.

In the case of an increasing function of $h(y)$ where $h(y)$ is very large, equation (4.4.108) takes the approximate form

$$A''(y) = k^2 A(y). \tag{4.4.109}$$

This admits the bounded general solution

$$A(y) = B \, \exp(-ky), \tag{4.4.110}$$

where B is a constant. For the case of uniform bottom slope β, $h(y) = \beta y$ so that (4.4.110) satisfies equation (4.4.108) everywhere provided that

$$\omega = (g\beta k)^{\frac{1}{2}}. \tag{4.4.111}$$

This is the classical *Stokes edge-waves* propagating along beaches of gradually varying slope $h'(y) = \beta$ with the wave speed $c(k) = \sqrt{\frac{g\beta}{k}}$, and the group velocity, $U(k) = \frac{d\omega}{dk} = \frac{1}{2} \, c(k)$. Indeed, the common observation reveals that ordinary beach waves caused by swell propagating towards the shoreline attains maximum amplitudes.

Lighthill concluded his book by adding an Epilogue with Part 1. A variety of waves in fluids, and Part 2. Nonlinear effects on dispersive wave propagation. In fact, Part 1. deals with inertial waves in a rotating fluid described by small changes in the undisturbed vorticity vector, $(0, 0, 2\Omega)$ where Ω is the angular velocity of rotation. This means that any small fluid velocity field $\mathbf{u}$ relative to the uniform rotation causes a rate of change of vorticity ($\boldsymbol{\omega} = \text{curl}\,\mathbf{u}$)

$$\frac{\partial \boldsymbol{\omega}}{\partial t} = \frac{\partial}{\partial t}\,(\text{curl}\,\mathbf{u}) = 2\Omega\,\frac{\partial \mathbf{u}}{\partial z} \tag{4.4.112}$$

proportional to the rate of change of $\mathbf{u}$ along a vortex line. Application of operation on the left-hand side a second time gives

$$-\frac{\partial^2}{\partial t^2}\left(\nabla^2 \mathbf{u}\right) = 4\Omega^2\,\frac{\partial^2 \mathbf{u}}{\partial z^2}, \tag{4.4.113}$$

in which curl curl $\mathbf{u}$ = grad div $\mathbf{u} - \nabla^2\mathbf{u}$, and div $\mathbf{u} = 0$ (for continuity equation for an incompressible fluid) are used. Equation (4.4.113) represents the wave equation for a uniformly rotating fluid. This equation admits plane wave solutions

$$\mathbf{u} = \mathbf{u}_0 \exp\left[i(\omega t - \boldsymbol{\kappa}\cdot\mathbf{x})\right], \tag{4.4.114}$$

provided the following dispersion relation is satisfied

$$\omega^2 = \frac{4\Omega^2 m^2}{k^2 + \ell^2 + m^2}. \tag{4.4.115}$$

Therefore, the frequency of these inertial waves is independent of the magnitude of the wavenumber vector, but it depends only on the direction, and can take any value less than 2Ω (just as in the internal waves with N replaced by 2Ω). In other words, there is a high-frequency cut off at the semi-diurnal frequency 2Ω. The frequency ω in (4.4.115) can be expressed in terms of the angle θ between the surface of constant phase and the z-axis of rotation in the form

$$\omega = 2\Omega\sin\theta, \tag{4.4.116}$$

which is similar to that for internal waves with N, and $\cos\theta$ replaced by 2Ω and $\sin\theta$ respectively.

The group velocity vector

$$\mathbf{U}(\boldsymbol{\kappa}) = \left(\frac{\partial\omega}{\partial k}, \frac{\partial\omega}{\partial \ell}, \frac{\partial\omega}{\partial m}\right)$$

can be written for positive (ω/m) in the form

$$\mathbf{U}(\boldsymbol{\kappa}) = \frac{2\Omega(k^2+\ell^2)}{\kappa^2}\left[\frac{(-km, -\ell m, k^2+\ell^2)}{\kappa(k^2+\ell^2)^{\frac{1}{2}}}\right], \tag{4.4.117}$$

which is perpendicular to the wavenumber vector (k, ℓ, m). Thus, the wave energy propagates with the group velocity $\mathbf{U}(\boldsymbol{\kappa})$ which is directed parallel to the surfaces of constant phase. The magnitude of the group velocity is given by

$$U(\kappa) = \left(\frac{2\Omega}{\kappa}\right)\cos\theta = \left(\frac{\lambda\Omega}{\pi}\right)\cos\theta, \tag{4.4.118}$$

where $\lambda = \left(\frac{2\pi}{\kappa}\right)$ is the wavelength.

The wave speed for inertial waves in a rotating fluid is

$$c(\kappa) = \left(\frac{\omega}{\kappa}\right) = \left(\frac{2\Omega}{\kappa}\right)\sin\theta = \left(\frac{\Omega\lambda}{\pi}\right)\sin\theta. \tag{4.4.119}$$

Results (4.4.118) and (4.4.119) are similar to those of internal waves with replacing 2Ω by N and with $\cos\theta$ and $\sin\theta$ interchanged. However, for inertial waves, the z-component of the group velocity and the wavenumber have the *same sign*. In rotating fluid, a source oscillating with a frequency $\omega < 2\Omega$ generates inertial wave energy moving at an angle $\theta = \sin^{-1}\left(\frac{\omega}{2\Omega}\right)$ to the z-axis. In the case of three-dimensional wave propagation, inertial waves form a double cone. Experimental observations reveal that the wave energy propagates at an angle $\frac{\pi}{2} - \theta = \cos^{-1}\left(\frac{\omega}{2\Omega}\right)$ to the axis of rotation by showing that waves due to a local source of frequency ω fill a circular cone with this semi-angle. When $\theta = \frac{\pi}{2}$ and $\omega = 2\Omega$, the group velocity is zero. Standing oscillations are possible at this frequency. In the other extreme limit $\theta \to 0 (\omega \to 2\Omega)$, the wave speed tends to zero, but there is steady propagation of wave energy along the axis of rotation at a group velocity $(2\Omega)/(k^2 + \ell^2)^{\frac{1}{2}}$. This can be considered as the speed of formation of the *Taylor column* which is observed when a solid body moves at right angles to the axis of rotation at a very low speed, small compared with the group velocity, $U = (\Omega\lambda/\pi)$ for waves of all wavelengths produced by a body of that size.

On the other hand, fluid motions in long waves are essentially horizontal with insignificant vertical variation. However, they allow small free surface elevation $\zeta(x, y, t)$. The linearized momentum equations for such long waves in a rotating fluid are

$$\frac{\partial u}{\partial t} - fv = -g\,\frac{\partial \zeta}{\partial x}, \qquad \frac{\partial v}{\partial t} + fu = -g\,\frac{\partial \zeta}{\partial y} \tag{4.4.120}$$

together with the continuity equation (4.4.103), where f is a Coriolis parameter.

If h and f are constant parameters, the additional vertical vorticity component due to wave motions can be calculated from the above equations in the form

$$\left(\frac{\partial v}{\partial x} - \frac{\partial u}{\partial y}\right) = \frac{f\zeta}{h}. \tag{4.4.121}$$

Consequently, equation (4.4.103) combined with (4.4.120) gives

$$\zeta_{tt} + f^2\zeta = gh\,\nabla_H^2\,\zeta. \tag{4.4.122}$$

This equation admits the plane wave solution

$$\zeta = A\,\exp\left[i(\boldsymbol{\kappa}\cdot\mathbf{x} - \omega t)\right] = A\,\exp\left[i(kx + \ell y - \omega t)\right], \tag{4.4.123}$$

where A is a constant.

This represents two-dimensional long wave solution in a rotating fluid. The associated isotropic dispersion relation is

$$\omega^2 = gh(k^2 + \ell^2) + f^2. \tag{4.4.124}$$

Thus, the wave velocity, $c(\kappa)$ and the group velocity, $U(\kappa)$ of long waves are given by

$$c(\kappa) = \left(gh + \frac{f^2}{\kappa^2}\right)^{\frac{1}{2}}, \qquad U(\kappa) = \sqrt{gh}\,\left(\omega^2 - f^2\right)^{\frac{1}{2}}\,\omega^{-1}, \tag{4.4.125}$$

where $\kappa^2 = \left(k^2 + \ell^2\right)$. These results show that long waves are dispersive in a rotating fluid, whereas in a non-rotating fluid ($f = 0$), they are non-dispersive in nature. So, there is a striking contrast between wave motion in a rotating fluid and that of non-rotating fluid.

In general, long waves of frequency ω much larger than f are often called *tsunamis* generated by earthquake motions inside the oceans. In any such situation, the effects of Earth's rotation on long wave propagation is negligible. However, these effects are significant because tide-generating forces excite motions in oceans at frequencies comparable to the Coriolis parameter, $f = 2\Omega\sin\phi$, where ϕ is the latitude. These are very important every fourteenth day (at new moon or full moon) when the tide-generating forces occur due to attraction of the Sun and the Moon. The solutions of long water waves equations have a major effect on the tides in shallow oceans. There is a significant increase in amplitude of these waves as their energy propagate into the shallower water.

There are other modes of propagation that are not described by the dispersion relation (4.4.124). They include a one-dimensional mode of waveguide propagation known as the *Kelvin wave* which is similar to that the

Stokes edge-wave modified by the Earth's rotation. On the other hand, the curvature of the earth causes the Coriolis parameter to vary with local latitude, the so called the "*beta effect*". Thus, the variation of the Coriolis parameter with latitude (that is, non-uniform rotation of the earth) is found to generate a new kind of waves known as the *Rossby waves* at much lower frequencies than those of gravity waves.

There is another new kind of wave motions that are found in electrically conducting fluids. In dynamics of a homogeneous fluid, both vortex lines and magnetic field lines moves with the fluid. The full nonlinear equation for the vorticity in an incompressible fluid is

$$\frac{\partial \boldsymbol{\omega}}{\partial t} = \text{curl}\,(\mathbf{u} \times \boldsymbol{\omega}) + \nu\,\nabla^2 \boldsymbol{\omega}, \tag{4.4.126}$$

where the last term on the right hand side of (4.4.126) represents the diffusion of vorticity and ν is the kinematic viscosity. Neglecting this term, equation (4.4.126) states that vortex lines move with the fluid. The linearized form of equation (4.4.126) with $\boldsymbol{\omega}$ replaced by $\boldsymbol{\Omega} = (0, 0, 2\Omega)$ on its right hand side leads to the equation (4.4.112).

Neglecting the displacement current field, the Maxwell's equations in electromagnetic theory are given by

$$\text{div}\,\mathbf{B} = 0, \quad \text{curl}\,\mathbf{B} = \mu \mathbf{j}, \quad \frac{\partial \mathbf{B}}{\partial t} = -\text{curl}\,\mathbf{E}, \tag{4.4.127}$$

where $\mathbf{B}$ is the magnetic field, μ is the magnetic permeability, $\mathbf{j}$ is the current density and $\mathbf{E}$ is the electric field. The current density $\mathbf{j}$ is determined by the equation

$$\tau \mathbf{j} = \mathbf{E} + \mathbf{u} \times \mathbf{B}, \tag{4.4.128}$$

where τ is the resistivity, and the second term, $\mathbf{u} \times \mathbf{B}$ on the right side of (4.4.128) is known as the *Lorentz force*. The presence of the velocity field $\mathbf{u}$ in equation for $\mathbf{j}$ incorporates the effect of fluid motions on the electromagnetic field.

In an electrically conducting fluid, the corresponding equation for the magnetic field $\mathbf{B}$ is

$$\frac{\partial \mathbf{B}}{\partial t} = \text{curl}\,(\mathbf{u} \times \mathbf{B}) + \left(\frac{\tau}{\mu}\right) \nabla^2 \mathbf{B}. \tag{4.4.129}$$

In the limit as $(\tau/\mu) \to 0$, equation (4.4.129) reduces to the form

$$\frac{\partial \mathbf{B}}{\partial t} = \text{curl}\,(\mathbf{u} \times \mathbf{B}). \tag{4.4.130}$$

It follows from a direct perturbation analysis that small variations of $\mathbf{B}$ from a uniform undisturbed magnetic field $\mathbf{B}_0 = (0, 0, B_0)$ satisfy the equation

$$\frac{\partial \mathbf{B}}{\partial t} = B_0 \frac{\partial \mathbf{u}}{\partial z}. \tag{4.4.131}$$

In view of a balance of fluid pressure and magnetic pressure, the fluid momentum equation is given by

$$\rho_0 \frac{\partial \mathbf{u}}{\partial t} = \left(\frac{B_0}{\mu}\right) \frac{\partial \mathbf{B}}{\partial z}, \tag{4.4.132}$$

where ρ_0 is the undisturbed density.

Eliminating $\mathbf{u}$ or $\mathbf{B}$ from equations (4.4.131) and (4.4.132) gives the Alfvén wave equations for $\mathbf{u}$ or $\mathbf{B}$ in the z-direction

$$\frac{\partial^2}{\partial t^2}\begin{pmatrix}\mathbf{u}\\ \mathbf{B}\end{pmatrix} = c_A^2 \frac{\partial^2}{\partial z^2}\begin{pmatrix}\mathbf{u}\\ \mathbf{B}\end{pmatrix}, \tag{4.4.133}$$

where $c_A = (B_0^2/\mu\rho_0)^{\frac{1}{2}}$ is the *Alfvén wave velocity.* Therefore, the velocity field $\mathbf{u}$ or the magnetic field $\mathbf{B}$ satisfies the one-dimensional wave equation which describes purely *magneto-hydrodynamic (or MHD) waves* in the z-direction with the Alfvén wave velocity, c_A. The remarkable feature of the above analysis is that they *do not* represent three-dimensional MHD waves due to a source of disturbance, instead, these waves propagate one-dimensionally along the z-direction, that is, along the magnetic field lines *without attenuation.*

For a compressible fluid in the presence of magnetic field, the linearized continuity and momentum equations combined with the linearized form of the magnetic force, $\mathbf{j} \times \mathbf{B} = \mu^{-1}(\text{curl}\, \mathbf{B}) \times \mathbf{B}$, that exerts force on the fluid (as the magnetic field lines are convected by the fluid)

$$\mu^{-1}\left[\mathbf{B}_0 \cdot \nabla \mathbf{B} - \nabla\left(\mathbf{B}_0 \cdot \mathbf{B}\right)\right] \tag{4.4.134}$$

can be used to study the interaction between the Alfvén waves and sound waves. The equation for the velocity component $w(\mathbf{x}, t)$ parallel to the magnetic field lines admits plane-wave solutions in the form

$$w(\mathbf{x}, t) = A\, \exp\left[i(\omega t - \boldsymbol{\kappa} \cdot \mathbf{x})\right], \tag{4.4.135}$$

where A is a constant, provided the dispersion relation

$$\left(k^2 + \ell^2\right)\left[\left(c_0^2 + c_A^2\right)\omega^2 - c_0^2\, c_A^2\, m^2\right] = \left(\omega^2 - c_0^2\, m^2\right)\left(\omega^2 - c_A^2\, m^2\right), \tag{4.4.136}$$

is satisfied, where c_0 is the velocity of sound. Equation (4.4.136) represents the wavenumber surface S for the MHD waves in a compressible fluid with the Alfvén velocity, c_A and sound velocity, c_0.

In the limit of large (c_A/c_0), the wavenumber surface S reduces to a sphere of radius (ω/c_A), corresponding to isotropic waves with the Alfvén velocity, c_A, together with two planes, $m = \pm\left(\frac{\omega_0}{c_0}\right)$ representing the one-dimensional waves with the velocity of sound c_0.

On the other hand, the equations for the velocity components u and v normal to the magnetic field lines are much more complicated. However, the two-dimensional divergence and curl of the two-dimensional velocity field

$$\Delta_2 = u_x + v_y \quad \text{and} \quad \Omega_2 = v_x - u_y \tag{4.4.137}$$

are relatively simple and provide unique solution for u and v that decay to zero at large distances.

The divergence Δ_2 admits solutions

$$\Delta_2 = iA_2\left(m - \omega^2 c_0^{-2}\, m^{-1}\right)\exp\left[i(\omega t - \boldsymbol{\kappa}\cdot\mathbf{x})\right] \tag{4.4.138}$$

so that its changes are directly proportional to those of $(\partial w/\partial z)$.

The wavenumber surface S given by (4.4.136) for the wave propagation of both w and Δ_2 can be drawn for various values of (c_A/c_0) between 0 and 1. For small values of (c_A/c_0), the wavenumber surface S represents a sphere of radius (ω/c_0) describing the isotropic propagation of sound waves and a pair of planes corresponding to one-dimensional MHD-wave propagation at the Alfvén velocity, c_A. However, except for the incompressible limit as $(c_A/c_0) \to 0$, the MHD waves propagating along the normals to the wavenumber surface S spreads out in the three-dimensional w and in two-dimensional Δ_2 signals radiated from the source of disturbance. When those waves have become fully separated from the one-dimensional waves of Ω_2, wave motions along the magnetic field lines are purely transverse and divergenceless.

The vorticity component Ω_2 satisfies the wave equation

$$\frac{\partial^2\Omega_2}{\partial t^2} = c_A^2\frac{\partial^2\Omega_2}{\partial z^2}. \tag{4.4.139}$$

This equation shows that the z-component of vorticity Ω_2 propagates one-dimensionally with the Alfvén speed c_A along the magnetic field lines (or, along the z-axis). In other words, compressibility does not change the nature of one-dimensional propagation; while it simply confines it to the z-component of vorticity distribution.

In wave motions in ionized gases dealing with a high temperature, low-density plasma of electrons and positive ions, the existence of various dissipative effects causes departures from the equation (4.4.130) for conducting fluids. One such effect, known as the *Hall current effect* , becomes significant as the frequency of waves increases. This effect is found to destroy the one-dimensional propagation of transverse wave motions along the magnetic field lines at the wave velocity c_A. In fact, it transforms it into a conical propagation confined within a cone of semi-angle $\sin^{-1}(\omega\rho_0/e\,nB_0)$, where n represents the number density of electrons of charge, $-e$. However, other one-dimensional longitudinal waves which arises at the sound velocity $c_0(< c_A)$ turns out to be unaffected by the Hall effect. Thus, the one-dimensional longitudinal waves may persist upto higher frequencies than the transverse waves.

Finally, Part 2 of the Epilogue is basically devoted to nonlinear effects on dispersive wave motions, in general and to equal influence of weak nonlinearity and weak dispersion on fairly long waves, in particular. Even quite *small* nonlinear terms in the equations of motion can, given a sufficiently long time, change ordinary waveforms into significantly different waveforms. In ordinary dynamics of homogeneous fluid, small nonlinear effects produce a significant cumulative action. However, the cumulative action of nonlinear effects whenever there is a single wave speed (zero for that vorticity field, or nonzero for a nondispersive system) is absent for a dispersive wave system. Indeed, dispersion can significantly limit the possibilities for such accumulative action of small nonlinear effects because different Fourier components in a wave system become more displaced from each other. This means that the dynamics of homogeneous fluid is really more complicated than that of a uniformly rotating homogeneous fluid, or that that of a stratified fluid because disturbance in these fluid systems, instead of staying together, are essentially dispersed three-dimensionally as the group velocity is different for each wavenumber vector.

Based on the averaged variational theory for dispersive waves in shallow water, the velocity potential may contain a slowly varying nonperiodic mean value ϕ whose gradient $\phi_x = s$ represents a mean horizontal stream velocity induced by the waves. The averaged Lagrangian density assumes the form $\mathcal{L}(\omega, k, \eta, s)$ where $\eta = -\phi_t$ leading to a system of two Euler equations

$$\frac{\partial}{\partial t}\left(\frac{\partial \mathcal{L}}{\partial \omega}\right) = \frac{\partial}{\partial x}\left(\frac{\partial \mathcal{L}}{\partial k}\right), \qquad \frac{\partial}{\partial t}\left(\frac{\partial \mathcal{L}}{\partial \eta}\right) = \frac{\partial}{\partial x}\left(\frac{\partial \mathcal{L}}{\partial s}\right). \tag{4.4.140}$$

The waves of moderate amplitude become *unstable* if and only if the wave steepness $kh > 1.365$ (or $\frac{\lambda}{h} < 4.6$). We next consider *fairly long waves* with $10 < (\lambda/h) < 20$ so that for these waves, dispersion is small but *not* negligible. Consequently, nonlinear effects seem to be moderately large. Such fairly long waves with $(\lambda/h) > 10$ can be well described by the first two terms

$$c(k) = c_0 \left(1 - \frac{1}{6}\, k^2\, h^2\right) \tag{4.4.141}$$

of the Taylor series of the linearized wave speed

$$c(k) = \left(\frac{\omega}{k}\right) = \left(\frac{g}{k}\tanh kh\right)^{\frac{1}{2}}, \tag{4.4.142}$$

where $c_0 = \sqrt{gh}$ is the shallow water speed.

In general, waves with *small dispersion* show such a k^2-term in the departure of the linearized value of the wave speed from a constant value c_0. The dispersion relation associated with (4.4.141) follows from the equation

$$v_t + c_0\, v_x + \sigma\, v_{xxx} = 0, \tag{4.4.143}$$

where $\sigma = \frac{1}{6}\, c_0\, h^2$ is constant.

The above analysis reveals how the nonlinear theory of one-dimensional waves can be modified for wave systems with small dispersion such that the linearized wave speed is $(c_0 - \sigma k^2)$ for wavenumber k. The wave speed due to nonlinear effects is denoted by v and a nonlinear equation

$$v_t + (c_0 + v)v_x = 0 \tag{4.4.144}$$

can be used to describe nonlinear nondispersive waves. For long waves $v = \frac{3}{2}u$, where u is the fluid velocity. Thus, it is possible to combine (4.4.143) and (4.4.144) to describe both nonlinear and dispersive effects so that $v(x,t)$ satisfies the *Korteweg and de Vries (KdV) equation*

$$v_t + (c_0 + v)v_x + \sigma\, v_{xxx} = 0. \tag{4.4.145}$$

This famous equation can be used to describe many nonlinear waves with small dispersion. Introducing a new coordinate $X = (x - c_0 t)$ moving with the basic wave speed c_0, equation (4.4.145) assumes the form

$$v_t + v\, v_X + \sigma\, v_{XXX} = 0. \tag{4.4.146}$$

The significance of the nonlinear term $v\,(\partial v/\partial X)$ compared with the dispersive term $\sigma\left(\partial^3 v/\partial X^3\right)$ in the KdV equation (4.4.146) depends on a non-dimensional parameter $\left(v_1\lambda^2/\sigma\right)$, where v_1 is the highest value of the excess wave speed and λ is the wavelength. Using the value of $\sigma = \frac{1}{6}\, c_0\, h^2$, and the fact that (v_1/c_0) is proportional to the ratio (a/h), the parameter $(v_1\lambda^2/\sigma)$ is proportional to $(a/h)\left(\lambda^2/h^2\right) = \left(a\lambda^2/h^3\right)$. This represents another *fundamental parameter* in the theory of water waves.

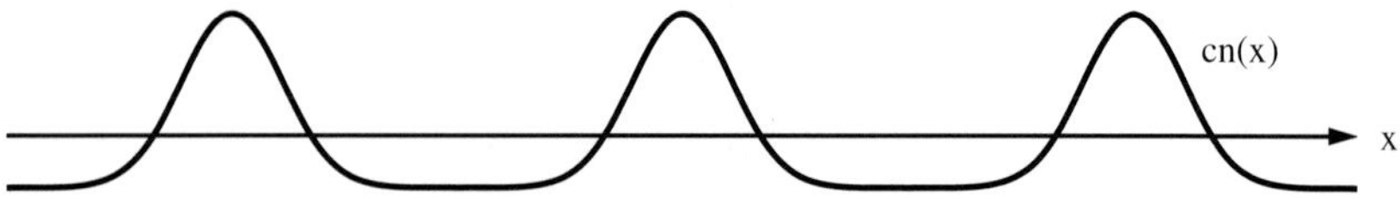

Fig. 4.6 A cnoidal wave solution.

For the smaller values of the parameter $\left(a\lambda^2/h^3\right) < 16$, nonlinear effects cause small deviation from the sinusoidal wave form. In other words, a balance between nonlinear steepening and dispersion is possible leading to periodic waves without change of form due to this balance. As the parameter $\left(a\lambda^2/h^3\right)$ increases, the first harmonic is obviously present initially, as the crests gradually become peakier and the tough shallower, followed by higher harmonics.

When the parameter, $\left(a\lambda^2/h^3 \approx 16\right)$ is close to the limiting value 16, the periodic waveform leads to a sequence of *isolating humps* with a flat water surface between them.

More precisely, substitution of $v_t = -U\; v_X$, where U is the wave speed in the (X, t) coordinate system in equation (4.4.146) leads to an ordinary differential equation. This equation admits a single family of periodic waves known as *cnoidal waves* because they are represented by the square of the Jacobian elliptic function, *cn* as shown in Figure 4.6 and one famous limiting solution in the form

$$v(X,t) = 3U\,\mathrm{sech}^2\left[\frac{1}{2}\left(\frac{U}{\sigma}\right)^{\frac{1}{2}}(X - U\,t)\right], \tag{4.4.147}$$

which is shown in Figure 4.7. This represents, *not* a periodic wave, but a *solitary wave* or *soliton* which describes the propagation of a single isolated symmetrical hump of unchanged form. In 1834, Scott Russell (1808-1882) experimentally discovered that such a solitary wave which propagate virtually unchanged except for a very gradual attenuation due to bottom friction. In 1895, based on nonlinear shallow water equations Korteweg and de Vries formulated a remarkable mathematical model equation to provide an explanation of the phenomenon observed by Scott Russell.

When the parameter $a\lambda^2/h^3\,(> 16)$ is large enough, nonlinear effects cannot be neglected. Consequently, the waveform is gradually modified by nonlinear effects and its front face continuously steepens to produce a hydraulic jump. Therefore, the waves behind the jump must be cnoidal. Indeed, they are observed to have the cnoidal waveforms and the cnoidal-wave relationships among wavelength, wave speed, and wave amplitude.

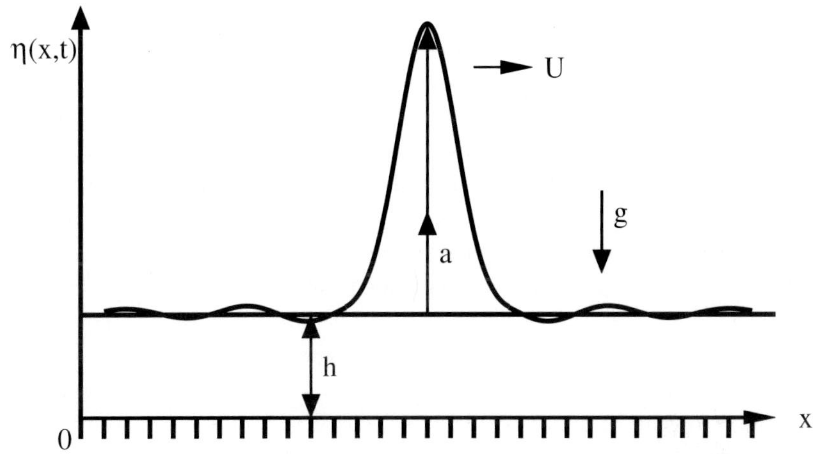

Fig. 4.7 A solitary wave solution.

In other words, the observed hydraulic jump exhibits rather foaming but is followed by a train of periodic waves with crests stationary relative to the jump. All these correspond to a loss of wave energy due to foaming or viscous action. However, foaming could be intense and is responsible for almost all energy loss. Observations show that hydraulic jumps are often found in nature as stationary phenomena in streams. A steady stream of speed exceeding the long-wave speed can slow down to a speed below the long-wave speed only by discontinuous deceleration at a jump.

It is fairly clear from the epilogue that to every property of surface water waves there are corresponding properties of internal waves. On the other hand, interfacial waves between two fluids of different densities also exist up to a certain limiting amplitude beyond which wave energy is rapidly dissipated due to interpenetration of the two fluids analogues to foaming. In continuously stratified fluids, it is possible that the steepening effect of the nonlinear term is counterbalanced by smoothening effect of dispersion so that a solitary wave can exist. Modern methods of functional analysis have been used to prove the existence of the solitary wave in a stratified fluid. The hydraulic jump is also observed in stratified fluids. When a stream of cold air flowing down the lee side of a mountain range acquires great speed and is then suddenly decelerated similar to a stationary hydraulic jump.

A self-contained treatment appropriate for advanced undergraduates, graduate and post-graduate students, this book offers general treatment of all kinds of wave propagation phenomena that occurs in fluids, and gives major views on many different phenomena of geophysical fluid dynamics from the standpoint of a unifying body of general theory. It also emphasizes the properties of boundary layers which, in general, constitute another class of unifying concepts, and there are many other kinds of fundamental ideas associated with wave motions in fluids. This remarkable and masterly treatise conveys the intellectual excitement of present-day wave phenomena in fluids from both mathematical and physical view points. It is nearly thirty years since this book was published and in the intervening period, considerable progress has been made in many aspects of knowledge of wave motions in fluids. Substantially, therefore, the fundamental nature of its contents remain unchanged. Not only it does offer an enormous source of essential information, much of which is difficult to find in other sources, but it is a fundamental work which virtually all other specialized studies build upon and refer to. The demand for this book continues to be high, both in the classroom and among research workers in the field. It is now available for the first time in a cheaper paperback edition for students.

4.5 *Newer Uses of Mathematics*, Penguin Books Ltd., Harmondsworth, U.K., 1978

Lighthill's expertise in the more abstruse realms of applied mathematics and fluid mechanics has not prevented him from writing an extremely lucid, popular and accessible introduction to a nontechnical book entitled *Newer Uses of Mathematics*. His main motivation in writing this book was to illustrate recent and diverse applications of mathematics in natural and social sciences, economics, finance, business, industry, world modeling, and to stimulate interest in mathematics among a very large number of the general public. This fascinating book was first published by Penguin, Harmondsworth, U.K., as a paperback edition in 1978. In the Preface of the book, Sir James Lighthill elucidates:

"Recently, mathematics has more and more been used to help with many of the things that really matter to us most. We care about the air we breathe, and its freshness and the climate that it brings us. We care about the water that quenches our thirst and cleans us and irrigates our crops. We care about the earth, as a source of food and of raw materials. We depend upon fuel for warmth and mobility. We want health and the conditions of life that promote it, and we want to see a balanced and healthy wildlife around us. All those things are what people mean when they speak about the *environment*. In the fight to improve it, mathematics is much used."

"By calling this book *Newer Uses of Mathematics* we admit that some good practical uses of mathematics which are well established and well known – for example, in all branches of engineering, and in statistics – are left out. Nor do we deal with applications of mathematics to pure science. Rather, we want to outline some of the many ways of using mathematics for significant practical purposes that have only become widespread in the past thirty years."

This extraordinary and popular book has six chapters:

Chapter 1. Mathematics and the Physical Environment includes the air, the analysis of weather forecasting, rainfall and runoff into rivers and streams, the fresh water and the sea, the earth's and its resources.

Chapter 2. Mathematics and Biological Environment deals with main topics including the introduction of mathematics into biology, genetics, populations, medicine and pharmacology.

Chapter 3. Methods of Operational Analysis include new methods of operational analysis, linear programming, nonlinear optimization, statistical decision theory, dynamic programming, queuing and simulation, combinatorics, catastrophic theory, stock control, game theory, replacement and renewal and economic forecasting.

Chapter 4. Networks include transportation networks, network techniques, telephone networks, project networks and network distribution.

Chapter 5. Finance deals with the new mathematics of finance that includes topics on accountancy, acturial work, banking, econometrics, and stock exchange.

Chapter 6. Planning. The major topics of this chapter include new mathematics of planning, boundary problem, the nature of the organization and its environment, linear deterministic models, manpower planning, the continuous approach, dynamic programming, competitive and world models.

Lighthill explained that weather patterns have a tremendous impact on each of us personally. The weather has also strong effects on many large-scale activities including production of food, fuel supply for heating and air conditioning, and transportation. Obviously, a reliable weather forecast has enormous value, not only to the general public but also to many different industries. In the first chapter of the book, he described, in nontechnical terms, the major use of mathematics for weather forecasting.

Most air contains water vapor that is not visible at all until it condenses. Condensation deals with conversion of water vapor into some visible form, such as fog, cloud, rain, hail or snow. Lighthill explained how mathematics is used to study condensation, pattern of rainfall, and large scale problems like weather prediction. On the other hand, both clean air and clean water are of great importance to human life, it is very important to maintain high quality standards of air and water by reducing pollution. There is a large number of uses of mathematics in pollution controls, and in predicting how the flow in a river or stream varies in response to varying amounts of rainfall. This information is vital for river management with the major goals of reducing pollution, minimizing danger of floods, and utilizing rivers and lakes for agricultural, industrial, recreational and navigational purposes.

According to Lighthill, the use of mathematics is essential for exploration of resources what lies on the earth's surface including the oceans and seas, and what lies beneath the earth's surface. About 71% of the earth's surface is covered by the oceans and seas which are important to us in many different ways. They are, indeed, are unlimited natural sources of food (fish and shellfish), oil, natural gas and other minerals. Naturally, the exploration, development and use of these tremendous resources would be extremely valuable for meeting the needs and demands of present day life. Available information and knowledge from various projects concerning coastal and marine life, marine vehicles and equipment, the design and building of coastal structures have already posed many new, challenging mathematical and experimental problems of geophysical, oceanographical and engineering interest. Considerable ocean and coastal research are also concerned with wave and current dynamics, generation and attenuation of energy, tidal phenomena in bays and rivers, the hydroelectric power development, food and energy resources. From a practical as well as scientific point of view, there are numerous incredible features of the oceans and seas which have a tremendous impact on life, coastal and marine environment, industrial and economic resources.

Of all natural hazards, hurricanes, cyclones, tsunamis, tornadoes, floods, thunderstorms, earthquakes, underwater explosions, landslides and volcanic eruptions are the most common and frequently observed to have catastrophic effects on life, economic and marine resources. Most of the hazards often occur without warning and cause various kinds of severe damage and loss. All coastal regions are highly susceptible to these destructive hazards. Topographically, low outer banks are, indeed, subject to erosion, flooding and considerable damage and loss during the occurrences of natural hazards. From a practical point of view, there is a critical need for mathematical study and research on numerous problems related to natural hazards and their impact on the coastal regions. Indeed, natural hazards remain the most serious mathematical problems facing man.

There are various mathematical methods that have been widely used in exploring the oceans and extensive offshore areas to discover deposits of oil and natural gas. One of the methods is to generate an explosion near the surface of the ocean at a given time and to record the signals received by a large number of instruments called *hydrophones* that automatically identify vibrations of sound in the ocean. Various rock layers beneath the ocean can be identified based on different travel times of sound in water and in different rocks. Similarly, mathematics has widely been used to identify minerals through their magnetic and electric properties.

In order to conclude Lighthill's discussion on the newer use of mathematics in exploring the earth and its various resources, it may be appropriate to quote his own statement:

"For centuries, people have been exploring the earth, looking for places to live and places to grow food; looking for sources of raw materials (such as metals) and of fuels; looking for good transport routes and terminals (such as harbours and, later, airports); looking for places to enjoy the beauties of nature. The exploring continues today – and is complicated by one widely recognized need; to plan for reducing conflict between all the different uses of the earth and its resources. We end this chapter with a brief outline of the contribution of mathematics to these activities."

This quotation illustrates a particular fact that the use of mathematics is essential for exploring the earth and its various resources of diverse nature. However, in general, the contents of this book reveal that at present there are exciting possibilities in quantifying and predicting a wide variety of outcomes from the real-world problems far more complex then hitherto. This book also revealed Sir James' lucid and delightful exposition and endearing interest in mathematical education. Lighthill has not only made a significant contributions to mathematical sciences, but also to a contribution of vital importance to the welfare of the human race.

4.6 *Monsoon Dynamics*, (Eds. Lighthill, J. and Pearce, R. B.), Cambridge University Press, Cambridge, 1981

In collaboration with Indian applied mathematicians, fluid dynamists, and other scientists, Lighthill organized a joint IUTAM/IUGG International Symposium on Monsoon Dynamics with applications to weather forecasting and flood prediction at Indian Institute of Technology (IIT), Delhi in the second week of December 1977. As a follow-up project, Lighthill and R. P. Pearce have published an edited volume as the Proceedings of this symposium. This volume has developed from manuscripts submitted by applied mathematicians, fluid dynamists and other scientists who have made some important contributions to oceanography, meteorology, geophysical fluid dynamics, hydrology, and to other related problems of water resources, river management, weather forecasting and flood prediction.

It is well-known that the atmosphere and the ocean have profound influence on each other. Many atmospheric phenomena including the tropical cyclones (known also as hurricanes or as typhoons) are frequently observed to have disastrous effects on life, economic and marine resources. The tropical cyclones are vastly bigger and more intense than mid-latitude cyclones, showing horizontal length scales of many hundreds of kilometers and surface wind speeds of the order of $50\,ms^{-1}$. They are formed when a wind pattern with substantial cyclonic vorticity comes in contact with a tropical ocean. It is essential to offer improved accuracies in forecasting the tracks of tropical cyclones for the human populations in the vulnerable coastlines. On the other hand, the problem of generating effective predictions of disastrous coastal flooding demands a special attention to study the combined tide and surge effects produced by a moving cyclonic pattern of winds, and of forecasting not just winds in general, but those features of cyclonic wind patterns that are most significant for wind-generated surges.

This symposium on Monsoon Dynamics had a great impact on Indian subcontinent for the following reasons. First, Indian agriculture depends critically on summer rainfall generated by the Southwest Monsoon which involves low-level winds beginning in the southern part of the Indian Ocean and blowing over East Africa and South Asia before traveling onwards across most of the Pacific. Second, the predictions of both water supply and energy demand depend on the accurate weather forecasting which indeed is extremely useful for most human activities. Third, it gives information about the extensive observational data that would enable many different aspects of monsoon behavior to be more accurately described and provide a much improved observational input into the study of monsoon dynamics. Fourth, this symposium provides information that puts the researcher at the forefront of current research on monsoon dynamics.

The Proceedings of the symposium have five parts with 46 research articles. Part I is devoted to the large-scale climatology of the tropical atmosphere. The summer monsoon over the Indian subcontinent and East Africa is discussed in Part II. Part III deals with the physics and dynamics of the Indian Ocean during the summer monsoon. Some important mathematical modelling techniques are described in Part IV. The final Part V is concerned with storm surges and flood forecasting.

Lighthill provided a unique leadership in organizing this international symposium in India with publication of Proceedings. This publication brought together current developments in monsoon dynamics with applications to weather forecasting, flood prediction and other related topics. At the same time, the subject of this symposium represents an unusually well-focused research effort on geophysical fluid dynamics of genuine importance to India.

In his introduction of Part V on Storm surges and flood forecasting, Lighthill described:

"From these discussions a consensus emerged among the participants from outside India that it would be of immense help for all purposes of giving reliable warning of dangers from intense cyclones if a suitable aircraft were available for use by the India Meteorological Department. When weather satellites indicates a strong cyclone in the Bay of Bengal, such an aircraft could fly directly into it to determine on the spot its actual strength (as indicated by the wind-velocity pattern, which could be quickly determined by an aircraft fitted with inertial navigation) and other parameters relevant to its subsequent path. This information could be immediately communicated to the synoptic and numerical forecasting teams to help make an early start on their predictions of the cyclone's effects."

In closing of his introduction of Part V, Lighthill concluded:

"The joint IUTAM/IUGG Symposium on Monsoon Dynamics not only increased the knowledge of those attending, but also established new contacts between workers in their own fields and in closely related areas. In particular, contacts were established between participants from geographically remote areas and from different spheres of activity (especially operational departments and universities). The future impact of material so multidisciplinary and multinational as that involved in the studies of monsoon dynamics and forecasting depends crucially upon the maintenance and growth of such new contacts."

4.7 *An Informal Introduction to Theoretical Fluid Mechanics*, Oxford University Press, Oxford, 1986

Sir James Lighthill used his extensive and effective teaching experience in fluid mechanics and applied mathematics while he was serving as the Beyer Chair of Applied Mathematics at the University of Manchester, as Royal Society Research Professor at Imperial College and then as Lucasian Professor of Applied Mathematics at the University of Cambridge to write another major book entitled "*An Informal Introduction to Theoretical Fluid Mechanics*" that was published by the Oxford University Press in 1986. The opening sentence of his Preface reads: "One of the major modern areas of successful application of mathematics is *fluid mechanics*." So, the study of fluid mechanics is important in many major contexts including locomotion through fluid media (as in animal swimming or flying or as in aircraft or ships), circulation systems including the circulation of the atmosphere and oceans, and the circulation of the blood, transfer of energy in engines, resistance of solid structures to wind and water, and wave motions in fluids. This is a remarkably well-written book which can be used as an introduction to theoretical fluid mechanics for undergraduate mathematics, science and engineering students.

Much of the modern work has been made possible by the remarkable advances in experimental studies and the development of powerful computer technology, combined with subsequent progress in computational techniques. In his preface of the book, Lighthill states: "... on the creative use of data from experimental studies and data from theoretical analyses to generate practically useful mathematical models (including manageable computer models) of a wide range of important fluid flows. Some of the analyses, as we shall see, involve mathematically exciting theories which, incidentally, are of a strikingly nonlinear character. The book's prime emphasis, however, is on the problem of how to use those as strong supports at one end of an effective bridge spanning the world of mathematics and the world of experiment and observation."

This relatively short but modern book has eleven chapters:

Chapter 1. Principles of mechanics applied to lumps of fluid addresses the study of fluid mechanics only as far as can be achieved by applying the basic principles of mechanics including Newton's laws of motion, momentum and energy to lumps of fluid on a large scale.

Chapter 2. Velocity fields and pressure fields deal with the development of mathematical model of the mechanics of a fluid, and critical reviews of the foundations of Leonhard Euler's (1707-1783) three-dimensional model of a frictionless fluid. Included are the basic properties of pressure fields and velocity fields.

Chapter 3. Equations of motion are concerned with the momentum principle that takes the form

$$\rho \frac{D\mathbf{u}}{Dt} = -\ \mathrm{grad}(p + \rho g H), \tag{4.7.1}$$

where the total derivative

$$\frac{D\mathbf{u}}{Dt} = \frac{\partial \mathbf{u}}{\partial t} + \mathbf{u} \cdot \nabla \mathbf{u}, \tag{4.7.2}$$

ρ is the density, $\mathbf{u} = (u, v, w)$ is the velocity field, p is the pressure field, ρg is the gravitational force per unit volume on a fluid particle directed vertically downwards and H represents the height above some reference level. Indeed, the total force per unit volume on a particle in the Euler model is $-\ \mathrm{grad}(p+g\rho H)$. The complete set of partial differential equations for the Euler model in an incompressible fluid are

$$\nabla \cdot \mathbf{u} = 0, \qquad \rho\left(\frac{\partial \mathbf{u}}{\partial t} + \mathbf{u} \cdot \nabla \mathbf{u}\right) = -\nabla(p + \rho g H). \tag{4.7.3ab}$$

This is a closed set of four equations for four unknowns, u, v, w and p.

On the other hand, the equations of motion of a compressible fluid of variable density ρ are

$$\frac{D\rho}{Dt} = -\rho\, \nabla \cdot \mathbf{u}, \qquad \rho \frac{D\mathbf{u}}{Dt} = -\nabla p - \rho\, \nabla(gH). \tag{4.7.4ab}$$

The first of these equations is the equations of continuity for a compressible fluid, and the second is the equation of motion. These equations constitute four scalar equations for five unknowns, ρ, p, u, v, and w so that a fifth equation expressing the compressibility properties of the fluid is to be added to make the system of equations closed. Included are boundary and initial conditions and some properties of steady flows in this chapter.

Chapter 4. Vorticity fields are introduced by $\boldsymbol{\omega} = \mathrm{curl}\,\mathbf{u}$. The vorticity and circulation and their relationship are discussed briefly.

Chapter 5. Vortex dynamics is the main topic of this chapter. Irrotational fluid flows (which are defined as flows with zero vorticity), structure of the boundary layer and boundary layer separation are discussed in some detail. The combined effects of viscous diffusion and convection (that is, the vortex lines are moving with the fluid) determine the thickness of the boundary layer around a solid body at rest. The effect of diffusion causes the boundary layer thickness to grow in proportion to $\sqrt{\nu t}$, where ν is the kinematic viscosity of the fluid. The vortex lines in the outer part of the boundary layer are carried away by the fluid moving with velocity U relative to the body of length ℓ so that they are swept away after a time $t = (\ell/U)$. This leads to the thickness of the boundary layer $\delta = \sqrt{\nu t} = \sqrt{(\nu\ell/U)}$. In other words, vorticity is diffused through a distance δ in time $t = (\ell/U)$ and it is then convected a distance $\ell = Ut$ in the same time t. In general, the thickness of the boundary layer is thin compared to the dimension ℓ of the body so that the ratio (ℓ/δ) can be large only if the quantity $(\ell/\delta) = \sqrt{(U\ell/\nu)}$ is very large. This is exactly the square root of the *Reynolds number* $R = (\frac{U\ell}{\nu})$, a non-dimensional quantity which measures the product of the flow speed and of the dimension of the body relative to the magnitude of viscous diffusion effects. It was Ludwig Prandtl (1875-1953) who first discovered the theory of boundary layer in fluid mechanics. On the other hand, Osborne Reynolds (1842-1912) first introduced the idea of the Reynolds number and demonstrated experimentally the significance of this number in the sense that many fluid flows become chaotic or turbulent when the Reynolds number exceeds a particular critical value. Although irrotational flows are immune to turbulence, many modern experiments confirmed that any boundary layer flow may become turbulent for a sufficiently large value of the Reynolds number.

Chapter 6. General properties of irrotational flows. Such flows are generated in a fluid from rest by the impulsive motion of a solid boundary. The existence of velocity potential ϕ ($\mathbf{u} = \nabla\phi$) for irrotational flows and its relation to the pressure field, Bernoulli's equation, and properties of irrotational flows in a simple or doubly connected region are discussed in this chapter.

Chapter 7. Three-dimensional examples of irrotational flows are presented with solution of the Laplace equation for the velocity potential ϕ. Included are spherically symmetric motions, axisymmetric fairings, bluff-body flows, and the concept of *dipole* (or *double-source*) which represents a combination of a source and a sink.

Chapter 8. Three-dimensional far fields. A study is made of the general properties of irrotational flows generated from rest by the impulsive motion of solid boundaries. Spherical means and Green's formula, irrotational flows with source and dipole far fields, impulse and added mass, steady flow past stationary solid bodies outside the boundary layer and associated wake are discussed.

Chapter 9. Two-dimensional irrotational flows are studied based on the theory of the complex potential and the method of conformal mappings in some detail. Examples of conformal mappings with a general introduction to aerofoil theory are given.

Chapter 10. Flows with circulation are developed in this chapter. It also includes flows around a rotating cylinders, flow due to motion of an aerofoil section, flow around an aerofoil held at rest due to an oncoming stream with given velocity, forces on aerofoils and in particular, the lift forces associated with the cambered Zhukovski aerofoil at zero incidence.

Chapter 11. Wing theory deals with the three-dimensional flows around wings and the analysis of conditions in which wings are aerodynamically efficient, in the sense that they can generate large lifts at the expense of very small drag forces. In addition to generation of irrotational flows with circulation around an aerofoil section, some residual vorticity, called the *trailing vorticity* is produced. The theory of vortex dynamics is developed to describe the properties of trailing vorticity. Special attention is given to lift and induced drag forces in two-dimensional flows around aerofoil sections and in three-dimensional flows around wings.

In this modern book, Lighthill has given an elegant first introduction to theoretical fluid mechanics and at the same time, provided the basic foundation of the theory of irrotational fluid flows based on theoretical and experimental studies to generate mathematical models. This book is sure to convince the reader that working in this area is challenging, enlightening and joyful. He concludes the book by adding the following statement in the end of Chapter 11:

"The purpose of this concluding section has been to indicate wide areas of applications of knowledge regarding wings and winglike surfaces able to achieve low ratios of drag to lift. All of them are areas where detailed theoretical studies of the types set out in this book, concerned with irrotational flows and boundary layers and shed vorticity, have been, and may still be, most fruitfully applied."

4.8 *Tropical Cyclone Disasters*, (Eds. Lighthill, J., Zhang, Z., Holland, G., and Emanuel, K.), Peking University Press, Beijing, China, 1993

When Sir James Lighthill was chairman of the International Council of Scientific Unions (ICSU) in 1991, he expressed deep and enduring concern on enormous loss of human lives and immense damage to property from natural disasters such as tropical cyclones (TC), hurricanes and tornadoes in many countries of the world. Coastal flooding by storm surges and river valley flooding both of TC origin are especially disastrous in many countries of southeast Asia including Bangladesh, China, India and Japan. He proposed an International Symposium on Tropical Cyclone Disasters in collaboration with the World Meteorology Organization (WMO), the Scientific Committee on Oceanic Research (SCOR), ICSU and the Chinese Academy of Sciences (CAS) for the benefit of developing and developed countries of the world. In collaboration with Zheng Zhemin of CAS, Lighthill organized probably the first international symposium in Beijing, China in October 12-16, 1992. Sponsored by 19 international and Chinese scientific organizations, this symposium turned out to be extremely successful in several aspects. A total of 80 scientists and engineers from 15 countries including Australia, Bangladesh, Barbados, France, Germany, Hong Kong, India, Japan, Macao, Madagascar, Malaysia, Russia, Switzerland, United Kingdom and the United States actively participated in the symposium. At the same time, this symposium has provided a unique opportunity to support the UN International Decade for Natural Disaster Reduction.

This volume serves as the Proceedings of the ICSU/WMO International Symposium that was edited by James Lighthill, Zheng Zhemin, Greg Holland and Kerry Emanuel. Published by Peking University Press, Beijing, China, this Proceedings has become a permanent and valuable document for a large public as it contributes in many ways toward advanced study and research of tropical cyclones and improving the timeliness and reliability of TC disaster warnings.

The Proceedings have eleven chapters and fifty nine research articles including Lighthill's first article on a broad introduction to the symposium's objectives and last article on final recommendations of the symposium. Included are also many theoretical, observational and computational research papers on different aspects of tropical cyclones. These studies may ultimately prove useful for development of more effective techniques producing reliable warnings of dangerous coastal inundations. The contents of this volume include:

A. General Introductory Material,

B. Observation, Data Collection and Assimilation,

C. Genesis of Tropical Cyclones,

D. Tropical Cyclone Forecasting from Numerical Models,

E. The Track Followed by a Tropical Cyclone,

F. Predicting Changes in Tropical Cyclone's Intensity,

G. Air-Sea Interaction and Surface Waves in Tropical Cyclones,

H. Storm Surges Generated by Tropical Cyclones,

I. Exceptionally Intense Precipitation and Associated Flooding,

J. Forecasting and Mitigation of Tropical Cyclone Disasters,

K. Concluding Paper.

In his introductory paper, Lighthill characterizes the tropical cyclone as a heat engine doing work by a classical Carnot cycle of heating at the warm ocean surface and cooling in the cold upper surface and cooling in the cold upper atmosphere. He also introduces Ertel's potential vorticity distributions as essential elements for understanding the development of intense wind fields and indicates its usefulness to study tropical cyclone mechanisms. In his broad introduction to the symposium's objectives, Lighthill states: "... Scientists have a major role in the work of the International Decade, especially by helping to improve the accuracy of techniques for detailed forecasting of Natural Disasters so that populations under threat may learn to rely on the predictions and consequently to follow the emergency measures recommended for their protection."

In his delightful simplified overview of generation of tropical cyclones, Lighthill states that oceans can be viewed as the major source of energy for intense tropical winds. This phenomenon depends on the detailed nature of very strong interactions between the atmosphere and oceans at high wind speeds. These interactions lead to extensive bubble and spray formation at the ocean surface with vigorously breaking of ocean waves. On the other hand, considerable research is needed for understanding of important problems which include:

(i) interpretation of extensive data on generation and intensification of tropical cyclones, and other similar natural hazards,
(ii) reliable prediction of the tropical cyclone tracks,
(iii) expected levels of storm-surge inundations,
(iv) any flooding as a result of heavy precipitation forecast for after the tropical cyclone moves over land.

To explain the mechanisms responsible for the generation of tropical cyclones, it is appropriate to quote Lighthill's own words:

"In particular, the energy source for intense TC winds depends on huge transfers of both water vapour and heat from ocean to atmosphere; transfers which also act, of course, to make the TC itself a source of dangerously extreme precipitation levels when it passes over land. Dissipation of the transferred energy, on the other hand, arises from the force F with which the winds are resisted by the ocean surface itself; this represents, needless to say, a transfer of momentum "back" atmosphere to ocean — which, in turn is responsible for generating storm-surge flooding when the region of momentum transfer approaches a coastline...."

The tropical cyclone problem basically consists of several major components associated with very strong interactions between the atmosphere and the ocean including generation and initial development, intensity and structure change, movement, landfall, extratropical transition, and seasonal and climate scale variations. The United Nations (UN) has announced the period from 1990-1999 as the International Decade for Natural Reduction (IDNDR) and has declared that tropical cyclones are among the greatest of all natural hazards. The research challenge for the IDNDR is to develop strategies that lead to a better tropical cyclone forecasting and warning system that minimizes catastrophic damage, loss of human lives and economic resources. Another challenge is to attract a multinational and multidisciplinary group of scientists to provide an integrated approach combining research, technique development, improved observation systems, and implementation of adequate community response procedures. Meeting these challenges require advanced study and research for improved understanding of all major components associated with tropical cyclone mechanisms. This may lead to the development of new techniques, major refinement of current methods and modern analysis of the diverse data available in many countries.

Sir James also provided special attention to some valuable new insights into tropical cyclones from a theoretical fluid mechanics point of view. According to Lighthill, a theoretical formulation of the dynamics of the atmosphere is beginning to enhance understanding of the formation and the subsequent development of tropical cyclones. In his own article jointly with Dr. Bryan Johns on Modelling of Storm Surges in the Bay of Bengal published in this Proceedings, they provided all detailed description of the major role of theoretical fluid mechanics in understanding storm-surge responses to intense tropical cyclones. The existence of very large amplitude of the ocean-surface elevation is primarily associated with the resonance of the atmospheric forcing effects. This resonance-type effect occurs when the phase velocity of the so-called *edge waves* coincides with the velocity of travel of the storm center along the coast. Such large amplitude wave motions cannot be explained by the linear theory. Consequently, it would be necessary to extend the theory taking nonlinear effects into account. In his presentation at the Air-Sea Interaction Symposium which was a part of the 8th International Congress of Theoretical and Applied Mechanics at Haifa in August 1992, Lighthill explained the development of nonlinear edge-waves in oceans generated by tropical cyclones, and described the impact of such large amplitude ocean waves on coastal areas with potential for devastating coastal inundations. He also indicated that his analysis can be applied to storm surges in the Bay of Bengal and the mechanisms responsible for their generation. In addition, he also made references to observations indicating that similar phenomena occur also in the coast of China and other coastal areas in the world.

In closing of his opening introduction to the symposium Lighthill states: "... should successfully achieve concensus on the right way forward both (a) in cost-effective new techniques for TC reconnaissance and (b) in new arrangements that may be needed to promote cooperative interdisciplinary research for Asia as a whole on Tropical Cyclone Disasters."

4.9 *Fluid Mechanics*, A Chapter in Twentieth Century Physics (Eds. Laurie M. Brown, Abraham Pais, and Sir Brian Pippard), Institute of Physics and American Institute of Physics, New York, Volume II, 1995

We close this chapter by adding Sir James Lighthill's famous and authorative 117-page chapter on Fluid Mechanics that was published in the *Encyclopedia of Twentieth Century Physics* in 1995. He considered fluid mechanics as a subdiscipline of physics and it is, indeed, another great success story of twentieth century physics. In this chapter, Lighthill described all major developments in fluid mechanics during the twentieth century with successful applications to a wide variety of subjects including the global climatology of the atmospheric and oceanic circulations, in general and to the understanding and prediction of natural disasters such as violent storms and floods, in particular. Many of these advances can be traced back with the revolutionary discovery of the boundary layer theory in fluid mechanics in 1904 by Ludwig Prandtl (1875-1953). He also indicated the great importance of the achievement of truely global weather forecasting of increased accuracy by means of computational fluid dynamics (CFD) during the last quarter of the twentieth century. The subject of CFD is a combination of three subjects: theoretical fluid dynamics, numerical mathematics, and computer science. John von Neumann (1903-1957) first realized the great importance of computational fluid dynamics research on the basis of the Euler or Navier-Stokes equations, and then recognized the fundamental nature of CFD for other problems in physics and engineering. So, John von Neumann may be called the founding father of computational fluid dynamics.

Describing the success story of the twentieth century fluid mechanics, Sir James emphasized its major roles in the service of mankind, and examined the endless interaction between many disciplines including applied mathematics, physics, geophysics, geology, mechanical engineering, aeronautical engineering, ocean engineering, environmental engineering, and biological sciences. Its completely interdisciplinary character is one of its powerful assets and it is noted for its combined strength in all of above disciplines. The twentieth century's technological revolution transformed human life on Earth specifically through aeronautical, ocean and environmental engineering. Research and development are involved in almost all types of service work to meet the major needs of the modern society and to improve the knowledge and understanding of natural environment. Quite recently, a very notable success in research and development in aeronautical and ocean engineering has been achieved that provided the fast and long – distance transport of people, high-value goods, ship-design technology, good 'seakeeping' (good performance in heavy seas) and economic activity, air-ocean interaction, wide spread use of offshore structures, oil and gas exploration, air and marine pollution leading to a matter of urgent public concern.

Although Lighthill made relatively less contribution to the understanding of the complex problem of turbulence except generation of noise by turbulent flows. However, he recognized that its occurrence is widespread in nature, especially in oceans and the atmosphere. In this article, Lighthill made some comments on turbulence and its recent developments in the twentieth century by stating that progress has been much more gradual except A. N. Kolmogorov's (1903-1987) universal equilibrium theory of isotropic turbulence based on the Navier-Stokes equations. However, there are certain major difficulties (see Debnath (1998)) associated with the three-dimensional Navier-Stokes equations as there are no general results for the Navier-Stokes equations on existence of solutions, uniqueness, regularity, and continuous dependence on the initial data. These are the challenging unsolved problems of the 21st century. According to Lighthill, "For turbulence itself, just as for transition, a key paper offering the subject increased coherence emerged from Russia in the early 1940's. Kolmogorov's 1941 paper [50] explained how, even though ideas of turbulence 'as a single integrated phenomenon', are erroneous (section 10.2.3), and although turbulence (like transition) exists in a variegated taxonomy of types, nonetheless all those types have their small-scale features in common at very high Reynolds number." He also stated that the Kolmogorov theory deals with only isotropic turbulence which is just one of more kind of turbulence, by no means *generic*. Thus, isotropic turbulence satisfies certain special rules and laws including the law for the Kolmogorov spectrum $E(k)$ of the kinetic energy of turbulence. This famous law of Kolmogorov implies that every kind of turbulence becomes isotropic at relatively high wavenumbers and obeys those special laws and rules. This has been one of his major concerns about the Kolmogorov theory.

Turbulence, is indeed, one of the most complex physical phenomena in all of fluid mechanics. It is a process where the main energy containing large eddies, while receiving energy directly, are gradually transformed into a cascade of small eddies which eventually dissipate into heat. At every stage of the cascade, nonlinear convective effects involved in the basic nonlinearity of momentum transport tend to generate *overtones* of somewhat higher wavenumbers. After several stages, the turbulent wavenumber spectrum $E(k)$ becomes statistically decoupled from the main-energy containing eddies, and attains a universal statistical equilibrium, independent of the kind of turbulence in which the small eddies are embedded. So, turbulence essentially deals with the interactions of eddies of various sizes or scales. Based on the assumption of the local similarity of small eddies, Kolmogorov and Oboukhov described the statistical properties of small eddies in turbulence at a very large Reynolds number and predicted the wavenumber spectrum of the turbulent energy. Their pioneering work provided the fundamental basis of all subsequent developments in modern theories of turbulence. No doubt, the complexity of the dynamics of turbulent flows still presents a number of new and challenging problems for the 21st century. Recent trends in turbulence research suggest that theoretical development alone, without experimental and computational work, can hardly solve the complex problem of turbulence.

Considerable progress has been made over the last three decades to gain understanding of turbulence through new developments of theory, experiment, and computation. More and more new evidence has been accumulated for the mathematical and physical descriptions of turbulent motions in both two and three dimensions. In spite of tremendous progress, there are still a number of open questions and unsolved problems. These include coherent structures and intermittency effects, existence, uniqueness, regularity and singularities of the Navier-Stokes equations, non-Gaussian statistics of turbulence flows, existence and uniqueness of vortices, vortex interactions and singularities of the vorticity distribution, perturbations to the small scale produced by non-isotropic, non-Gaussian and inhomogeneous large-scale motions, precise measurements and computations of small-scale turbulence. No complete theory is yet available for the problem how eddy structure of turbulence evolves under the action of mean distortion and even during the mutual random interaction of eddies of different sizes or scales.

No doubt, the complexity of turbulent flows still presents a number of new and challenging problems. In recent years, considerable research on various aspects of turbulence has been carried out by using wavelet transforms, fractals and multifractal analysis. These new methods are found to be useful in studying self-similarity, intermittency and local nature of the dynamics and other aspects of turbulent flows. Even though the above methods have not yet been fully successful, they still have great potential. Some researchers raised new questions as to whether the structure of velocity fluctuations in turbulence resembles a distribution of accumulating oscillations or has fractal or multifractal structure. In order to resolve these questions and other open questions and unsolved problems, the wavelet transform, and/or multifractal analysis either modified or combined with other theoretical, experimental and/or computational methods and algorithms is likely to play a major role in future turbulence research during the twenty first century.

Describing enormous interactions between the atmosphere and oceans and their significant impact on everyday life, Lighthill expressed his views as follows:

"The Earth's fluid envelope is a mixture of air and water. In the atmosphere air predominates, yet water (transferred from the ocean) plays a spectacular role. In oceans, rivers and lakes water predominates yet they teem with life because of dissolved oxygen and CO_2. Not only mass exchange between the different components but also transfers of heat and of momentum are major influences on their dynamics.

Study of that dynamics, besides its scientific interest, has crucially important applications to meeting the requirements of countless industries — agriculture, shipping, aviation, energy-supply, coastal engineering, river-management, etc — for improved weather forecasting; see section 10.5.2, where some account of climate-change forecasting is also given. Practically none of the dynamical or forecasting problems involve only one component of the Earth's fluid envelope in isolation from others; nevertheless, this section 10.5.1 is devoted to certain features of ocean and river dynamics which, while influenced by atmospheric input, assume something of a life of their own because patterns of water flow propagate in a wave-like manner."

Sir James concluded this chapter by stating: "The fluid mechanics of just air and water were seen in this chapter to have yielded 'a fair share' out of many exciting developments in twentieth century physics; those mentioned having been chosen for their general importance and interest to non-specialists readers.... Above all he hopes that the chapter will extend the interests of many generally 'physics-minded' readers into fluid mechanics."

As global advocate, Lighthill described many major theoretical, experimental and computational progress made on the subject during the twentieth century, and cited many unsolved problems and open questions in this rapidly growing field for the 21st century. This is one of his most influential articles ever written on fluid mechanics which is really another success story of the twentieth century physics as Sir James indicated. For a complete information on the historical introduction to turbulence, the reader is referred to the list of large number of major references cited in the end.

Chapter 5

Supersonic and Subsonic Aerodynamic Flows

> "The first process ... in the effectual study of science must be one of simplification and reduction of the results of previous investigations to a form in which the mind can grasp them."

James Clerk Maxwell

> "As a *science*, mathematics has been adapted to the description of natural phenomena, and the great practitioners in this field, such as von Kármán, Taylor and Lighthill, have never concerned themselves with the logical foundations of mathematics, but have boldly taken a pragmatic view of mathematics as an intellectual machine which works successfully. Description has been verified by further observation, still more strikingly by prediction ...."

George Temple

5.1 Introduction

Even a glance at the chronological list of his research papers and books appended to *Collected Papers of Sir James Lighthill* (1997) will show that his whole life was packed with research and publication activities. His creativity was simply phenomenal, and his deep involvement in teaching, academic and professional service was enormous, and has many ramifications. I have no intention of discussing all of his contributions in many areas which include aerodynamics, fluid dynamics, aeroacoustics, waves in fluids, biofluid mechanics, geophysical fluid dynamics, boundary layer theory, vorticity dynamics, magnetohydrodynamics, atmospheric and ocean dynamics. In each of these areas, he made major and unique contributions with new mathematical results and profound physical significance. His mathematical research always dealt with vivid imagination that responded to the 1983 quotation of V. I. Arnold: "In every mathematical investigation, the question will arise whether we can apply our mathematical results to the real world."

Sir James Lighthill's research career began at National Physics Laboratory (NPL) with publication of several reports and memoranda of the Aeronautical Research Committee (ARC) on supersonic airfoil theory in various configurations during 1943-1947. Based on research on the foundations of the supersonic airofoil theory of Prandtl and his students, Ackeret (1925) and Busemann (1935), Lighthill published a series of technical reports on the two-dimensional supersonic flow past the general thin sharp-nosed aerofoil, and development of shock waves. The early days of aerodynamics became James' favorite subject for a lecture of reminiscences to a scientific audience in his subsequent years.

5.2 Supersonic Airfoil Theory

In 1943, Lighthill's first research paper was published in the form of an ARC Report when he was 19 years old. Subsequently, he published many more ARC Reports and Memoranda that contained the deflection of a supersonic stream at the trailing edge of a two-dimensional wing and a general theory of supersonic flow past wings of finite span.

Based on the theory of compressible fluid flow and small disturbances to the original uniform stream with velocity V along the x-axis, Lighthill formulated the equation of continuity and the equation of motion in the form

$$\text{div}\,\mathbf{q} = \frac{V^2}{c^2}\,\frac{\partial u}{\partial x}, \tag{5.2.1}$$

$$\nabla^2\phi = M^2\,\frac{\partial^2 u}{\partial x^2}, \tag{5.2.2}$$

where $\mathbf{q} = \nabla\phi$, $q^2 = (u+V)^2 + v^2 + w^2$, u, v, w are the excess velocities in the x, y, z directions, $\frac{dp}{d\rho} = c^2$ is the square of the speed of sound and $M = (V/c)$ is the Mach number.

When $M = 0$, equation (5.2.2) reduces to the Laplace equation for the incompressible flow. For $0 < M < 1$, it is the fundamental basis of the Prandtl-Glauert theory of compressible fluid flow in which correction factor of the form $(1 - M^2)^{r/2}$ is applied to the incompressible fluid flow theory.

When $M > 1$, equation (5.2.2) assumes the form

$$\frac{\partial^2 \phi}{\partial y^2} + \frac{\partial^2 \phi}{\partial z^2} - \alpha^2 \frac{\partial^2 \phi}{\partial x^2} = 0, \tag{5.2.3}$$

where $\alpha^2 = M^2 - 1$.

Equation (5.2.2) admits a fundamental solution of the form

$$\phi(x, y, z) = \frac{1}{\sqrt{(x^2 + y^2 + z^2) - M^2(y^2 + z^2)}}, \tag{5.2.4}$$

which, when $M < 1$, describes the motion due a source in uniform stream. It can readily be obtained from the incompressible source-potential $(x^2 + y^2 + z^2)^{-\frac{1}{2}}$ by a transformation $x = x'\sqrt{1 - M^2}$. When $M > 1$, (5.2.4) still represents a solution of equation (5.2.2), and of equation (5.2.3) and it can be written as

$$\phi(x, y, z) = \frac{1}{\sqrt{x^2 - \alpha^2(y^2 + z^2)}}. \tag{5.2.5}$$

It would be more accurate to define the potential $\phi = 0$ when $x < \alpha\sqrt{y^2 + z^2}$ and by (5.2.5) when (x, y, z) lies on the cone $x > \alpha\sqrt{y^2 + z^2}$. On the other hand, as $x \to \alpha\sqrt{y^2 + z^2}$, $\phi \to \infty$ which has no physical significance. When ϕ is independent of y, equation (5.2.3) reduces to the form

$$\frac{\partial^2 \phi}{\partial z^2} = \alpha^2 \frac{\partial^2 \phi}{\partial x^2}. \tag{5.2.6}$$

The solution of this equation describes the plane supersonic flow. In an earlier ARC report, Lighthill also described a theory of plane supersonic flow to orders higher than the first. Von Kármán and Moore (1932) first applied solution (5.2.5) to describe flow past bodies of revolution using cylindrical polar coordinates (x, r, θ) where $r^2 = y^2 + z^2$. If the single source is replaced by a linear superposition of sources and sinks along the axis $r = 0$, he obtained a physically realistic velocity field. With the strength of an element dx' replaced by $f(x')dx'$, a velocity potential exists and is given by

$$\int_0^{x-\alpha r} \frac{f(x')dx'}{\sqrt{(x - x') - \alpha^2 r^2}} \tag{5.2.7}$$

which satisfies the boundary conditions for flow past a body of revolution for certain function f.

Integrating the Kármán solution (5.2.7) with $r^2 = (y-y')^2 + z^2$ over any range of y', Lighthill obtained a new solution of equation (5.2.3) in the form

$$\phi(x,y,z) = \frac{\pi}{\alpha}\int_0^{x-\alpha|z|} f(x')dx'. \tag{5.2.8}$$

Similarly, solution of (5.2.7) with a doublet distribution can be obtained in the form

$$\phi(x,y,z) = \pi\ sgnz \int_0^{x-\alpha|z|} g(x')dx'. \tag{5.2.9}$$

Thus, the sum of (5.2.8) and (5.2.9) represents the most general two dimensional supersonic flow.

In particular, he showed that there is no induced drag and, for a rectangular wing of aspect ratio A at Mach number M, the effective angle of incidence is reduced by a factor

$$1 - \frac{0.5}{A\sqrt{M^2-1}}. \tag{5.2.10}$$

Based on the work of Kármán and Moore (1932), Sir James developed the theory of supersonic flow past bodies of revolution. For slender bodies of revolution pointed at both ends, the slope of whose meridian section is continuous, he obtained the velocity distribution over the entire field correct to the first and second orders in the thickness of the body that was placed in a uniform supersonic flow. He also derived the asymptotic formula for the drag as the thickness tends to zero in the form

$$Drag = \frac{1}{2}\,\rho\,V^2 \cdot \frac{1}{2\pi}\int_0^{\ell}\int_0^{\ell} \log\left(\frac{\ell}{|x-y|}\right) S^{''}(x)\,S^{''}(y)\,dx\,dy, \tag{5.2.11}$$

where $S(x)$ is the cross section of the body at a distance x along its length ℓ. The drag coefficient $C_D = (drag)/\left(\frac{1}{2}\rho V^2\right)\left(\frac{\pi}{4}t^2\right)$ varies as $(t/\ell)^2$ and is independent of the Mach number $M(=V/c)$ of the main stream.

Much more remarkable, however, were the series of four papers published in the *Proceedings of the Royal Society, London* in 1947 on the hodograph transformation for the solution of transonic (both subsonic and supersonic) flow problems. Lighthill developed the hodograph transformation as a new approach for the study of compressible flow problems with regions of both subsonic and supersonic flows. With independent flow variables τ (square of velocity) and θ (direction of motion), he derived the partial differential equation for the stream function of the compressible flow in the form

$$PQ\psi_{\tau\tau} + PQ_\tau\psi_\tau - \psi_{\theta\theta} = 0, \tag{5.2.12}$$

where P and Q are functions of τ only so that

$$P(\tau) = \left[\frac{2\tau(\gamma-1)}{(\gamma-1)\tau-(\gamma-1)}\right](1-\tau)^{\gamma/\gamma-1}, \tag{5.2.13}$$

$$Q(\tau) = \frac{2\tau}{(1-\tau)^{1/(\gamma-1)}}, \tag{5.2.14}$$

γ is a adiabatic constant.

He also obtained the alternative form of the equation for ψ in the case of supersonic flow with t in place of τ as

$$\psi_{tt} + S(t)\,\psi_t - \psi_{\theta\theta} = 0, \tag{5.2.15}$$

where

$$\frac{dt}{d\tau} = (PQ)^{-\frac{1}{2}}, \qquad S(t) = \frac{1}{2}(PQ)^{\frac{1}{2}}\left(Q_\tau Q^{-1} - P_\tau P^{-1}\right), \tag{5.2.16}$$

and

$$t = \left(\frac{\gamma+1}{\gamma-1}\right)^{\frac{1}{2}} \tan^{-1}\left[\left\{\frac{(\gamma+1)\tau-(\gamma-1)}{(\gamma+1)(1-\tau)}\right\}^{\frac{1}{2}}\right] - \tan^{-1}\left[\left\{\frac{(\gamma+1)\tau-(\gamma-1)}{(\gamma-1)(1-\tau)}\right\}^{\frac{1}{2}}\right]. \tag{5.2.17}$$

For subsonic flow, he presented the equation in the similar form

$$\psi_{ss} - T(s)\,\psi_s + \psi_{\theta\theta} = 0, \tag{5.2.18}$$

where

$$s = \sigma + \left(\frac{\gamma+1}{\gamma-1}\right)^{\frac{1}{2}} \tanh^{-1}\left[\left\{\frac{(\gamma-1)-(\gamma+1)\tau}{(\gamma+1)(1-\tau)}\right\}^{\frac{1}{2}}\right] - \tanh^{-1}\left[\left\{\frac{(\gamma-1)-(\gamma+1)\tau}{(\gamma-1)(1-\tau)}\right\}^{\frac{1}{2}}\right], \tag{5.2.19}$$

$$\sigma = \frac{1}{2}\log\{2(\gamma-1)\} - \left(\frac{\gamma+1}{\gamma-1}\right)^{\frac{1}{2}} \tanh^{-1}\left[\left\{\frac{\gamma-1}{\gamma+1}\right\}^{\frac{1}{2}}\right], \tag{5.2.20}$$

and

$$T(s) = -\frac{1}{2}\left[(-PQ)^{\frac{1}{2}}\left(Q_\tau Q^{-1} - P_\tau P^{-1}\right)\right]. \tag{5.2.21}$$

It is noted that the constant σ is to make τ to be asymptotically $\exp(2s)$ as $\tau \to 0$, and $s \to -\infty$, σ is the value of s at the sonic speed, $\tau = (\gamma-1)/(\gamma+1)$.

This work was not only mathematically rigorous, but also provided some justification of appropriate methods that have been applied to the general theory of supersonic flow past bodies of revolution. He first introduced the hodograph transformation for the development of a general theory of compressible flow applicable to problems with regions both of subsonic and supersonic flows. He obtained the solutions by the hydrograph method in the complex plane of velocity q and direction θ for the linear problem, while these solutions in the physical plane are not even approximately linear for transonic flows. In his first paper, he used the transonic problems in the steady plane adiabatic flow of gas in symmetrical channels in which the velocity increases from zero to infinity on the left to a supersonic value at infinity on the right. He also calculated the value of the stream functions $\psi_n(\tau)$ to help design the flow problem in symmetrical channels. This was followed by a mathematical study of the properties of functions $\psi_n(\tau)$ in this second paper in which he also developed asymptotic formula for both supersonic and subsonic τ. In the third paper of this series, Lighthill investigated how the problem of finding a plane steady adiabatic compressible flow round a body can be reduced to a given incompressible flow when the Mach number $M \to 0$. In his fourth paper, he constructed tables of functions $F_n(\tau)$, $F_n'(\tau)$, $\psi_n(\tau)$ and $\psi_n'(\tau)$ where $\psi_n(\tau) = \tau^{n/2} F_n(\tau)$ and $F_n(\tau)$ represents a hypergeometric function defined by $F(a_n, b_n; n+1; \tau)$, $a_n + b_n = n - (\gamma - 1)^{-1}$, and $a_n b_n = -n(n+1)/2(\gamma - 1)$. This series of four papers represents his significant contribution to fluid mechanics.

During 1947-1949, Lighthill published a series of significant research papers on supersonic flow past slender bodies of revolution at yaw and the position and strength of shock waves in aerodynamic problems of physical interest in Volume 1 of *Quarterly Journal of Mechanics and Applied Mathematics (QJMAM)*. He extended the linearized theory of supersonic flow around slender bodies of revolution, yawed or unyawed, with pointed or open bows to the case when the meridian section of the outer surface has discontinuous slope. His theory is fairly general so that it can be applied to flow outside slender ducts whose thickness is small compared with their lengths, but whose radius is not approximately constant. For the yawed flow at an angle ψ to the axis, the disturbance-velocity potential can be written $\phi_0(r, x) + \psi\, \phi_1(r, x) \cos\theta + O(\psi^2)$, where ϕ_1 approximately satisfies the equation

$$\frac{\partial^2 \phi_1}{\partial r^2} + \frac{1}{r}\frac{\partial \phi_1}{\partial r} - \frac{1}{r^2}\phi_1 - \alpha^2 \frac{\partial^2 \phi_1}{\partial x^2} = 0, \qquad \alpha = \sqrt{M^2 - 1}\,. \quad (5.2.22)$$

The solution of this equation is

$$\phi_1(r,x) = \frac{1}{r}\int_{c_0}^{x-\alpha r} \frac{(x-y)\,g(y)\,dy}{\left\{(x-y)^2-\alpha^2 r^2\right\}^{\frac{1}{2}}} = \alpha K_1(\alpha pr)\,g(x)\,, \quad (5.2.23)$$

where $K_1(z) = -K_0'(z) \sim z^{-1}$ as $z \to 0$, $K_0(z)$ is the Bessel function, c_0 is any positive constant, p^{-1} is $\int_0^x$ (in Heaviside notation). He also obtained correct second approximations of the solution of the yawed flow past a body of revolution. It is shown that the drag coefficient is no longer independent of Mach number, and tends to zero more slowly than square of the thickness of the body. The approximate expressions for the pressure field, drag, lift and moment. He also demonstrated the importance of his results in the theory of cylindrical sound-wave generated by the contraction and expansion of a small cylinder. His result for the drag of a given body of revolution gives the work done in contracting and expanding the small cylinder in a prescribed manner. In 1948, he wrote another paper on the position of the shock-wave in certain aerodynamic problems in QJMAM. In this paper, Lighthill obtained a first approximation to the strength and position of the shock-wave for small disturbances where the motions are generated by the uniform expansion of a cylinder or sphere into still air and for steady symmetrical supersonic flow past a cone. When air flows past any body of revolution whose tip is conical but which curve back convexly behind the tip, shock waves exist near the tip and have the same strength as if the entire body were conical. Based on a rigorous mathematical analysis, he provided a solid justification of approximate methods which have been employed in the general theory of supersonic flow past bodies of revolution.

In his two 1949 significant research papers published in the *Proceedings of the Royal Society of London*, Sir James investigated mathematically the theory of the diffraction of blast as well as the behavior of a plane shock wave of any strength, with a solid corner angle $\pi - \delta$, where δ is small. His two other 1949 papers dealt with the prediction of the positions and strengths of shock waves induced by three-dimensional bodies or other disturbances. He also considered the complete flow field by calculating the positions of purely cylindrical, spherical and conical shocks that are obtained analytically to a good approximation from the exact nonlinear equations. However, for more general situations, any regular perturbation expansion method breaks down in the far field since shock deviates increasingly from the Mach cone as it propagates in the medium. Lighthill then formulated a new mathematical method to obtain an approximate solution that is uniformly valid everywhere in the flow field. His series of famous papers on a technique for rendering approximate solutions to physical nonlinear problems uniformly valid is an important step forward in shock dynamics. In these papers, he developed a new method which is known as the *Method of Strained Coordinates*. In this paper and in other papers, Lighthill made some significant contributions to dynamics of shock waves which is an integral part of supersonic flow past a solid body or in a duct of nonuniform width. Shocks are treated as surfaces of discontinuity in inviscid aerodynamics.

Citing only one paper of Henri Poincaré, Lighthill published another important paper on "A Technique for Rendering Approximate Solutions to Physical Problems Uniformly Valid" in the *Philosophical Magazine*, **40** (1949) 1179-1201. In this paper, he suggested a new and important approximate method in finding solution of nonlinear partial differential equations near a singular characteristic of the approximate solution with several examples of interest. His new method of solution can be applied to any wave propagation problem with initially large disturbances. He demonstrated that the critical behavior of the expansion of the unknown flow variable, v is associated with a singularity (algebraic or logarithmic) of the first approximate solution v_0. When such a singularity is present in the first approximate solution of a partial differential equation, it propagates along a characteristic of the linear equation for v_0. In short, Lighthill developed a method of solution of partial differential equations in two variables near a singular characteristic of the approximate solution and then applied his method to an equation of mixed type of the form

$$\left(1-r^2\right)\frac{\partial^2 v}{\partial r^2}+\left(r^{-1}-2r\right)\frac{\partial v}{\partial r}+\frac{1}{r^2}\frac{\partial^2 v}{\partial \theta^2}=0. \tag{5.2.24}$$

Such equation arises in aerodynamics and it is elliptic for $r < 1$ and singular at $r = 1$ which is a characteristic. Whitham (1952) discovered a general theory of an axisymmetric supersonic flow. This theory provided all information similar to that from the Friedrichs theory of two-dimensional supersonic flow past airfoils, but only to a first, not to a second approximation. However, Whitham's theory holds without modification in more general flows.

This is followed by Lighthill's another remarkable paper published in the *Aeronautical Quarterly*, Vol. **III** (1951) 193-210 on "A New Approach to Thin Aerofoil Theory" in which he applied his general method of approximate solutions to the problem of improving the theory of thin wings near leading edge due to Goldstein (1948). The main reason of writing this paper is the following. The lifting theory of flow over wings was not completely satisfactory, particularly near the leading edge because this theory neglected the squares of the disturbances to the main stream. From the mathematical and physical points of view, the theory is not realistic as for no wings, however, thin, the neglect of the square of the disturbances near the leading edge stagnation point cannot be logically justified. Obviously, Lighthill took a significant step to modify the existing theory by developing a simple lifting-surface theory so that it becomes a correct approximation even near to the leading edge. He made some important arguments against the existing theory of the two-dimensional incompressible potential fluid flow about a thin aerofoils due to Goldstein (1948) and others. He stated that the existing theory is satisfactory for practical purpose of obtaining a sufficiently accurate results for a short time, but it does not offer any possibility of extension to three-dimensional problems or to compressible fluid flows because the theory is based on the transformation of conformal mapping in complex analysis.

With $C(x)$ and $F(x)$ as functions defining the shape of the aerofoil, Lighthill obtained the complete solution for the velocity distribution for the general case of flow about the aerofoil by a direct series expansion in powers of the disturbances, upto and including either the first or the second power. His 1951 formula for the velocity field is given by

$$\begin{aligned} u_1 - i\,v_1 = {} & \frac{1}{\pi}\int_0^c \frac{F'(w)dw}{x+iy-w} + \left(\frac{c-x-iy}{x+iy}\right)^{\frac{1}{2}} \\ & \times \left[\alpha + \frac{1}{\pi}\int_0^c \frac{C'(w)}{x+iy-w}\left(\frac{w}{c-w}\right)^{\frac{1}{2}} dw\right], \end{aligned} \tag{5.2.25}$$

where $0 < x < c$ and $(\varepsilon\alpha)$ is the angle of incidence and the parameter ε is the size of the disturbances. This formula represents Goldstein's Approximation I for the complete velocity distribution. Lighthill's method is fairly general in the sense that the velocity field (u_n, v_n), for $n = 1, 2, 3, \ldots$ can be calculated, and the velocity field given by the first or second order approximation is uniformly valid, even near the leading edge, if it is shifted bodily downstream through a distance equal to half the leading edge radius of curvature. Lighthill also obtained the formulas for the fluid speed $q = (u^2 + v^2)^{\frac{1}{2}}$ on the surface of the aerofoil and the distribution (q/U), the ratio of fluid speed to its value at infinity, over the surface of the aerofoil. The upshot of Lighthill's theory is that it is a significant work which made possible to verify the theory against Goldstein's Approximations I, II and

III. Lighthill's theory also gave a new formula for the lift coefficient

$$C_L = 2\pi\varepsilon\alpha - \frac{4}{c}\int_0^c \{\varepsilon C'(w) + \varepsilon^2 U'(w)\}\left(\frac{w}{c-w}\right)^{\frac{1}{2}} dw, \quad (5.2.26)$$

where $U(x) = \alpha\left(\frac{c-x}{x}\right)^{\frac{1}{2}} F(x)$ and $V(x) = \alpha\left(\frac{c-x}{x}\right)^{\frac{1}{2}} C(x)$. This formula represents a second order approximation to the no-lift angle.

Finally, Lighthill concluded his paper by formulating the second order approximation to the surface velocity field. With the equation of the aerofoil surface as $y = C(x) \pm F(x)$, $0 \leq x \leq c$, where $C(0) = C(c) = 0$ and $F(x) \sim (2\rho_L x)^{\frac{1}{2}}$ as $x \to 0$, so that ρ_L is the leading edge of radius of curvature, and if α is the angle of incidence, the ratio (q/U) of the fluid speed on the aerofoil surface to its value at infinity is given by the Lighthill formula

$$\begin{aligned}\left(\frac{q}{U}\right) = \left(\frac{x}{x+\frac{1}{2}\rho_L}\right)^{\frac{1}{2}} \Bigg[1 &- \frac{1}{2}\alpha^2 + \frac{1}{\pi}\int_0^c \frac{F'(w)+V'(w)}{(x-w)} dw \\ &\pm \left(\frac{c-x}{x}\right)^{\frac{1}{2}} \left\{\alpha + \frac{1}{\pi}\int_0^c \frac{C'(w)+U'(w)}{(x-w)}\left(\frac{w}{c-w}\right)^{\frac{1}{2}} dw\right\} \\ &+ \{C(x) \pm F(x)\}\,\{C''(x) \pm F''(x)\} \\ &+ \frac{1}{2}\{C'(x) \pm F'(x)\}^2 + \frac{1}{4}\left(\frac{\rho_L}{x}\right)\Bigg], \end{aligned} \quad (5.2.27)$$

where the $\pm$ signs refer to the upper and lower surfaces respectively, $U(x)$ and $V(x)$ are defined by

$$U(x) = u_{11}C(x) + u_{12}F(x), \qquad V(x) = u_{11}F(x) + u_{12}C(x), \quad (5.2.28)$$

$$u_{11} = \frac{1}{\pi}\int_0^c \frac{F'(w)dw}{(x-w)}, \quad (5.2.29)$$

$$u_{12} = \left(\frac{c-x}{x}\right)^{\frac{1}{2}}\left[\alpha + \frac{1}{\pi}\int_0^c \left(\frac{w}{c-w}\right)^{\frac{1}{2}} \frac{C'(w)dw}{(x-w)}\right]. \quad (5.2.30)$$

Lighthill's solution (5.2.27) is uniformly valid provided singular part like x^{-1} is subtracted and the remainder is multiplied by $\left\{x/(x+\frac{1}{2}\rho_L)\right\}^{\frac{1}{2}}$.

Lighthill's analysis here deals with the two-dimensional incompressible irrotational flow problem of correcting the theory of thin wings near a rounded leading edge. However, it has great potential for extension to three-dimensional compressible flow problems.

Chapter 6

Aeroacoustics and Nonlinear Acoustics

"It seems to be one of the fundamental features of nature that fundamental physics laws are described in terms of great beauty and power. As time goes on, it becomes increasingly evident that the rules that the mathematician finds interesting are the same as those that nature has chosen."

Paul Dirac

"Our present analytical methods seem unsuitable for the solution of the important problems arising in connection with nonlinear partial differential equations, and, in fact, with virtually all types of nonlinear problems in pure mathematics. The truth of this statement is particularly striking in the field of fluid dynamics. Only the most elementary problems have been solved analytically in this field..."

John von Neumann

6.1 Introduction

During his stay in Manchester from 1946-1959, Lighthill first initiated two new and major fields of modern fluid mechanics that included aeroacoustics and nonlinear acoustics. Lighthill's pioneering work on sound generated aerodynamically appeared in the *Proceedings of the Royal Society of London*, Vol. **A211** (1952) 564-587. This work along with his many other papers on high-speed aerodynamics, how flows generate sound, and the theory of turbulent jet noise are considered as classic, and led to a totally new branch of fluid mechanics which became known as *aeroacoustics*, and has immense industrial and technological importance. Indeed, he is considered as the founder of the modern theory and applications of aeroacoustics that basically deals with the mutual interaction of sound with fluid flows and turbulence as a source of sound. He was probably the first applied mathematician who developed the whole of modern aeroacoustics from unified mathematical and physical viewpoints.

6.2 Lighthill's Theory of Sound Generated Aerodynamically

Based on the classical Newton's law of motion and conservation of mass, Lighthill supposed that sound originates in a limited region of disturbed air, such as the turbulent jet behind a jet engine. Outside that region where air is at rest, sound field is governed by the classical wave equation. Inside the disturbed region, fluid motions can be considered as the main source of sound field that is radiated according to law of acoustics. If the source region does not contain any sources of mass, or of momentum, the far field would be dominated by quadrupole radiation field. Hence, the propagation of sound in a uniform medium at rest due to externally applied fluctuating stresses is governed by the equations

$$\frac{\partial \rho}{\partial t} + \frac{\partial}{\partial x_i}(\rho v_i) = 0, \tag{6.2.1}$$

$$\frac{\partial}{\partial t}(\rho v_i) + a_0^2 \frac{\partial \rho}{\partial x_i} = -\frac{\partial T_{ij}}{\partial x_j}, \tag{6.2.2}$$

$$\frac{\partial^2 \rho}{\partial t^2} - a_0^2 \nabla^2 \rho = \frac{\partial^2 T_{ij}}{\partial x_i \partial x_j}, \tag{6.2.3}$$

where the instantaneous applied stress field at any point is

$$T_{ij} = \rho v_i v_j + p_{ij} - a_0^2\, \rho\, \delta_{ij}, \tag{6.2.4}$$

where ρ is the density, v_i is the velocity in the x_i direction, p_{ij} is the compressive stress tensor, a_0 is the speed of sound in the uniform medium outside the flow field and δ_{ij} is the Kronecker delta symbol. The stress T_{ij} represents the difference between the effective stresses in the real flow and the stresses in the uniform acoustic medium at rest. In fact, this is a static distribution of acoustic quadrupoles whose instantaneous strength per unit volume is T_{ij}.

Equations (6.2.1)–(6.2.3) are the basic equations of the theory of aerodynamic sound generation. It is noted that in many situations with small temperature variation and negligible viscous terms (large Reynolds numbers), only nonlinear term $\rho_0 v_i v_j$ become important in the stress field T_{ij} with a proportional error of the square of the Mach number, M which is equal to the fluid velocity divided by the *atmospheric* speed of sound a_0.

If the sources of matter are concentrated into a point, where the total rate of mass distribution is $q(t)$ in a unbounded medium, then the density field is given by the formula

$$\rho - \rho_0 = \frac{1}{4\pi a_0^2}\frac{1}{r}\, q'\left(t - \frac{r}{a_0}\right), \tag{6.2.5}$$

where ρ_0 is the undisturbed value of the density field and r is the distance from the source. In the case of the sources of matter are not so concentrated, the density distribution is represented by a volume integral of terms such as (6.2.5) in the form

$$\rho(\mathbf{x}) - \rho_0 = \frac{1}{4\pi a_0^2}\int \frac{\partial}{\partial t} Q\left(\mathbf{y}, t - \frac{|\mathbf{x}-\mathbf{y}|}{a_0}\right)\frac{d\mathbf{y}}{|\mathbf{x}-\mathbf{y}|}, \tag{6.2.6}$$

where the integral is taken over whole space, and $Q(\mathbf{x}, t)$ represents the mass per unit volume per unit time at position $\mathbf{x}$ at time t.

Indeed, the sound field generated is a dipole field. If the force is concentrated at a point with value $f_i(t)$ in an unbounded medium, the density distribution is given by

$$\rho(\mathbf{x}) - \rho_0 = -\frac{1}{4\pi a_0^2}\frac{\partial}{\partial x_i}\left[\frac{1}{r} f_i\left(t - \frac{r}{a_0}\right)\right]. \tag{6.2.7}$$

This follows from (6.2.5) when q' is equal to $\pm\, \varepsilon^{-1} f_i$ placed at $(0, 0, 0)$ and $(-\varepsilon, 0, 0)$, and similarly for f_2 and f_3 with $\varepsilon \to 0$.

It follows from (6.2.7) that the density field for the general volume distribution of dipole field is given by

$$\rho(\mathbf{x}) - \rho_0 = -\frac{1}{4\pi a_0^2}\frac{\partial}{\partial x_i}\int F_i\left(\mathbf{y}, t - \frac{|\mathbf{x}-\mathbf{y}|}{a_0}\right)\frac{d\mathbf{y}}{|\mathbf{x}-\mathbf{y}|}, \tag{6.2.8}$$

where a volume distribution of dipoles whose strength per unit volume is F_i. Hence, at large distances from the flow field, the density fluctuations can be represented by

$$\rho(\mathbf{x}) - \rho_0 \sim \frac{1}{4\pi a_0^2}\int \frac{(x_i - y_i)}{|\mathbf{x}-\mathbf{y}|^2}\frac{1}{a_0}\frac{\partial}{\partial t} F_i\left(\mathbf{y}, t - \frac{|\mathbf{x}-\mathbf{y}|}{a_0}\right) d\mathbf{y}. \tag{6.2.9}$$

This shows explicitly how the sound radiated to large distances. It depends only on the rate of change of dipole strength with time t.

Finally, for a concentrated quadrapole field in an unbounded medium, the density field similar to (6.2.8) can be written in the exact form

$$\rho(\mathbf{x}) - \rho_0 = \frac{1}{4\pi a_0^2} \frac{\partial^2}{\partial x_i\, \partial x_j} \int T_{ij} \left(\mathbf{y}, t - \frac{|\mathbf{x}-\mathbf{y}|}{a_0}\right) \frac{d\mathbf{y}}{|\mathbf{x}-\mathbf{y}|}, \quad (6.2.10)$$

where the stress tensor T_{ij} is the strength of the quadrupole field per unit volume.

In his famous 1952 paper, Sir James gave entirely new and novel derivation of a formula for the sound radiation field at large distances outside the source region in terms of an integral of T_{ij} over the source region in the form

$$\rho(\mathbf{x}) - \rho_0 \sim \frac{1}{4\pi a_0^2} \int \frac{(x_i - y_i)(x_j - y_j)}{|\,\mathbf{x}-\mathbf{y}\,|^3}\, a_0^{-2} \frac{\partial^2}{\partial t^2} T_{ij} \left(\mathbf{y}, t - \frac{|\,\mathbf{x}-\mathbf{y}\,|}{a_0}\right) d\mathbf{y}. \quad (6.2.11)$$

This is known as the *Lighthill formula* for the sound radiation field that gives an exact estimate for the total energy radiated and its directional distribution. This lead him to formulate a new *fourth power law* for the density variation $\rho - \rho_0$ at a distance x that is proportional to the product

$$\frac{1}{a_0^2} \cdot \frac{1}{x} \cdot \frac{1}{a_0^2} \left(\frac{U}{\ell}\right)^2 \rho_0 U^2 \ell^3 = \rho_0 \left(\frac{U}{a_0}\right)^4 \cdot \left(\frac{\ell}{x}\right), \quad (6.2.12)$$

where U is the typical velocity correlated over a length scale ℓ.

This formula gives a remarkable description of the dependence of the density changes in the sound radiation field on the *fourth power* of the Mach number $M = (U/a_0)$. He then derived the sound intensity field $I(\mathbf{x})$ which is proportional to (a_0^3/ρ_0) times the square of (6.2.12). More precisely, $I(\mathbf{x})$ is given by

$$I(\mathbf{x}) = \frac{a_0^3}{\rho} \overline{(\rho - \overline{\rho}_0)^2(\mathbf{x}, t)} = \frac{a_0^3}{\rho} \sigma^2 \{\rho(\mathbf{x}, t)\}, \quad (6.2.13)$$

where $\sigma^2 \rho(\mathbf{x}, t)$ is the variance representing the mean value at a point of the square of (6.2.12).

In the other words, Lighthill proved a remarkable formula

$$I(\mathbf{x}) \sim \rho_0 \left(\frac{U^8}{a_0^5}\right) \left(\frac{\ell}{x}\right)^2. \quad (6.2.14)$$

The significant impact of this law was realized in the 1950s in the theory of airfoils and the design of aircraft and missile engines.

Thus, he established a precise formula for the total acoustic power output that is proportional to

$$\rho_0\, U^8\, a_0^{-5}\, \ell^2 \quad (6.2.15)$$

which gives an eighth-power dependence on the flow velocity. In this case, the acoustic efficiency is proportional to $(U/a_0)^5 = M^5$.

This is the famous *Lighthill eighth* power law that has had almost simultaneous experimental confirmation. Thus, Lighthill's theory of the sound generated aerodynamically has not only remarkable applications to the noise made by jet aircraft, but also to other areas of fluid mechanics. On the other hand, a solid body which is subjected to a fluctuating aerodynamic force scaling as $\rho U^2 \ell^2$ (at frequency scaling as U/ℓ), radiates as an acoustic dipole of strength F. The associated acoustic power is proportional to $\rho U^6 \ell^2/a_0^3$ which depends on the sixth power of flow speed. So, the acoustic efficiency, defined by the ratio of acoustic power to a rate of delivery of energy to the flow, is proportional to $(U/a_0)^3 = M^3$. It turns out that the quadrupole radiation, though often significant, may become negligible near a solid body when dipole radiation due to fluctuating body force with its sixth-power dependence is also present as suggested by Curle (1955) and Ffowcs Williams and Hawkings (1969).

In order to describe a broad scientific background to several aspects of the interaction between airflows and atmospheric sound, Sir James once said: "Near not necessarily compact bodies a more refined calculation — using Green's functions not for free space but internally bounded space — leads in general to same conclusion: that quadrupole radiation with its eighth-power dependence is negligible alongside the sixth-power dependence of dipole radiation due to fluctuating body forces; but important exceptions to this rule include sharp-edged bodies, where features of the relevant Green's function imply a fifth-power dependence on flow speed of acoustic radiation from turbulence (Ffowcs Williams and Hall, 1970; Crighton and Leppington, 1970; Crighton and Leppington, 1971; and Crighton 1981). In his 1961 Bakerian Lecture to the Royal Society of London, Lighthill emphasized that the reduction of jet noise in civil aircraft is evident from his eighth power of the jet velocity. Designers of aircraft engines could realize at once that it was essential to reduce the fluid speed in their jets which they achieved by increasing the bypass ratio of the aircraft engines. Indeed, his new theory and its extensive experimental study provided the practical impetus behind the expansion of aeroacoustic in the middle of the twentieth century. According to Lighthill (1962): "The practical impetus behind the expansion, then has been the need to understand aircraft noise, including propeller noise, boundary-layer noise, supersonic bangs, and above all, the noise of jets, in order to discover new methods in both the design and operation of aircraft which may reduce these noises to levels tolerably by those exposed to them. We aim, in fact, to reap the benefits of high-speed travel while retaining within bearable limits the resulting noise nuisance."

Subsequently, in collaboration and cooperation with his students and research associates including Curle (1955), Ffowcs Williams (1963, 1977), Ffowcs Williams and Hawkings (1969), Goldstein (1976) and Crighton (1981), Sir James was actively involved in further research and advanced study in aeroacoustics that provided a continuous development in acoustic power output of jets, from subsonic jets through supersonic jets (including turbojet engine exhausts) to jets of speeds many times the speed of sound (such as rocket exhausts). Interestingly, the research of Ffowcs Williams was devoted to a good understanding not only of the initial rise in acoustic efficiency with jet speed, but also of its subsequent leveling off and of the associated changes in directional distribution and frequency spectrum. He also showed that, for subsonic motion, Lighthill's formula (6.2.14) would be modified by a factor $f(\theta)(1 - M_c \cos\theta)^{-5}$, where θ is the angle between the line from the source to the observer and the direction of mean motion and M_c represents the convection Mach number defined by the eddy convection velocity divided by the speed of sound a_0. The factor $f(\theta)$ depends on the orientation of the dominant quadrupoles and $f(\theta) = 1$ for purely longitudinal quadrupoles. Some of these works were concerned with applications of Lighthill's ideas to the sound field induced by turbulent flow near a rigid boundary.

At distances large compared with the dimensions of the fluid flow, it is appropriate to approximate $x_i - y_i$ by x_i in (6.2.11) provided that the origin is taken within the flow. Any terms of order $x^{-1}(x = |\,\mathbf{x}\,|)$ can be retained in (6.2.11) so that a simple representation for the sound radiation field can be written as

$$\rho(\mathbf{x}) - \rho_0 \sim \frac{1}{4\pi a_0^2}\left(\frac{x_i x_j}{x^3}\right)\int \frac{1}{a_0^2}\,\frac{\partial^2}{\partial t^2}T_{ij}\left(\mathbf{y}, t - \frac{|\,\mathbf{x}-\mathbf{y}\,|}{a_0}\right)d\mathbf{y}. \quad (6.2.16)$$

He then extended his theory of sound generated aerodynamically by taking into account the statistical properties of turbulent airflows, from which the sound radiated is called *aerodynamic noise*. His remarkable physical explanation of the theory of aerodynamic noise generation was simultaneously verified by several experimental works on both subsonic and supersonic jets. His mathematical work was so fundamental that it remains today the only practical means for computing noise in complex flow configurations. So his deep and intricate research on aeroacoustics influenced all subsequent work on new aircraft engine designs in the second half of the twentieth century. In the early nineties, Lighthill delivered many special lectures on different aspects of the aeroacoustics of high-speed jets by describing a new dynamical theory of nonlinear propagation of statistically random assemblages of conical shock waves which are radiated by eddies convected at supersonic speed. He showed, by both a general theoretical analysis and by computational method, how their propagation is dominated by a characteristic "bunching process"—a name given by Lighthill himself. This marked process is associated with a strong tendency of early local unions of shock waves that act to stimulate further unions with neighboring shock waves. His completely general theoretical study led to the demonstration of the *universal* tendency towards bunching process. Later on, Ffowcs Williams and Hawkings (1968) made some significant extension of Lighthill's theory to arbitrary time dependent source motion. Subsequently, Lighthill delivered several major lectures on aeroacoustics and its developments.

His 1961 Bakerian Lecture to the Royal Society of London was devoted to his new and original theory of sound generated aerodynamically, and its experimental study of pulse-jet-noise, hydrodynamic sound generation, aeolian tones, propeller noise, and boundary-layer noise as well as for an extensive discussion of the noise of turbulent jets, both stationary and in flight. In order to provide his definite motivation of this outstanding lecture, it is a delight to quote his magnificent opening statement:

"Before the end of the nineteenth century Lord Rayleigh (1896) set down in mellifluously lucid fashion in his *Theory of sound* the fundamentals of acoustic theory and vibration theory, and applied these to the study of such harmonious phenomena as the tuning-fork, the violin, the kettledrum, the clarinet, whispering galleries, organ pipes, church bells, singing flames, Rijke's tube, Sondhauss's bird-call and Trevelyan's rocker. The widespread continued use of this book as a compendium of fundamental knowledge on acoustics and vibrations not only bears witness to the outstanding qualities of the scientific writing, but also reminds us that acoustics has, in its fundamentals, changed far less since Rayleigh's time than most other sciences. Applied acoustics has made great strides, notably in the study of sound attenuation and insulation, in the techniques of sound measurement and reproduction, and in many matters bearing on subjective reaction to sound, including concert-hall design. But basic theory, in the meantime, was advancing slowly.

This, indeed, is the first Bakerian lecture on an acoustical topic since 1893. My aim in it is to describe the major revival and expansion of acoustics which has characterized the middle of the twentieth century, and which has involved extensive new progress in basic theory as well as even more extensive experimental study, public interest, and industrial expenditure. The subject matter of this expansion is, perhaps, characteristic of the present age, both in a scientific sense, for it is concerned with non-linear problems, and specifically with how the inherently non-linear flow fields of aerodynamics act on the more typically linear motions of acoustics; and also in a social sense, for Rayleigh's world of church bells, violins and bird-calls has been stormed, even though defended by the niceties of acoustic insulation and high-fidelity reproduction; it has been invaded from the air by the cacophonous sequence of whines and roars and bangs, which, I must admit, is the main subject-matter of Sound Generated Aerodynamically."

His lecture included convected quadrupoles and jet noise, radiation fields of moving sources, dipoles and quadrupoles, the noise of supersonic jets, and dependence of jet noise on velocity. Special attention was given to methods of reducing aircraft noise based on mathematical as well as physical knowledge. The most effective approach to the reduction of jet noise is evident from its dependence on the eighth power of the jet velocity, which means that, for a given thurst, a reduction in jet speed, with reciprocal increase in diameter, will be raised to the sixth power in the consequent reduction in noise. He also discussed recent improved knowledge of space-time correlations in turbulent flows which was used to throw new light on the noise radiated by turbulent boundary layers, as well as by jets at high Mach numbers. Supersonic bangs and the scattering of both sound and shock waves by turbulent flow are described briefly. In conclusion, he described the methods used for the reduction of jet aircraft noise based on his theory. His concluding statement is also a delight to quote:

"This has been a lecture in which I have principally tried to expound the science of sound generated aerodynamically, while including glimpses of its practical applications. I hope at least to have earned gratitude from my audience by resisting all temptation to bring aircraft noise tapes and play them within a few wavelengths of our founder's portrait! More presumptuously, I hope both to have exemplified the subtle modes of operation of mathematical theory that are necessary in practical spheres, and to have established that, on the frontiers between acoustics and aerodynamics, there exists this rather clearly defined new science, in which Lord Rayleigh himself might be taking an interest were he alive today."

He delivered the Wright Brothers Memorial Lecture on Jet Noise at the 31st Annual Meeting of the Institute of Aerospace in January 1963. In this expository and survey lecture, Lighthill presented the current developments in generation and reduction of jet noise. Included are theoretical analysis and experimental evidence of how the shearing motions in a turbulent jet manage to shed some of their energy as sound radiation. He described three major difficulties of the simple-source theory which deals with the noise generated in various regions of a jet.

His 1972 Fourth Annual Fairey Lecture in the Institute of Sound and Vibration Research at University of Southampton dealt with the propagation of sound through moving fluids with application to aero-engine compressor noise.

In 1978, Lighthill presented his general lecture on the modern developments in acoustic streaming that is forced by the action of the Reynolds stress field at the Institute of Acoustic Spring Meeting in Cambridge, England. His focus was to explain the significance of a variety of attenuation mechanisms of sound waves involved in various acoustic streaming motions due to a Reynolds stress.

In 1994, Sir James delivered The Inaugural Theodorsen Lecture in honor of a renowned NASA Langley aerodynamist on Some Aspects of the Aeroacoustics of High-Speed Jets that is mainly deal with nonlinear propagation of noise from extreme-speed jets. It is very appropriate to read Sir James' opening remarks:

"Fundamental investigations of turbulent flows had, of course, been yet another of the major interests pursued by Theodore Theodorsen. I have accordingly felt that a lecture concerned with turbulent jets and with one of the gravest environmental problems posed by their use for aircraft propulsion purposes — namely, the noise they generate — would form a fitting memorial tribute to that great aerodynamicist.

Furthermore, a lecture on this theme is particularly appropriate to an important occasion here at Langley; where early pioneering researches on aeroacoustics, including some crucial experiments on jet noise, were carried out by Harvey Hubbard and others (Lassiter and Hubbard, 1952). Last year, moreover, ICASE and NASA Langley Research Center joined forces to host an exceptionally fine workshop, which gave an authoritative view of the past, present, and future of aeroacoustics, with special emphasis on computational aeroacoustics. This workshop looked forward towards "a second golden age of aeroacoustics," in which both new theoretical and new computational approaches, closely linked to new experimental techniques, would be employed in response to the challenge of those toughened environmental requirements that are now to be imposed on all new aircraft designs (Hussaini and Hardin, 1993).

I was delighted to participate in this workshop by giving an introductory lecture looking back to some of the earlier fundamental discoveries in aeroacoustics, as well as by chairing the Final Panel Discussion which looked forward to likely future developments as aeroacoustics enters its second golden age. In between I was stimulated by hearing many brilliant contributions concerned with meeting some of those new challenges to which I just referred.

One of the most exciting of these is posed by the U.S. High Speed Civil Transport project (HSCT), an extremely promising plan for a supersonic transport aircraft ingeniously designed to minimize the level of supersonic-boom annoyance. However, the corresponding problems of reducing engine noise for such an aircraft to within acceptable limits raise some thorny questions and may, in particular, demand that a fundamental aeroacoustic analysis of jets at relatively high speed be undertaken."

He gave a detailed study of the nonlinear acoustic propagation of a thickly packed random assemblage of conical shock waves once they have been generated. In this lecture, he described his own remarkably new dynamical theory of the nonlinear acoustic propagation of statistically random assemblages of conical shock waves. This study shows how the inherent tendency of the shock strengths to decay through conical spreading and internal dissipation is counteracted by increases in strength that occur when adjacent shocks unite. A certain bunching process tend to arise, because union of two adjacent shocks is found to enhance the likelihood of further union with other neighboring shocks. The high-frequency part of the noise spectrum is made more intense than it would otherwise be by these bunching tendencies. Using Whitham's (1956) supersonic-boom theory, a transformation of coordinates is introduced to reduce conical wave problem into plane wave problem where a temporal waveform varies with distance r along the Mach direction. In the latter problem the time t replaces $\sqrt{r}$ to discuss how a given spatial waveform varies with time. In the former conical wave problem, Lighthill used t_c to denote the time variable that would avoid confusion with the use of t as the time for the plane wave problem.

With the dependent flow variable v_c defined by

$$v_c = c_0^{-1} - (u+c)^{-1}, \tag{6.2.17}$$

where c_0 is the undisturbed sound velocity, c is the local sound velocity, and u is the fluid velocity.

This dependent variable varies as $r^{-\frac{1}{2}}$ along the characteristics satisfying

$$dt_c = (u+c)^{-1}dr = (c_0^{-1} - v_c)dr, \tag{6.2.18}$$

can be written as

$$\left[\frac{\partial}{\partial r} + \left(c_0^{-1} - v_c\right)\frac{\partial}{\partial t_c}\right]\left(\sqrt{r}\, v_c\right) = 0. \tag{6.2.19}$$

Then a simple transformation of coordinates

$$2\sqrt{r} = t, \qquad \frac{r}{c_0} - t_c = X, \tag{6.2.20}$$

transforms the operator in the square bracket in (6.2.19) into the form

$$\frac{1}{\sqrt{r}}\frac{\partial}{\partial t} + \frac{1}{c_0}\frac{\partial}{\partial X} - \left(\frac{1}{c_0} - v_c\right)\frac{\partial}{\partial X}. \tag{6.2.21}$$

Thus, $v = \sqrt{r}\, v_c$ satisfies the familiar nonlinear partial differential equation in the form

$$\frac{\partial v}{\partial t} + v\frac{\partial v}{\partial X} = 0, \tag{6.2.22}$$

with its characteristics of slope v in the (t, X) plane. This equation describes the waveform shearing at a uniform rate that is associated with the formation of shock waves and propagation in nonlinear plane-wave acoustics. From physically relevant solutions of (6.2.22), there exists famous N-wave solution which is generated by an initial signal (such as an aircraft's passage through air) that is first compressive and then expansive. However, all results for the plane wave problem can be directly applied to the conical wave problem with v and t replaced by $\sqrt{r}\, v_c$ and $2\sqrt{r}$ respectively. The temporal evolution as t increases of the spatially unbounded waveform in $-\infty < X < \infty$ for plane waves is then found from (6.2.20) as a representation of the spatial evolution as $r \to \infty$ of the temporal unlimited waveform in $-\infty < t_c < \infty$ for conical waves. Lighthill introduced a nondimensional quantity σ as an appropriate measure of times at which unions of shock waves occur, where

$$\sigma = (t - T)/2t, \qquad t = T/(1 - 2\sigma), \tag{6.2.23}$$

where the quantity T denotes the time when a plane wave of random sawtooth type is established so that $T = 2\sqrt{R}$ for the conical wave problem with R as the corresponding propagation distance from the axis at which such a wave has become established. Consequently, the times for unions of shocks in the plane wave problem can be expressed in terms of σ corresponding to conical wave propagation in the form

$$\sigma = (\sqrt{r} - \sqrt{R})/2\sqrt{r}, \qquad r = R/(1 - 2\sigma)^2. \tag{6.2.24}$$

Lighthill (1994) also first introduced an expression for a new time-like variable σ at which the very first union of nth and $(n-1)$th shock waves occurs by

$$\sigma = \min_{t \leq n \leq N} \left[H_n / \left(Y_n + Y_{n-1} \right) \right] . \tag{6.2.25}$$

He also introduced a general expression for σ as a weighted mean of the σ values σ_n, σ_{Am} and σ_{Bm} in the form

$$\sigma = \frac{(Y_{n-1} + Y_n)\, \sigma_n + \sum_{m=1}^{m_{n-1}} Y_{Am}\, \sigma_{Am} + \sum_{m=1}^{m_n} Y_{Bm}\, \sigma_{Bm}}{(Y_{n-1} + Y_n) + \sum_{m=1}^{m_{n-1}} Y_{Am} + \sum_{m=1}^{m_n} Y_{Bm}}, \tag{6.2.26}$$

where Y_{Am} and Y_{Bm} are the weights of σ_{Am} and σ_{Bm} respectively, the weight attached to σ_n is the combined initial Y-values of the nth and $(n-1)$th shocks which would have united at $\sigma = \sigma_n$, and the denominator of (6.2.26) denotes the sum of all these weights.

Thus, expressions for the values of σ at which shock waves unite, like (6.2.25) for the first such union, or (6.2.26) for a quite general subsequent union as given in terms of Y-values of shocks and the initial values of H_n can be interpreted by (6.2.24), as specifying the distance r required for merger of adjacent shocks. All these results in turn can be used effectively to quantify tendencies to bunching process in conical wave propagation. This process is found to play a fundamental role for the analysis of how aspects of aerodynamic noise are changed in the region of conical wave propagation that surrounds of an extreme-speed jet.

Finally, Sir James suggested an algorithm which is very useful for the quantitative study of the bunching process through numerical experiments. His numerical experiments indicate a general tendency for ΣY_n^2 to increase in approximate proportionality to $t = 2\sqrt{r}$. This has interesting implications for the quantity

$$\sum_{n=1}^{N} z_n^2 = \sum_{n=1}^{N} (Y_n / 2r)^2, \tag{6.2.27}$$

which associates with high-frequency part of the noise spectrum. The inherent tendency of z_n to decay at r^{-1} would force the sum (6.2.27) to vary as r^{-2}. Thus, bunching tendencies tend to modify this decay law for high frequency noise into a variation as $r^{-3/2}$. In summary, Lighthill's general mathematical analysis and illustrative computational study confirmed the fact that the propagation of conical shock waves is dominated by the characteristic bunching process. This process associated with a tendency for shock waves which have already formed unions of other shock waves to acquire an increased proneness to generate further unions acts so as to enhance the high frequency part of the spectrum of noise emitted from jets at these high exit speeds. It is a delight to quote the concluding statement of his lecture:

"Then general nonlinear acoustics (Lighthill, 1978) tells us that waveform modifications due to nonlinear propagation effects are enormously greater in a conically spreading wave than in a spherically spreading wave; specifically, the square-root transformation (A.4) derived in the Appendix below is replaced by a logarithmic transformation for a spherical wave which, accordingly, undergoes far less modification from nonlinear propagation. The conical field is specially important, then, as the region of occurrence of most of the spectrum-modification effects of nonlinear propagation.

However, theoretical studies may have even more importance for the problem of how aeroengine installations are to be designed so as to reduce ground levels. In this context a jet noise field where propagation is directionally concentrated (say, in the Mach direction) may offer engine-installation designers a rather special opportunity to incorporate shielding devices aimed at limiting community noise. This, then, is yet another good reason why, in this inaugural Theodorsen Lecture, I have presented to my distinguished audience quite a detailed analysis of how the noise spectrum in that conical field with its "random assemblage of shocks" may be influenced by the general tendency on which 1 have affectionately ventured to bestow the name "bunching"."

Chapter 7

Boundary Layer Theory and Vorticity Dynamics

"It was Helmholtz who demonstrated rather remarkable characteristics of the vortex motion ... starting from the properties of the vortex rings Sir W. Thomson suggested a new version of atomic theory ...."

Ludwig Boltzmann

"The enormous usefulness of mathematics in the natural sciences is something bordering on the mysterious and there is no rational explanation for it. It is not at all natural that "laws of nature" exist, much less that man is able to discover them. The miracle of the appropriateness of the language of mathematics for the formulation of the laws of physics is a wonderful gift which we neither understand nor deserve."

Eugene Wigner

7.1 Introduction

Historically, in 1761, Leonhard Euler formulated the continuity equation and the equation of motion for incompressible inviscid fluid flows in the exact form

$$\operatorname{div}\mathbf{u} = 0, \qquad \frac{\partial \mathbf{u}}{\partial t} + (\mathbf{u}\cdot\nabla)\,\mathbf{u} = -\frac{1}{\rho}\,\nabla p, \tag{7.1.1ab}$$

where $\mathbf{u} = (u, v, w)$ is the velocity field, ρ is a constant density and p is the pressure field. These closed system of equations constitute four scalar nonlinear partial differential equations for four scalar variables u, v, w and p.

The relationship of the Euler flow model, which neglects all effects of viscosity, to real fluid flows with very thin boundary layers (layers whose thickness tends to zero with the viscosity) is that a flow determined by the Euler model equations is a good representation of the flow outside the boundary layer. The large-scale motions of the inviscid fluid can be described well by this model and involve substantial tangential motion at the solid boundary. However, this model does not identify the extremely small scale of the thin boundary layer separating those tangential motions from the real solid boundary surface.

On the other hand, L. M. H. Navier (1785-1836) in 1822 and George G. Stokes (1819-1903) in 1845 formulated the continuity equation and the equation of motion for incompressible viscous fluid flows with no external forces that are in the form

$$\text{div}\,\mathbf{u} = 0, \qquad \frac{\partial \mathbf{u}}{\partial t} + (\mathbf{u}\cdot\nabla)\mathbf{u} = -\frac{1}{\rho}\,\nabla p + \nu\,\nabla^2\mathbf{u}, \tag{7.1.2ab}$$

where $\nu = \frac{\mu}{\rho}$ is the kinematic viscosity and μ is the dynamic viscosity of the fluid.

Unfortunately, the Navier-Stokes equations are very difficult nonlinear partial differential equations and so we only deal with the very easy and special cases. No useful general solution is known and existence, and uniqueness, and regularity theorems are only available in very special situations.

For fluids of small viscosity, including water and air, many flows do involve only thin boundary layers, so that all of their large scale features can be determined precisely by the Euler equation model. For many flow models, the associated boundary layers do *not* remain thin, however small viscosity may be. For certain types of fluid motion, the mechanics of the boundary layer itself causes the whole flow to separate from the solid boundary surface. Consequently, this separation yields a flow very different from that of the Euler flow model for any value of viscosity.

In 1905, Ludwig Prandtl first discovered the boundary layer theory in order to resolve the inherent difficulty that irrotational (potential) flow solutions obtained by neglecting the viscosity ($R = (U\ell/\nu) \to \infty$) do not satisfy the boundary condition at a solid boundary surface, where R stands for the Reynolds number at a typical flow speed U and length scale ℓ. This nondimensional number, R gives the measure of the fluid velocity and scale of the body in relation to the magnitude of viscous diffusion effects. Rather, the potential flow solutions require slip over the surface that seems to be impossible. Prandtl made a major assumption that within a thin layer of fluid adjacent to the boundary the relative fluid velocity increases very rapidly from zero at the solid surface to the actual value at the edge of the layer. This thin layer is called the *boundary layer* where the velocity gradient is high, viscosity within the layer would be important, no matter how small it is, and flow becomes turbulent in this thin boundary layer. The kinetics effects, due to the individual fluid particles, cause drag and heat dissipation to occur in this layer, and the real understanding of the layer made clear the process of aerodynamic flow over a wing and led to the engineering design of more efficient airfoils. However, for high-altitude aerodynamical problems, we use what are called *slip conditions* as no such slip condition holds. Experiments support the correctness of Prandtl's assumptions so that, not only the boundary layer exists, but also the velocity distribution across it develop in accordance with the laws of fluid mechanics.

Diffusion plays a significant role in determining the detailed structure of boundary layers. If U is the velocity of a body relative to the undisturbed fluid, then U also represents a typical value of the velocity of the fluid at the outer edge of the boundary layer. The vortex lines in the outer part of the boundary layer are carried alone by the fluid, then at a velocity whose value is U, and for a body of typical length ℓ, they swept clear of the body after a time $t = (\ell/U)$. This leads to set the limit of the growth of the thickness of the boundary layer to a value proportional to

$$\delta = \left(\frac{\nu\ell}{U}\right)^{\frac{1}{2}}, \tag{7.1.3}$$

because vorticity generated at the solid boundary diffuses outward to a distance δ. Thus, the ratio of the body length ℓ to the boundary layer thickness δ can be large provided the quantity

$$\frac{\ell}{\delta} = \sqrt{\frac{U\ell}{\nu}} \equiv \sqrt{R}, \tag{7.1.4}$$

is very large, that is, if Reynolds number R itself is very large.

Although irrotation fluid flows are susceptible to become turbulent, subsequent experimental findings confirmed that any boundary layer flow may become turbulent at a sufficiently large value of the Reynolds number. The required value of R depends on various factors including the shape of boundary, but it is very close to 10^6.

7.2 Lighthill's Contributions to Boundary Layer Theory

Lighthill published at least seven research papers on the boundary layer theory and its numerous aspects describing significance of boundary layers in fluid flows. In his paper published in the *Proceedings of the Royal Society*, **200A** (1950) 359-377, Lighthill developed a new method to obtain an approximation of the heat transfer rate across an incompressible laminar boundary layer for arbitrary distribution of main stream velocity and of wall temperature distributions using the von Mises energy equation without the frictional heating term as

$$\frac{\partial T}{\partial x} = \frac{\mu\rho}{\sigma}\frac{\partial}{\partial \psi}\left(u\,\frac{\partial T}{\partial \psi}\right), \tag{7.2.1}$$

where T is the temperature, $\nu/\sigma = \kappa$ is the diffusivity of heat ($= k/\rho c_p$ in terms of thermal conductivity k, density ρ and specific heat at constant pressure c_p) so that σ is the Prandtl number (0.71 for Air), ψ is the stream function as an independent variable. An approximate value of the velocity field u is inserted into equation (7.2.1) which represents the most accurate value near the surface, where $u = (\tau(x)/\mu)\,y$, $\tau(x)$ is the skin friction at a distance x along the surface from the beginning of the boundary layer. This means that

$$\psi = \rho\int_0^y u\,dy = [\rho\,\tau(x)/2\mu]\,y^2, \tag{7.2.2}$$

so that

$$u = \left[\frac{2\tau(x)\psi}{\mu\rho}\right]^{\frac{1}{2}}. \tag{7.2.3}$$

Then equation (7.2.1) becomes

$$\frac{\partial T}{\partial x} = \frac{1}{\sigma}\{2\mu\rho\tau(x)\}^{\frac{1}{2}}\frac{\partial}{\partial \psi}\left(\sqrt{\psi}\;\frac{\partial T}{\partial \psi}\right). \tag{7.2.4}$$

This equation was solved by Lighthill under the boundary conditions

(i) $T = 0$ at $x = 0$ and $y \to \infty$,
(ii) $T = T_0(x)$ and $k\frac{\partial T}{\partial y} = Q(x)$ at $y = 0$,

and obtained a relationship between $Q(x)$, $T_0(x)$ and $\tau(x)$ in the form

$$Q\,(x) = -k\left(\frac{\sigma\rho}{9\mu^2}\right)^{\frac{1}{3}}\frac{\sqrt{\tau\,(x)}}{\left(\frac{1}{3}\right)!}\int_0^x\left[\int_{x_1}^x\sqrt{\tau\,(z)}\,dz\right]^{-\frac{1}{3}} d\,T_0(x_1). \tag{7.2.5}$$

This is the relation between local heat transfer rate to the wall $Q(x)$, skin friction $\tau(x)$, and excess $T_0(x)$ of wall temperature over that of the stream.

Using the Heaviside operator p corresponding to $\frac{\partial}{\partial t}$ with $T = 0$ at $t = 0$, equation (7.2.4) reduces to the form

$$pT = \frac{\partial}{\partial \psi}\left(\sqrt{\psi}\,\frac{\partial T}{\partial \psi}\right). \tag{7.2.6}$$

This equation admits two linearly independent solutions in terms of the Bessel functions I_λ, and $I_{-\lambda}$ so that

$$T_1 = \left(\frac{2}{3}\,p^{\frac{1}{2}}\psi^{\frac{3}{4}}\right)^{\frac{1}{3}}\left(-\frac{1}{3}\right)!\,I_{-\frac{1}{3}}\left(\frac{4}{3}p^{\frac{1}{2}}\psi^{\frac{3}{4}}\right), \tag{7.2.7}$$

and

$$T_2 = \psi^{\frac{1}{2}}\left(\frac{2}{3}\,p^{\frac{1}{2}}\psi^{\frac{3}{4}}\right)^{-\frac{1}{3}}\left(\frac{1}{3}\right)!\,I_{\frac{1}{3}}\left(\frac{4}{3}\,p^{\frac{1}{2}}\psi^{\frac{3}{4}}\right). \tag{7.2.8}$$

In view of the boundary condition (ii) as $y \to 0$, the combination of solutions T_1 and T_2 is required so that

$$\begin{aligned} T = {} & \left(\frac{2}{3}\,p^{\frac{1}{2}}\right)^{\frac{1}{3}}\psi^{\frac{1}{4}}\left(-\frac{1}{3}\right)!\,I_{-\frac{1}{3}}\left(\frac{4}{3}\,p^{\frac{1}{2}}\psi^{\frac{3}{4}}\right)T_0(x) \\ & + \left(\frac{2}{3}\,p^{\frac{1}{2}}\right)^{-\frac{1}{3}}\psi^{\frac{1}{4}}\left(\frac{1}{3}\right)!\,I_{\frac{1}{3}}\left(\frac{4}{3}\,p^{\frac{1}{2}}\psi^{\frac{3}{4}}\right)\frac{Q(x)}{k}\left(\frac{2\mu}{\rho\,\tau(x)}\right)^{\frac{1}{2}}. \end{aligned} \tag{7.2.9}$$

Thus the total heat transfer rate over an area of surface between $x = 0$ and $x = l$ per unit breadth is given by

$$\int_0^l Q(x)dx = -\frac{\left(\frac{1}{2}k\right)}{\left(\frac{1}{3}\right)!}\left(\frac{3\,\sigma\rho}{\mu^2}\right)^{\frac{1}{2}}\int_0^l\left[\int_z^l \sqrt{\tau(z)}\;dz\right]^{\frac{2}{3}} d\,T_0(x), \tag{7.2.10}$$

where $\left(\frac{1}{3}\right)! = 0.8930$.

It turns out that formula (7.2.10) is very accurate for large Prandtl number σ. Therefore, the thermal boundary layer is only a small fraction of the velocity boundary layer, and in it the approximation $u = (\tau(x)/\mu)\,y$ is very accurate.

In this paper published in *Quarterly Journal of Mechanics and Applied Mathematics*, **III** (1950) 303-325, Lighthill considered a laminar boundary layer of a weak steady disturbance to a supersonic stream without viscosity and heat conduction. In a steady two-dimensional flow perturbed only slightly from a basic flow parallel to a solid plane wall, the Mach number of the basic flow varies continuously and monotonically from zero at the wall to a supersonic value at a distance δ from the wall, beyond which it assumes this supersonic value uniformly. He also investigated the disturbance due to a weak stationary simple plane wave incident on the boundary layer from outside. It is shown that the streamline pattern and pressure coefficient distribution resulting from a given incident wave depend only on the Mach number in the basic flow, and are independent of the individual velocity and temperature distributions.

The pressure coefficient $q = (p - P)/\gamma P$ and the streamline pattern depend not on the separate velocity and temperature distributions in the boundary layer, but only on the Mach number $M(y)$. The pressure coefficient $q(x, y)$ satisfies the equation

$$\frac{\partial^2 q}{\partial y^2} - \frac{2M'(y)}{M(y)}\frac{\partial q}{\partial y} + \left\{1 - M^2(y)\right\}\frac{\partial^2 q}{\partial x^2} = 0. \tag{7.2.11}$$

This equation was solved under the boundary condition at the wall at $y = 0$ appropriate to an inviscid fluid. It follows from $q \to 0$ at infinity upstream that $(q)_{y=0} = 0$ and $q = 0(y^3)$ as $y \to 0$.

Using the inverse Fourier transform $q(x, y)$ of $Q(k, y)$ defined by

$$q(x, y) = \int_{-\infty}^{\infty} e^{ikx} Q(k, y)dk, \tag{7.2.12}$$

equation (7.2.11) can be reduced to the form

$$\frac{d^2 Q}{dy^2} - \frac{2M'(y)}{M(y)}\frac{dQ}{dy} + k^2\left\{M^2(y) - 1\right\}Q = 0, \tag{7.2.13}$$

with the transformed boundary conditions.

If $Q(k, y)$ is any solution of (7.2.13) with $Q(k, 0) = 0$, it turns out that $Q(k, y)$ satisfies the asymptotic result

$$Q(k, y) \sim y^3 \quad \text{as } y \to 0 \text{ for all } k. \tag{7.2.14}$$

This is followed by a discussion of any singularity in the incident wave field which is propagated along the characteristics in the supersonic term of the boundary layer and reflected back from the modified sonic line. This led to upstream influence of the disturbance. The upstream influence of the disturbance on the boundary layer is found to be completely insignificant at a distance of one layer thickness ahead of the wavefront on the outside of the boundary layer. It is interesting to point out that Lighthill suggested four radically different physical mechanisms for the interaction of complex nature of this problem and upstream influence of the shock. They are:

(i) It is appropriate to neglect the subsonic layer and to include the results associated with viscous diffusion in the supersonic layer; or

(ii) Lighthill's theory could be improved by taking viscosity into account after considering the boundary layer stability, especially near the wall where the full viscous boundary condition could then be applied; or

(iii) the present theory is correct for a weak disturbance without viscosity, but the solution pattern changes as the shock amplitude is increased so that to predict the effect the full nonlinear adiabatic equations would be required; or

(iv) there is a bubble of dead air at the wall near the base of the shock of sufficient length but small breadth, this has been formed by repeated separation until the pressure field was such as to allow attachment of the main laminary layer.

None of the above possible mechanisms can be disregarded in the light of experimental and numerical data available. There is no significant upstream influence or boundary layer separation, and thus, Lighthill's predictions agree with observational findings.

In 1953, Lighthill published two major papers on "On Boundary Layers and Upstream Influence, I. A Comparison between Subsonic and Supersonic Flows"; *Proceedings of the Royal Society, London*, (*PRSL*) **A217** (1953) 334-357; "On Boundary Layers and Upstream Influence, II. Supersonic Flows without Separation"; *Proceedings of the Royal Society, London*, (*PRSL*) **A217** (1953) 478-507. In Part I of his paper, Lighthill suggested two separate mechanisms for upstream influence through the boundary layer in supersonic flow, and one of these involving separation acts also in subsonic flow. He also described the main differences between subsonic and supersonic flows as follows:

(i) The boundaries of dead-air regions are almost straight in supersonic flow, but are highly curved in subsonic flow.

(ii) Seperation occurs at a much lower pressure coefficient in supersonic flow.

(iii) For a given pressure coefficient in the dead-air region, the distance of upstream influence is somewhat greater in the subsonic flow except at the highest pressures.

Part II of his paper essentially deals with interaction of between a boundary layer and a supersonic main stream without separation. Such an interaction is investigated mathematically by a method of perturbation of a parallel flow and neglecting the disturbances of the viscous forces. He showed that these are not negligible in an inner viscous sublayer in which the Mach number remains small, and which for the turbulent boundary layer is inside the so-called *laminar sublayer*. If disturbances are analyzed longitudinally by the Fourier transform, the effect of the inner viscous sublayer on the nature of each harmonic component outside it is exactly as if there were a solid wall at some position in the stream, with no flow across it and inviscid flow outside it. The position depends on the wavenumber k and on the skin friction in the undisturbed boundary layer. Solutions are obtained for small and large wavenumber k and used to derive the following: (i) the main local behaviors of both the wall-pressure distribution and the flow outside the boundary layer, and (ii) the extent of upstream influence, and the comparison of the upstream influence with boundary layer thickness. Further, he made an attempt to predict the complete form of the reflected wave when a shock is incident upon a turbulent boundary layer. His theory was found to be in reasonable agreement with the experimental result. Finally, he completed his paper with a discussion of the skin friction and its significant influence on the onset of nonlinearity. The importance of nonlinear effects depends on the sign of a disturbance, and the overall pressure change (leading to large amplification only when this is positive) is a common feature in nonlinear phenomena.

Lighthill's another major contribution to the unsteady-boundary layer theory deals with the problem of how a boundary layer responds to fluctuations of the external flow about a steady mean value. He published a long paper on the "Response of Laminar Skin Friction and Heat Transfer to Fluctuations in the Stream Velocity", *Proceedings of the Royal Society, London*, (*PRSL*) **A224** (1954) 1-23. In this paper, he developed a new mathematical theory of the laminar boundary layer in two-dimensional flow about a fixed cylindrical body, when the velocity of the oncoming flow relative to the body oscillates in magnitude, but not in direction.

He formulated the boundary layer equations for two-dimensional incompressible viscous flow over a fixed cylindrical body given by

$$\frac{\partial u}{\partial t} + u\,\frac{\partial u}{\partial x} + v\,\frac{\partial u}{\partial y} = -\frac{1}{\rho}\,\frac{\partial p}{\partial x} + \nu\,\frac{\partial^2 u}{\partial y^2}, \tag{7.2.15}$$

$$\frac{\partial u}{\partial x} + \frac{\partial v}{\partial y} = 0, \tag{7.2.16}$$

$$u = v = 0 \text{ at } y = 0, \quad u \to U(x,t) \text{ as } y \to \infty, \tag{7.2.17}$$

$$\frac{\partial U}{\partial t} + U\,\frac{\partial U}{\partial x} + \frac{1}{\rho}\,\frac{\partial p}{\partial x} = 0, \tag{7.2.18}$$

where x is along the surface from the front stagnation point, y represents the normal distance from the surface, u, v are the corresponding velocities, $p = p(x,t)$ is the pressure, ρ is constant density and $\nu = \mu/\rho$ is the kinematic viscosity. The main stream velocity $U = U(x,t)$ outside the boundary layer is that given by the theory of irrotational flow with the displacement of the external flow by the boundary layer and wake.

It is assumed that the quantities u, v, and U are all performing small oscillations about a steady mean value so that they assume the form

$$\left.\begin{aligned} u &= u_0(x,y) + \varepsilon\, u_1(x,y)\exp(i\omega t) \\ v &= v_0(x,y) + \varepsilon\, v_1(x,y)\exp(i\omega t) \\ U &= \quad U_0(x) + \varepsilon\, U_1(x)\exp(i\omega t) \end{aligned}\right\}. \tag{7.2.19}$$

Substituting (7.2.19) into (7.2.15)–(7.2.18), retaining only terms of order ε and neglecting the squares of the perturbation gives

$$i\omega u_1 + u_0\frac{\partial u_1}{\partial x} + u_1\frac{\partial u_0}{\partial x} + v_0\frac{\partial u_1}{\partial y} + v_1\frac{\partial u_0}{\partial y} = i\omega U_1 + \frac{d}{dx}(U_0\,U_1) + \nu\frac{\partial^2 u_1}{\partial y^2}, \tag{7.2.20}$$

$$\frac{\partial u_1}{\partial x} + \frac{\partial v_1}{\partial y} = 0, \quad u_1 = v_1 = 0 \text{ at } y = 0, \; U_1 \to U_1(x) \text{ as } y \to \infty. \tag{7.2.21}$$

When the fluctuations in the main stream velocity $U(x,t)$ are produced by fluctuations in the magnitude of the oncoming flow velocity which at time t is

$$V\left(1 + \varepsilon\, e^{i\omega t}\right). \tag{7.2.22}$$

The external flow about the body will fluctuate by the same factor so that at time t it is given by

$$U(x,t) = U_0(x)\left[1 + \varepsilon\, e^{i\omega t}\right], \tag{7.2.23}$$

where we deal with the case $U_0 = U_1$ when (7.2.23) is compared with (7.2.19).

The solution of equations (7.2.20)–(7.2.21) for the limiting case $\omega \to 0$ represents the quasi-steady which is denoted by (u_s, v_s) so that they are the coefficient of ε in the velocity field for steady flow with an incident stream velocity $V(1+\varepsilon)$.

For general ω, we write

$$(u_1, v_1) = (u_s, v_s) + i\omega(u_2, v_2), \tag{7.2.24}$$

so that (u_2, v_2) represents the acceleration - dependent part of the velocity field in the boundary layer.

Substituting (7.2.24) in (7.2.20)–(7.2.21) leads to the equations for (u_2, v_2) in the form

$$i\omega u_2 + u_0\frac{\partial u_2}{\partial x} + u_2\frac{\partial u_0}{\partial x} + v_0\frac{\partial u_2}{\partial y} + v_2\frac{\partial u_0}{\partial y} = \nu\frac{\partial^2 u_2}{\partial y^2} + (U_0 - u_s)\,, \tag{7.2.25}$$

$$\frac{\partial u_2}{\partial x} + \frac{\partial v_2}{\partial y} = 0, \quad u_2 = v_2 = 0 \quad \text{at} \quad y = 0, \quad u_2 \to 0 \quad \text{as} \quad y \to \infty. \tag{7.2.26}$$

For small frequency parameter $(\omega\ell/V)$, where ℓ is a characteristics length scale of the body, equations (7.2.25)–(7.2.26) can be solved by the Kármán-Pohlhausen method. For large frequency parameter, we retain the terms involving ω, and the derivative of highest order so that

$$i\omega\,(u_1 - U_1) = \nu\frac{\partial^2 u_1}{\partial y^2}. \tag{7.2.27}$$

This is identical with the equation for *shear waves* confined to the shear-wave layer near the wall whose thickness is of the order $\sqrt{\nu/\omega}$.

Thus, the solution of (7.2.27) with $u_1 \to U_1$ as $y \to \infty$ is

$$u_1 = U_1\left[1 - \exp\left(-y\sqrt{\frac{i\omega}{\nu}}\right)\right]. \tag{7.2.28}$$

This is valid for large ω regardless of the nature of the main stream fluctuations. The skin friction intensity is given by

$$\mu\left(\frac{\partial u_0}{\partial y}\right)_{y=0} + \varepsilon e^{i\omega t}\,(\mu U_0)\sqrt{\frac{i\omega}{\nu}}. \tag{7.2.29}$$

The amplitude of its fluctuations increases with frequency, and its phase is ahead of that of the fluctuations in the stream velocity by $\frac{\pi}{4}$.

Following Lighthill's analysis, we conclude that there exists a critical frequency ω_0 such that for frequencies $\omega > \omega_0$, the oscillations are to a close approximation ordinary *shear waves* unaffected by the mean flow, and the phase advance in the skin friction is $\pi/4$. For frequencies $\omega < \omega_0$, the oscillations are closely approximated as the sum of component proportional to the instantaneous velocity and acceleration of the oncoming stream, the phase advance in the skin friction is then $\tan^{-1}(\omega/\omega_0)$.

We next consider the fluctuations in heat transfer, and in the temperature distribution in the boundary layer, when the temperatures of both the body surface and the oncoming stream remain constant. The equations for the temperature distribution $T(x, y, t)$ in the boundary layer, measured from that of the stream as zero, are

$$\frac{\partial T}{\partial t} + u\frac{\partial T}{\partial x} + v\frac{\partial T}{\partial y} = \kappa\,\frac{\partial^2 T}{\partial y^2}, \tag{7.2.30}$$

$$T = \Theta \quad \text{at} \quad y = 0, \qquad T \to 0 \quad \text{as} \quad y \to \infty, \tag{7.2.31}$$

where Θ is the constant excess temperature of the body surface over that of the stream, and κ is the thermal diffusivity.

Using u and v from (7.2.19) and introducing

$$T = T_0\,(x, y) + \varepsilon\, e^{i\omega t}\, T_1\,(x, y)\,, \tag{7.2.32}$$

where T_0 is the unperturbed temperature, and retaining only terms of order ε in equation (7.2.30), we obtain the equation for $T_1(x, y)$ as

$$i\omega T_1 + u_0\frac{\partial T_1}{\partial x} + u_1\frac{\partial T_1}{\partial x} + v_0\frac{\partial T_1}{\partial y} + v_1\frac{\partial T_0}{\partial y} = \kappa\,\frac{\partial^2 T_1}{\partial y^2}, \tag{7.2.33}$$

$$T_1 \to 0 \quad \text{as} \quad y \to 0, \quad \text{and} \quad y \to \infty, \tag{7.2.34}$$

Using a form similar to that of (7.2.24), we write T for general ω as

$$T = T_s + i\omega\, T_2 \tag{7.2.35}$$

so that for oscillations of relatively small frequency $(-T_2/T_s)$ represent the time lag in temperature over its quasi-steady state value.

In view of (7.2.24) and (7.2.35) and the fact that T_s is a solution of (7.2.33)–(7.2.34) for $\omega = 0$, we obtain the equations for T_2 in the form

$$i\omega T_2 + u_0\frac{\partial T_2}{\partial x} + v_0\frac{\partial T_2}{\partial y} - \kappa\frac{\partial^2 T_2}{\partial y^2} = -\left(T_s + u_2\frac{\partial T_0}{\partial x} + v_2\frac{\partial T_0}{\partial y}\right), \tag{7.2.36}$$

with the conditions

$$T_2 = 0 \quad \text{at} \quad y = 0, \quad \text{and} \quad T_2 \to 0 \quad \text{as} \quad y \to \infty. \tag{7.2.37}$$

Equation (7.2.36) is suitable for small values of frequency ω, but we adopt another approach to solve it for large frequency ω. Thus, the significant terms of equation (7.2.33) are

$$-i\omega T_1 + \kappa\,\frac{\partial^2 T_1}{\partial y^2} = u_1\frac{\partial T_0}{\partial x} + v_1\frac{\partial T_0}{\partial y}. \tag{7.2.38}$$

Using the solution for u_1 in (7.2.28) and the continuity equation we determine both u_1 and v_1 in the form

$$u_1 = U_1\left[1 - \exp\left(-y\sqrt{\frac{i\omega}{\nu}}\right)\right], \tag{7.2.39}$$

$$v_1 = -\left(\frac{dU_1}{dx}\right)\left[y - \frac{1}{\sqrt{\frac{i\omega}{\nu}}}\left\{1 - \exp\left(-y\sqrt{\frac{i\omega}{\nu}}\right)\right\}\right]. \tag{7.2.40}$$

Thus, the solution for large ω is obtained far away from the wall for not rapidly varying u_1 and v_1 so that the term $\left(\kappa\,\frac{\partial^2 T_1}{\partial y^2}\right)$ can be neglected.

It turns out that

$$-i\omega T_1 = \left(u_1\frac{\partial T_0}{\partial x} + v_1\frac{\partial T_0}{\partial y}\right). \tag{7.2.41}$$

However, for small y, the exponential terms in u_1 and v_1 make them change very rapidly for this approximation to be valid, but to these can be assumed a simple form $\Theta + y\left(\frac{\partial T_0}{\partial y}\right)_{y=0}$ on the right hand side of (7.2.38). This is a reasonable approximation near the wall, since by (7.2.30), $\left(\frac{\partial^2 T_0}{\partial y^2}\right) = 0$ on $y = 0$. Consequently, the solution of T_1 can be obtained in the form

$$\begin{aligned} T_1 = {} & \left(\frac{i}{\omega}\right) U_1 \left(\frac{\partial^2 T_0}{\partial x \partial y}\right)_{y=0} \left[y + \left(\frac{\sigma y}{1-\sigma}\right)\exp\left(-y\sqrt{\frac{i\omega}{\nu}}\right)\right. \\ & \left. + \frac{2\sigma}{(1-\sigma)^2}\left\{\frac{\exp\left(-y\sqrt{\frac{i\omega}{\nu}}\right) - \exp\left(-y\sqrt{\frac{i\omega}{\kappa}}\right)}{\sqrt{\left(\frac{i\omega}{\nu}\right)}}\right\}\right] \\ & - \left(\frac{i}{\omega}\right)\left(\frac{dU_1}{dx}\right)\left(\frac{\partial T_0}{\partial y}\right)_{y=0} \\ & \times \left[y - \frac{1-\sigma+\sigma\exp\left(-y\sqrt{\frac{i\omega}{\nu}}\right) - \exp\left(-y\sqrt{\frac{i\omega}{\kappa}}\right)}{(1-\sigma)\sqrt{\frac{i\omega}{\nu}}}\right], \end{aligned} \tag{7.2.42}$$

where $\sigma = (\nu/\kappa)$ is the Prandtl number.

The solution (7.2.42) fits well into (7.2.41) as the exponentials are small. Both are valid for any changes $\varepsilon U_1(x)\exp(i\omega t)$ in the stream velocity (7.2.19), but only the case $U_1 = U_0$ will be studied.

Finally, the heat transfer rate per unit area is given by

$$\begin{aligned} -k\left(\frac{\partial T_0}{\partial y}\right)_{y=0} + \left(\frac{i}{\omega}\right)(\varepsilon k)\, e^{i\omega t}\left[\left(\frac{dU_0}{dx}\right)\left(\frac{\partial T_0}{\partial y}\right)_{y=0}\frac{1}{(1+\sqrt{\sigma})}\right. \\ \left. -U_0\left(\frac{\partial^2 T_0}{\partial x \partial y}\right)_{y=0}\frac{(1+2\sqrt{\sigma})}{(1+\sqrt{\sigma})^2}\right]. \end{aligned} \tag{7.2.43}$$

This result shows that its amplitude decreases with frequency and its phase is behind that of fluctuations in stream velocity by $(\pi/2)$.

We close this discussion by adding the following comments. When the body is hot, the maximum value of the heat transfer from it tend to lag behind those of the stream velocity, as a result of thermal inertia, but this is counteracted to some extent by the influence of convection by the phase-advanced velocities near the wall. For boundary layers with a favorable gradient in the mean flow, it turns out that the tendency to lag predominates. On the other hand, for the Blasius boundary layer, two effects cancel out fairly closely, and layers with adverse pressure gradient in the main stream, there is a phase advance at lower frequencies. At high frequencies $\omega \gg \omega_0$, there is always a phase lag of $\frac{\pi}{2}$, but the amplitude of heat-transfer fluctuations is then much decreased, even though that of the skin friction fluctuations is increased.

In one case, the exact solution of equations (7.2.36)–(7.2.37) with neglected ω can be obtained, given the right hand side of (7.2.36). This is the case of the Hiemenz layer $U_0 = \beta x$ and the predicted time lag is constant, $0.63\beta^{-1}$.

It is appropriate to verify at one case whether the regions of validity of low- and high-frequency approximations overlap. The Hiemenz boundary layer is used for this purpose as the process is impossible for the Blasius boundary layer. Using a more integration procedure, the resulting temperature T_2 would have non-negligible negative values near the wall and positive ones far away for $\omega = 0$, and the variation with ω would make $(T_s + i\omega T_2)$ change in the direction of the asymptotic value of T_1.

It is relevant to include the application of the results to calculate the fluctuations of heat transfer from a heated circular wire in a fluctuating stream. The heat transfer distribution for the unperturbed state is given in an edited volume of Howarth (1953, page 798). Assuming heat-transfer rate from the stagnation point at $\theta = 0$ as unity, it reduces to the value, 0.8 at $\theta = 45°$, but reduces very rapidly to 0.25 at separation ($\theta = 0.82°$). There is considerable heat transfer at the back due to the recirculating turbulent vortex pair behind the wire, and rate increases to 1 again at $\theta = 180°$.

In most fluctuating fluid flows, it can be stated that the heat transfer due to recirculating turbulent vortex pair would not fluctuate much, and that the fluctuations in heat transfer would occur mainly from the boundary layer on the front of the wire through which about half the unperturbed heat transfer takes place. Of the quasi-steady state fluctuation in heat transfer through the boundary layer, two-third arises from the front quadrant ($\theta < 45°$) in which the modulation obtained for the Hiemenz layer may be used. The modulation for the rest of the layer is relatively less.

In case of very low frequency ($\omega \ll \omega_0$) fluctuations, the time lag is $0.7\beta^{-1}$. This will be decreased somewhat by the greater effect of convection by the acceleration-dependent velocity profile, in the rear part of the boundary layer.

In a joint paper with M. B. Glauert on the axisymmetric boundary layer on a long thin cylinder, published in *Proceedings of the Royal Society*, **A230** (1955) 188-203, Lighthill investigated the axisymmetric laminar boundary layer flow of a uniform stream of velocity U over a long thin cylinder of radius a. They solved the problem by a 'Pohlhaussen' method based on a velocity profile chosen to represent conditions near the boundary surface as accurately as possible. They also used the second method which deals with an asymptotic series solution valid far enough downstream from the nose for the boundary-layer thickness to have become large compared with the radius of the cylinder. This asymptotic solution is valid for very large $(\nu x/Ua^2)$.

The boundary layer equations for this problem are

$$\frac{\partial u}{\partial x}+\frac{\partial v}{\partial y}+\frac{v}{a+y}=0, \quad u\frac{\partial u}{\partial x}+v\frac{\partial u}{\partial y}=\nu\left(\frac{\partial^2 u}{\partial y^2}+\frac{1}{a+y}\cdot\frac{\partial u}{\partial y}\right), \tag{7.2.44ab}$$

where y is the distance from the solid boundary and $y+a$ is therefore the distance from the axis. These equations (7.2.44ab) are to be solved under the boundary conditions

$$u=v=0, \text{ on } y=0; \qquad u\to U \quad \text{as} \quad x\to 0 \quad \text{or as} \quad y\to\infty. \tag{7.2.45}$$

Substituting $y=0$ in (7.2.44b) yields

$$\frac{\partial^2 u}{\partial y^2}+\frac{1}{a}\frac{\partial u}{\partial y}=0, \quad \text{on} \quad y=0. \tag{7.2.46}$$

Differentiating this equation with respect to y and putting $y=0$ again gives

$$\frac{\partial^3 u}{\partial y^3}+\frac{1}{a}\frac{\partial^2 u}{\partial y^2}-\frac{1}{a^2}\frac{\partial u}{\partial y}=0, \quad \text{on} \quad y=0, \tag{7.2.47}$$

with no contribution from the left side of equation (7.2.44b).

Consequently, equations (7.2.46)–(7.2.47) show that as $y\to 0$

$$u=A\left[\frac{y}{a}-\frac{y^2}{2a^2}+\frac{y^3}{3a^3}+O\left(\frac{y}{a}\right)^4\right], \tag{7.2.48}$$

for some values of A which may depend on x. Equation (7.2.48) can be written in the form

$$u=A\left[\ln\left(1+\frac{y}{a}\right)+O\left(\frac{y}{a}\right)^4\right]. \tag{7.2.49}$$

It follows from this solution that, as the boundary layer thickness increases, the velocity profile departs more and more from that of the Blassius solution. The shear stress $\mu\left(\frac{\partial u}{\partial y}\right)$ is maximum on the boundary, from which it falls off initially as the reciprocal of the distance from the axis. This is due to the fact that the shearing force on the cylinder of fluid of unit length is equal to the shear stress, $\mu\left(\frac{\partial u}{\partial y}\right)$ multiplied by C, where $C = 2\pi(a+y)$ is the circumference of the cylinder. This force must be independent of y in the region near the solid boundary, where the acceleration of the fluid is negligible. It turns out that the skin friction and other features of the boundary layer profile near the solid boundary decrease like the reciprocal of $\beta = \ln\left(4\nu x/Ua^2\right)$ for large x so that they vary very slowly with x. In other words, Lighthill's profile which is fairly good for small x and gets progressively better as x increases. The error for small x can be determined by comparing with the Seban-Bond-Kelly solution. It is expected to form an upper bound for the error for general x which can be confirmed by the asymptotic theory.

In order to make a match between the solution profile and the constant U, where $u \to U$ as $y \to \infty$, we replace the constant A above by U/α, where α is a non-dimensional quantity and write

$$u = \begin{bmatrix} \frac{U}{\alpha}\ln\left(1+\frac{y}{a}\right), & y \le \delta \\ \\ U, & y \ge \delta \end{bmatrix}, \tag{7.2.50}$$

where $\delta = a\left(e^{\alpha}-1\right)$ is the thickness of the boundary layer involved in the Pohlhausen method of solution.

Based on the Pohlhausen solution, they first examined the accuracy of three quantities: (i) the skin friction,

$$\tau = \mu\left(\frac{\partial u}{\partial y}\right)_{y=0} = \frac{\mu U}{a\alpha},$$

(ii) the non-dimensional displacement area, $\left(\Delta/\pi a^2\right)$ and (iii) the nondimensional momentum defect area, $\left(\Theta/\pi a^2\right)$.

They have shown that

$$\frac{a\tau}{\mu U} = \frac{1}{\alpha} \sim \frac{2}{\beta} - \frac{6}{\beta^3} + \cdots, \tag{7.2.51}$$

$$\frac{\Delta}{\pi a^2} \sim \frac{4\nu x}{Ua^2}\left\{\frac{1}{\beta} + \frac{3}{\beta^2} + \frac{5}{\beta^3} + \cdots\right\}, \tag{7.2.52}$$

$$\frac{\Theta}{\pi a^2} \sim \frac{4\nu x}{Ua^2}\left\{\frac{1}{\beta} + \frac{1}{\beta^2} - \frac{1}{\beta^3} + \cdots\right\}. \tag{7.2.53}$$

They then compared these asymptotics results (7.2.51)–(7.2.53) with those for a turbulent boundary layer on a flat plate, for which also the velocity profile in logarithmic close to the wall. This profile occurs in both problems because the ratio of viscous force to velocity gradient increases linearly with distance from the wall. In both problems, a consequence of this logarithmic profile is that the viscous stress decreases only very slowly in the reciprocal - logarithm manner suggested by (7.2.51). On the other hand, both the displacement area and the momentum defect area increases almost linearly as in (7.2.52) and (7.2.53).

Glauert and Lighthill then presented the asymptotic series solution of the full boundary layer equations which remain valid far down the cylinder, where the radius a of the cylinder is much smaller than the thickness of the boundary layer. With $r = a + y$ is the distance from the axis, the boundary layer equation (7.2.44b) becomes

$$u\frac{\partial u}{\partial x} + v\frac{\partial u}{\partial r} = \nu\left(\frac{\partial^2 u}{\partial r^2} + \frac{1}{r}\frac{\partial u}{\partial r}\right). \tag{7.2.54}$$

This has to be solved with the boundary conditions $u = v = 0$ at $r = a$ and $u \to U$ as $r \to \infty$. The equation of continuity is satisfied by the Stokes stream function ψ so that

$$\frac{\partial \psi}{\partial r} = ru, \quad \text{and} \quad \frac{\partial \psi}{\partial x} = -rv. \tag{7.2.55}$$

It is convenient to introduce a nondimensional form of the new independent variable

$$\xi = \frac{Ur^2}{4\nu x},$$

and to write $\psi = \nu x\, f(\xi)$ so that $u = \frac{1}{2}Uf'$ and $v = \frac{\nu}{r}(\xi f' - f)$.

Consequently, equation (7.2.54) reduces to the form

$$\xi f''' + f'' + \frac{1}{2}ff'' = 0. \tag{7.2.56}$$

The transformed boundary conditions are

$$f' \to 2 \quad \text{as} \quad \xi \to \infty,$$

and

$$f = f' = 0 \quad \text{at} \quad \xi = e^{-\beta},$$

where β is defined before.

We seek a series expansion in the form

$$\psi = \nu x f(\xi, x) \sim \nu x\left[f_0(\xi) + \frac{f_1(\xi)}{\beta} + \frac{f_2(\xi)}{\beta^2} + \cdots\right]. \tag{7.2.57}$$

Then

$$u = \frac{1}{2}Uf' \quad \text{and} \quad v = \frac{\nu}{r}\left(\xi f' - f - x f_x\right), \qquad (7.2.58\text{ab})$$

where a prime represents differentiation with respect to ξ, but a suffix is used to denote differentiation with respect to x.

Without detailed analysis, the skin friction, displacement area and momentum defect area are given as follows

$$\left(\frac{a\tau}{\mu U}\right) \sim \frac{2}{\beta} + \frac{2\gamma}{\beta^2} + \frac{\left(2\gamma^2 - \frac{1}{2}\pi^2 - 4\ln 2\right)}{\beta^3} + O\left(\frac{1}{\beta^4}\right). \qquad (7.2.59)$$

Comparing this result (7.2.54) with the series (7.2.51) obtained by the Pohlhausen method and that given by the Rayleigh method

$$\left(\frac{a\tau}{\mu U}\right) \sim \frac{2}{\beta} + \frac{2\gamma}{\beta^2} + \frac{\left(2\gamma^2 - \frac{1}{3}\pi^2\right)}{\beta^3} + O\left(\frac{1}{\beta^4}\right), \qquad (7.2.60)$$

where γ is Euler's constant.

The Rayleigh solution is correct in predicting the second term because the Rayleigh method uses $u = U$ as the first approximation to the velocity. The coefficient of third term in both (7.2.59) and (7.2.60) seems to be incorrect as $2\gamma^2 - \frac{1}{3}\pi^2 \sim -2.62$ and $2\gamma^2 - \frac{1}{2}\pi^2 - 4\ln 2 \sim -7.04$.

Using the physical meaning of the Stokes stream function ψ in terms of stream tube flux, the displacement area is found as

$$\frac{\Delta}{\pi a^2} = \frac{4\nu x}{Ua^2}\left(\frac{1}{\beta} + \frac{1 + \gamma + 2\ln 2}{\beta^2} + \cdots\right). \qquad (7.2.61)$$

The momentum defect area is given by

$$\frac{\Theta}{\pi a^2} = \left(\frac{4\nu x}{Ua^2}\right)\left[\frac{1}{\beta} + \frac{1+\gamma}{\beta^2} + \frac{2\,(1+\gamma) + \gamma^2 - \frac{1}{4}\pi^2 - 2\ln 2}{\beta^3} + \cdots\right]. \qquad (7.2.62)$$

Once again, results (7.2.61) and (7.2.62) can be compared with those given by the Pohlhausen solutions, (7.2.52) and (7.2.53), and with the Rayleigh solution in the form

$$\frac{\Delta}{\pi a^2} = \frac{\Theta}{\pi a^2} = \left(\frac{4\nu x}{Ua^2}\right)\left[\frac{1}{\beta} + \frac{1+\gamma}{\beta^2} + \frac{2\,(1+\gamma) + \gamma^2 - \frac{1}{6}\pi^2}{\beta^3} + \cdots\right]. \qquad (7.2.63)$$

The asymptotic series solution (7.2.61) is valid for large $\left(\frac{\nu x}{Ua^2}\right)$ and the first three terms for the skin friction seems to give good accuracy for $\left(\frac{\nu x}{Ua^2}\right) > 100$. It is important to note that only the third term differs from that predicted by the Rayleigh method, which is rather accurate for these large values of $\left(\frac{\nu x}{Ua^2}\right)$. Finally, the asymptotic series solution enables us to bridge the gap $\left(0.04 < \left(\nu x/Ua^2\right) < 100\right)$ between the Seban-Bond-Kelly series and the asymptotic series by drawing a curve to join them up which is closely parallel to the Pohlhausen curve, or in the case of the displacement area, identical with it. This interpolation leads to tables of skin friction, displacement area and momentum defect area for all values of $(\nu x/Ua^2)$ which are correct to within 2%.

Chapter 8

Linear and Nonlinear Waves in Fluids

"Our present analytical methods seem unsuitable for the solution of the important problems arising in connection with nonlinear partial differential equations and, in fact, with virtually all types of nonlinear problems in pure mathematics. The truth of this statement is particularly striking in the field of fluid dynamics...."

John Von Neumann

"Everything is vague to a degree you do not realize till you have tried to make it precise."

Bertrand Russell

"I believe there is no philosophical high-road in science, with epistemological signposts. No, we are in a jungle and find out way by trial and error, building our road behind us as we proceed."

Max Born

8.1 Sound Waves and Shock Waves

In his 1956 famous paper "On Viscosity Effects in Sound Waves of Finite Amplitude" written in honor of the seventieth birthday of Sir G. I. Taylor, Lighthill initiated the theory of nonlinear acoustics. According to the linearized theory of sound, pressure waves in gases initially at rest are propagated with constant wave velocity and constant energy. Sinusoidal acoustic waveforms are likely to be altered or attenuated by various thermodynamically irreversible processes (or viscosity), or other diffusive effects including heat conduction, relaxation effects, and nonlinear effects which emerge from the theory of plane waves of finite amplitude. In view of dominant effects of nonlinearity, not only does the velocity of sound waves change, being greater in the compression phase of a sound wave than in the expansion phase, but also the waves are convected with the fluid, which is propagating in the direction of propagation in the compression part and in the opposite direction in the expansion part. Both effects significantly distort the waveform by causing the compression part to move forward faster than the expansion part. Thus, the second effect, known as *convection*, is greater so that it tends to produce parts of the waveform, grow steeper and steeper compared with other effects mentioned above. So, Lighthill described the opposite effect on sound wave propagation of convection on one hand and of diffusion and relaxation on the other. The opposite effect leads to the formation of shock waves due to a balance between the steepening tendency of the nonlinear (convective) terms in the equations of motion and smoothening tendency of the diffusion term. The effect of steepening in the absence of dissipation was successfully described for plane waves by many authors including Riemann who presented the most general approach in 1859. In this paper, Lighthill provided a fairly complete account of the development and propagation of shock waves with decay. However, other important effects of high pressure and temperature, gas imperfection, dissociation, ionization and radiation have not been described by Lighthill in this paper.

Lighthill began his formulation of the equations of mass, momentum and energy in the form

$$\frac{D\rho}{Dt} + \rho\, div\, \mathbf{u} = 0, \tag{8.1.1}$$

$$\frac{D\mathbf{u}}{Dt} + \frac{1}{\rho}\, \nabla p = 0, \tag{8.1.2}$$

$$\frac{De}{Dt} + p\, \frac{D}{Dt}\left(\frac{1}{\rho}\right) = 0, \tag{8.1.3}$$

where

$$\frac{D}{Dt} = \frac{\partial}{\partial t} + \mathbf{u} \cdot \nabla,$$

is the total derivative, ρ is the density, p is the pressure, $\mathbf{u}$ is the velocity of fluid, and e is the internal energy per unit mass.

Equations (8.1.1)–(8.1.2) constitute five equations with six unknowns ρ, p, $\mathbf{u}$ and e. Thus, this system of equations is supplemented by another equation (8.1.4) stated below. To a close approximation, air and other gases in non-uniform motion pass through a successive equilibrium states with a fixed relation

$$e = (\gamma - 1)^{-1}\left(\frac{p}{\rho}\right), \tag{8.1.4}$$

where $\gamma = \left(\frac{c_p}{c_v}\right)$ is the ratio of the specific heats, and typical values of γ range from a maximum of 5/3 for monotonic gases through 7/3 for di-atomic gases. Thus, equations (8.1.2)–(8.1.4) constitute six equations for six unknowns, and they are known as the equations of thermodynamically reversible fluid flow.

On the other hand, the equations for irreversible gas flow, when departure from equilibrium are taken into account to a first approximation are given by

$$\frac{D\rho}{Dt} + \rho\, div\, \mathbf{u} = 0, \tag{8.1.5}$$

$$\frac{Du_i}{Dt} + \frac{1}{\rho}\frac{\partial p_{ij}}{\partial x_j} = 0, \tag{8.1.6}$$

$$\frac{De}{Dt} = -\frac{1}{\rho}\, p_{ij}\, e_{ij} + \frac{1}{\rho}\frac{\partial}{\partial x_i}\left(k\frac{\partial T}{\partial x_i}\right), \tag{8.1.7}$$

where p_{ij} is the compressive stress tensor, e_{ij} is the pure rate of strain represented by

$$e_{ij} = \frac{1}{2}\left(\frac{\partial u_i}{\partial x_j} + \frac{\partial u_j}{\partial x_i}\right), \tag{8.1.8}$$

k is a constant and T is the temperature. Thus, equations (8.1.5)–(8.1.7) constitute five equations for five unknowns p, ρ and $\mathbf{u}$.

With linear approximation of the effects of diffusion, the Riemann equations of one-dimensional sound waves of finite amplitude under thermodynamically reversible conditions are given by

$$\frac{\partial u}{\partial t} + u\,\frac{\partial u}{\partial x} + \frac{2}{\gamma - 1}\,c\,\frac{\partial c}{\partial x} = \delta\,\frac{\partial^2 u}{\partial x^2}, \tag{8.1.9}$$

$$\frac{\partial c}{\partial t} + u\,\frac{\partial c}{\partial x} + \frac{1}{2}(\gamma - 1)\,c\,\frac{\partial u}{\partial x} = 0, \tag{8.1.10}$$

where c is the speed of an isentropic infinitesimal sound wave at the local values of pressure and density, and δ is called the *diffusivity of sound* given by

$$\delta = \frac{4}{3}\nu + \frac{\mu_u}{\rho} + (\gamma - 1)\,\frac{k}{\rho\, c_p} = \nu\left[\frac{4}{3} + \frac{\mu_u}{\mu} + \frac{\gamma - 1}{\sigma}\right], \quad \sigma = \frac{\mu c_p}{k}, \tag{8.1.11}$$

We consider the solution of equations (8.1.9)–(8.1.10) for waves of infinitesimal amplitude, where the nonlinear terms of order (U/c) is much less than dissipative terms of order $(\nu\omega/c^2)$, ν is the kinematic viscosity and ω is the frequency so that the governing equations (8.1.9)–(8.1.10) become

$$\frac{\partial u}{\partial t} + \left(\frac{2}{\gamma - 1}\right) c_0 \frac{\partial c}{\partial x} = \delta \frac{\partial^2 u}{\partial x^2}, \tag{8.1.12}$$

$$\frac{\partial c}{\partial t} + \frac{1}{2}(\gamma - 1) c_0 \frac{\partial u}{\partial x} = 0. \tag{8.1.13}$$

We seek a solution for $u(x,t)$ and $c(x,t)$ proportional to $\exp(i\omega t)$ so that

$$\left(c_0^2 + i\omega\delta\right) \frac{d^2 u}{dx^2} + \omega^2 u = 0 \tag{8.1.14}$$

and a progressive wave solution must be of the form

$$u(x,t) = A \exp\left[i\omega \left\{t - x/\left(c_0^2 + i\omega\delta\right)^{\frac{1}{2}}\right\}\right], \tag{8.1.15}$$

where A is a constant.

We next expand the square root involved in (8.1.15) with $\left(\nu\omega/c_0^2\right)^2$ neglected so that we obtain the solution

$$u(x,t) = A \exp\left(-\frac{1}{2}\delta\, \omega^2/c_0^3\right) \exp\left[i\omega\left(t - \frac{x}{c_0}\right)\right]. \tag{8.1.16}$$

This solution decays like $\exp(-\alpha x)$, $\alpha = \left(\delta\omega^2/c_0^3\right)$. Thus, the sinusoidal waveform is unaffected by the damping effect of diffusion.

However, other waveforms may be affected by diffusion as different harmonic components are differently attenuated. Thus, if, at $x = 0$, and at time t,

$$u = u_0(t) = \int_{-\infty}^{\infty} e^{i\omega t} F(\omega)\, d\omega \tag{8.1.17}$$

then, at a distance x, and time t,

$$u(x,t) = \int_{-\infty}^{\infty} \exp(-\frac{1}{2}\alpha\, x) \exp\left[i\omega\left(t - x/c_0\right)\right] F(\omega)\, d\omega. \tag{8.1.18}$$

Using the Fourier inversion theorem leads to the solution

$$u(x,t) = \left[2\pi\left(\frac{\delta x}{c_0}\right)\right]^{-\frac{1}{2}} \int_{-\infty}^{\infty} u_0(\xi) \exp\left[-\frac{\left(x + c_0\xi - c_0 t\right)^2}{2\delta\left(x/c_0\right)}\right] c_0\, d\xi. \tag{8.1.19}$$

This represents the time average of $u_0(\xi)$ weighted by a Gaussian distribution centered at $\xi = \left(t - \frac{x}{c_0}\right)$ with standard deviation $\frac{1}{c_0}\left(\frac{\delta x}{c_0}\right)^{\frac{1}{2}}$ as simple diffusion gives a spatial standard deviation of $\sqrt{\delta t}$ after time t. In an extreme situation, a step function waveform leads to an error-integral waveform that will spread the discontinuity over a length increasing in proportion to $\left(\frac{\delta x}{c_0}\right)^{\frac{1}{2}}$ as the wave progresses.

This is followed by a discussion of Lighthill on the attenuation of sound waves due to relaxation effects and the theory of relaxation time.

According to the most general approach of Riemann, the equations (8.1.9)–(8.1.10) of motion with zero right hand yield the equations for Riemann invariants (see Debnath 2005, page 316)

$$\frac{\partial r}{\partial t} + (c+u)\frac{\partial r}{\partial x} = 0, \qquad \frac{\partial s}{\partial t} - (c-u)\frac{\partial s}{\partial x} = 0, \tag{8.1.20ab}$$

where r and s are the Riemann invariants given by

$$r = \frac{c}{\gamma - 1} + \frac{1}{2}u, \qquad s = \frac{c}{\gamma - 1} - \frac{1}{2}u. \tag{8.1.21ab}$$

These equations signify that

$$r = \text{constant along positive wavelet on } \Gamma_1 : \frac{dx}{dt} = c + u, \tag{8.1.22}$$

$$s = \text{constant along negative wavelet on } \Gamma_2 : \frac{dx}{dt} = c - u. \tag{8.1.23}$$

Furthermore, $u = r - s = \text{constant}$ and $c = \frac{1}{2}(\gamma - 1)(r + s) = \text{constant}$ confirming that u and c are separately constant.

For a disturbance initially limited to $x < 0$, all negative wavelets beginning from initial positions with $x > 0$ must carry a value of s equal to $c_0/(\gamma - 1)$, its value in the undisturbed fluid. In the regions filled by these wavelets,

$$c = c_0 + \frac{1}{2}(\gamma - 1)u, \qquad r = \frac{c_0}{\gamma - 1} + u. \tag{8.1.24ab}$$

This part of the fluid flow which is being bounded on the left by a negative wavelet, contains continually more and more fluid, and is called a *traveling wave of finite amplitude*. The value of u is constant on each positive wavelet, and so is that of

$$c = c_0 + \frac{1}{2}(\gamma - 1)\, u, \tag{8.1.25}$$

$$p = p_0 \left[1 + \frac{1}{2c_0}(\gamma - 1)\, u\right]^{2\gamma/(\gamma-1)}, \tag{8.1.26}$$

$$\rho = \rho_0 \left[1 + \frac{1}{2c_0}(\gamma - 1)\, u\right]^{2/(\gamma-1)}. \tag{8.1.27}$$

The speed of the wave is

$$c + u = c_0 + \frac{1}{2}(\gamma - 1)\, u + u = c_0 + \frac{1}{2}(\gamma + 1)\, u \tag{8.1.28}$$

which is also constant in time. All these quantities are increasing function of u. The wavelets carrying the maximum pressures travel fastest both relative to the fluid and in space. Evidently, any acoustic waveform is continually changing. The compressive part of the wave becomes shorter and hence, it becomes steeper. On the other hand, the expansive part of the wave becomes longer. This may be broadly described as due to the convection of sound.

It follows from (8.1.22)–(8.1.23) that the characteristics $\frac{dx}{dt} = c \underline{+} u =$ constant, which represent two families of straight lines in the (x, t)-plane. Mathematically, the characteristics exist because the governing quasi-linear equations (8.1.9) and (8.1.10) with zero right hand side are hyperbolic. Physically, they represent the *wavelets* propagating in both directions with speed c while also connected by the fluid with local speed u. If, as often happens, one of the Riemann invariants is constant everywhere (that is, it takes the *same* constant value on every characteristic line of its family), the resulting solution is said to form a *simple wave*.

The solution of the equations of motion in the above simple wave assumes the form

$$u(x, t) = f(\xi), \qquad \xi = x - (u + c)\, t, \tag{8.1.29}$$

where $u(x, 0) = f(x)$ represents the initial data. If $f(x) \geq 0$, $u + c$ will be everywhere positive, and hence, the family of characteristics carrying the simple wave has positive slope in the (x, t)-plane, and the wave is said to be *forward progressing*. The speed of propagation of the forward progressing wave is given by

$$u + c = \frac{1}{2}(\gamma + 1)\, u + c_0, \tag{8.1.30}$$

where a slightly different form of (8.1.21a), that is, $c = \frac{1}{2}(\gamma+1)\,u +$ constant, is used, and c_0 is the sound speed of a one-dimensional compressive wave which cannot remain continuous indefinitely, whereas a similar expansion wave can. Differentiating (8.1.29) combined with (8.1.30) gives

$$\frac{\partial u}{\partial x} = f'(\xi)\,\xi_x = \frac{f'(\xi)}{1+\frac{1}{2}(\gamma+1)\,t\,f'(\xi)}, \tag{8.1.31}$$

$$\frac{\partial u}{\partial t} = f'(\xi)\,\xi_t = -\frac{1}{2}(\gamma+1)\left[\frac{f'(\xi)\,f(\xi)}{1+\frac{1}{2}(\gamma+1)\,t\,f'(\xi)}\right]. \tag{8.1.32}$$

Evidently, both u_x and u_t tend to infinity as

$$t \to -\frac{2}{(\gamma+1)\,f'(\xi)}, \tag{8.1.33}$$

which is positive provided that $f'(\xi) < 0$, corresponding to the compressive part of the wave profile. The discontinuity occurs first at the instant

$$t = \frac{2}{(\gamma+1)\{-f'(\xi)\}_{\max}}. \tag{8.1.34}$$

The solution develops a discontinuity in the form of *shock waves*. The gas is at rest. From the constant value in this simple wave of the Riemann invariant, it is evident that the greater the speed u the greater the value of c and, therefore, of the pressure p. Hence the largest velocities and pressures move most rapidly through the gas. The peaks of velocity or pressure tend to overtake the troughs, thus, leading to the breakdown of the uniqueness of the solution whenever the initial profile contains a compressive part. This has already been illustrated in Debnath's (2005) book (see Chapter 5, Figure 5.4) for a similar situation observed in the case of a general, nonlinear initial-value problem. As time increases, the initial velocity (or pressure) profile is progressively distorted, and the part of the wave profile that was initially expansive (velocity increases in the direction of motion) tends to elongate. Ultimately, the compressive part of the wave profile develops multiple-valuedness and then breaks at the point when it becomes vertical. This simple analysis confirms that, in inviscid flow, a shock wave develops in one-dimensional unsteady flow as exhibited in the photographs included in Figure 8.1, which is due to Dr. W. Bleakney and Dr. Wayland Griffith at the Palmer Physical Laboratory, Princeton University. This photograph shows that in (a) the density gradient is moderate, in (b) it becomes steeper, and in (c) it is clearly discontinuous. The steepening effect of the waveform for the case of one-dimensional flow is equally true for flows with cylindrical and spherical symmetry. This fundamental result was obtained long ago

(a) (b) (c)

Fig. 8.1 Development of shock waves at successive instants. From Lighthill (1956).

by Burton (1893) by intuitive arguments and has been confirmed by Pack (1960) by the method of characteristics.

Example 8.1. (*The Piston Problem in a Polytropic Gas*). The problem is to determine how a simple wave is produced by the prescribed motion of a piston in the closed end of a semi-infinite tube filled with gas.

This is a one-dimensional unsteady problem in gas dynamics. We assume that the gas is initially at rest with a uniform state $u = 0$, $\rho = \rho_0$, and $c = c_0$. The piston starts from rest at the origin and is allowed to withdraw from the tube with a variable velocity for a time t_1, after which the velocity of withdrawal remains constant. The piston path is shown by a dotted curve in Figure 8.2. In the (x, t)-plane, the path of the piston is given by $x = X(t)$ with $X(0) = 0$. The fluid velocity u is equal to the piston velocity $\dot{X}(t)$ on the piston $x = X(t)$, which will be used as the boundary condition on the piston.

The initial state of the gas is given by $u = u_0$, $\rho = \rho_0$, and $c = c_0$ at $t = 0$ in $x \geq 0$. The characteristic line Γ_0 that bounds it and passes through the origin is determined by the equation

$$\frac{dx}{dt} = (u + c)_{t=0} = c_0$$

so that the equation of the characteristic line Γ_0 is $x = c_0 t$.

In view of the uniform initial state, all of the Γ_2 characteristics start on the x-axis so that the Riemann invariants s in (8.1.21b) must be constant and have the form

$$\frac{2c}{\gamma - 1} - u = \frac{2c_0}{\gamma - 1}, \tag{8.1.35}$$

or, equivalently,

$$u = \frac{2(c - c_0)}{\gamma - 1}, \qquad c = c_0 + \frac{\gamma - 1}{2} u. \tag{8.1.36ab}$$

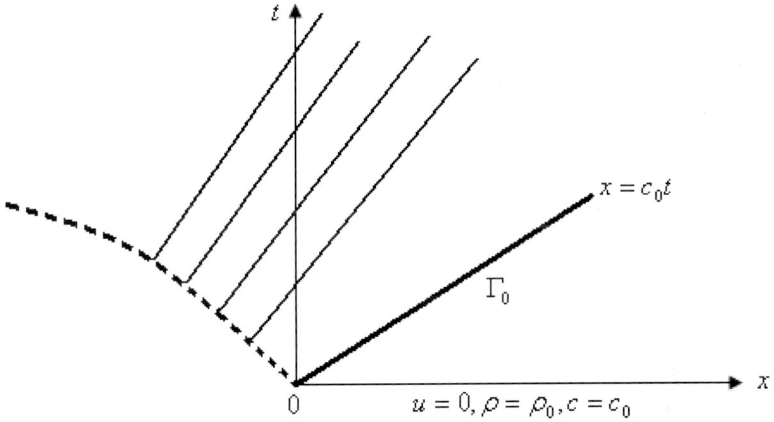

Fig. 8.2 Simple waves generated by the motion of a piston.

The characteristics Γ_1 meeting the piston are given by

$$\frac{2c}{\gamma - 1} + u = 2r \quad \text{on each} \quad \Gamma_1 : \frac{dx}{dt} = u + c, \tag{8.1.37}$$

which is, by (8.1.36ab) which holds everywhere,

$$u = \text{constant on } \Gamma_1 : \frac{dx}{dt} = c_0 + \frac{1}{2}(\gamma + 1)\, u. \tag{8.1.38}$$

Since the flow is continuous with no shocks, $u = 0$ and $c = c_0$ ahead of and on Γ_0, which separates those Γ_1 meeting the x-axis from those meeting the piston. The family of lines Γ_1 through the origin has the equation $(dx/dt) = \xi$, where ξ is a parameter with $\xi = c_0$ on Γ_0. The Γ_1 characteristics are also defined by $(dx/dt) = u + c$, so that $\xi = u + c$. Hence, eliminating c from (8.1.36b) gives

$$u = \left(\frac{2}{\gamma + 1}\right)(\xi - c_0). \tag{8.1.39}$$

Substituting this value of u in (8.1.36b), we obtain

$$c = \left(\frac{\gamma - 1}{\gamma + 1}\right)\xi + \frac{2c_0}{\gamma + 1}. \tag{8.1.40}$$

It follows from $c^2 = \gamma\, k\, \rho^{\gamma - 1}$ and (8.1.36b) with the initial data $\rho = \rho_0$ and $c = c_0$ that

$$\rho = \rho_0 \left[1 + \frac{\gamma - 1}{2c_0}\, u\right]^{2/(\gamma - 1)}. \tag{8.1.41}$$

With $\xi = x/t$, results (8.1.39) through (8.1.41) give the complete solution of the piston problem in terms of x and t.

Finally, the equation of the characteristic line Γ_1 is found by integrating the second equation of (8.1.38) and using the boundary condition on the piston. When a line Γ_1 intersects the piston path at time $t = \tau$, then $u = \dot{X}(\tau)$ along it, and the equation becomes

$$x = X(\tau) + \left\{ c_0 + \frac{\gamma + 1}{2} \dot{X}(\tau) \right\} (t - \tau). \tag{8.1.42}$$

Note that the family Γ_1 represents straight lines with slope (dx/dt) increasing with velocity u. Consequently, the characteristics are likely to overlap on the piston, that is, $\dot{X}(\tau) > 0$ for any τ. If u increases, so do c, ρ, and p, so that instability develops. This shows that shocks will be formed in the compressive part of the disturbance.

We next consider more general theory of plane shock-wave formation by using equation of progressive sound waves with convection and diffusion allowed for to first approximation. If U is the maximum fluid velocity in a flow, and the maximum velocity gradient is written $(U\omega/c)$, by analogy with sound waves, then weak shock adjusts itself so that (U/c) and $(\nu\omega/c^2)$ are small quantities of the same order. So, it is reasonable to equations (8.1.9) and (8.1.10) and simplify them by neglecting terms of order $(U/c)^2$. To do this we write equations (8.1.9)–(8.1.10) in terms of the Riemann invariants (8.1.21ab), but including the right hand sides. This leads to the following equations

$$\frac{\partial r}{\partial t} + (c + u) \frac{\partial r}{\partial x} = \frac{1}{2} \delta \left(\frac{\partial^2 r}{\partial x^2} - \frac{\partial^2 s}{\partial x^2} \right), \tag{8.1.43}$$

$$\frac{\partial s}{\partial t} - (c - u) \frac{\partial s}{\partial x} = \frac{1}{2} \delta \left(\frac{\partial^2 s}{\partial x^2} - \frac{\partial^2 r}{\partial x^2} \right), \tag{8.1.44}$$

where

$$c + u = \frac{1}{2} (\gamma + 1) r + \frac{1}{2} (\gamma - 3) s. \tag{8.1.45}$$

We consider waves propagating in the undisturbed fluid in the positive x-direction, with the aid of the simplified equation (8.1.43) by replacing s by s_0, its value in the undisturbed fluid, so that

$$\frac{\partial r}{\partial t} + \left\{ \frac{1}{2} (\gamma + 1) r + \frac{1}{2} (\gamma - 3) s_0 \right\} \frac{\partial r}{\partial x} = \frac{1}{2} \delta \frac{\partial^2 r}{\partial x^2}, \tag{8.1.46}$$

In view of this assumption, and since $\frac{\partial}{\partial t}\left(\frac{\partial r}{\partial x}\right) = -c_0 \frac{\partial}{\partial x}\left(\frac{\partial r}{\partial x}\right)$ plus terms of higher order in (U/c) and $(\nu\omega/c^2)$ and equation (8.1.44) can be solved to a first approximation by

$$s = \left(\frac{\delta}{4c_0}\right)\frac{\partial r}{\partial x}, \tag{8.1.47}$$

In fact, $(r - r_0)/c_0 \sim \frac{U}{c}$ and $(s - s_0)/c_0 \sim (\nu\omega/c^2)(U/c)$ and hence can be neglected in (8.1.43) so that (8.1.46) is the approximate equation of the desired order.

Substituting $X = x - c_0 t$ and

$$v = \frac{1}{2}(\gamma+1)(r - r_0) = \frac{1}{2}(\gamma+1)r + \frac{1}{2}(\gamma-3)s_0 - c_0 \approx c + u - c_0 \tag{8.1.48}$$

in the equation (8.1.46), we obtain the celebrated Burgers equation for the excess wave velocity v

$$\frac{\partial v}{\partial t} + v\,\frac{\partial v}{\partial X} = \frac{1}{2}\,\delta\,\frac{\partial^2 v}{\partial X^2}, \tag{8.1.49}$$

where $X = (x - c_0 t)$ is a coordinate measured in a frame of reference moving in the same direction as the waves at the undisturbed sound velocity c_0.

This is the simplest nonlinear model equation representing the balance between time evolution, convection and diffusion. Burgers (1948) first developed this equation primarily to shed light on the study of turbulence which describes the interaction of the two opposite effects of convection and diffusion. However, turbulence is more complex in the sense that it is three dimensional and statistically random in nature. The Burgers equation arises in many other physical problems.

Replacing v by u, X by x and $\frac{1}{2}\delta$ by ν, we write the Burgers equation in standard form (see Debnath (2005) page 374) and we first solve standard form of (8.1.49) for two simple cases: (i) a linearized Burgers equation and (ii) an equation with the linearized convective term $c\,u_x$, where c is a constant.

In the first case, the linearized Burgers equation

$$u_t = \nu\,u_{xx}, \qquad -\infty < x < \infty, \quad t > 0, \tag{8.1.50}$$

with the initial conditions at $t = 0$

$$u(x,0) = \begin{Bmatrix} u_1, & x < 0 \\ \\ u_2, & x > 0 \end{Bmatrix}, \tag{8.1.51ab}$$

can readily be solved by applying the Fourier transform with respect to x. The final solution of (8.1.50) with (8.1.51ab) for $t > 0$ is

$$u(x,t) = \frac{1}{2}(u_1 + u_2) - \frac{1}{2}(u_1 - u_2)\,\mathrm{erf}\left(\frac{x}{2\sqrt{\nu t}}\right). \tag{8.1.52}$$

This shows that the presence of the diffusion term νu_{xx} is to smooth out the initial distribution like $(\nu t)^{-\frac{1}{2}}$. The solution (8.1.52) tends to constant values u_1, as $x \to -\infty$, and u_2, as $x \to +\infty$. The absence of the diffusion term in (8.1.49) leads to gradual nonlinear steepening, and eventually breaking. Indeed, equation (8.1.49) combines the two opposite effects of nonlinearity and diffusion. In the absence of the diffusion term in (8.1.49), the resulting equation reduces to the first-order, nonlinear wave equation which admits a progressively distorted wave profile as a solution. Eventually, this solution develops a discontinuity as a shock wave.

In the second case, equation (8.1.49) reduces to the linear parabolic equation

$$u_t + c\,u_x = \nu u_{xx}. \tag{8.1.53}$$

We seek a plane wave solution of (8.1.53) in the form

$$u\,(x,t) = a \exp\left\{i\,(k\,x - \omega t)\right\}, \tag{8.1.54}$$

where $\operatorname{Im}\,\omega = -\nu k^2 < 0$, since $\nu > 0$.

Thus, the solution (8.1.54) becomes

$$u\,(x,t) = a\,e^{-\nu k^2 t} \exp\left[i\,k\,(x - ct)\right]. \tag{8.1.55}$$

This represents a diffusive wave with wavenumbers k and phase velocity $c = (\omega/k)$. The amplitude of the wave decays exponentially with time t, and the decay time $t_0 = \left(\nu k^2\right)^{-1}$, which becomes smaller as k increases with fixed ν. Thus, the waves of smaller wavelengths decay faster than the waves of longer wavelengths. On the other hand, for a fixed wavenumber k, the decay time decreases as ν increases so that waves of a given wavelength attenuate faster in a medium with a larger ν. This quantity ν may be regarded as a *measure of diffusion*. Finally, after a sufficiently long time $(t \gg t_0)$, only disturbances of long wavelength will survive, whereas all waves of short wavelength will decay very rapidly. For $\nu < 0$, solution (8.1.55) tends to infinity as $t \to \infty$, and hence, it becomes unstable.

Traveling Wave Solutions and Shock-Wave Structure

To investigate the effects of nonlinear steepening and diffusion, we seek a traveling wave solution of the standard Burgers equation (8.1.49) in the form

$$u\,(x,t) = u\,(\xi)\,, \qquad \xi = x - Ut, \tag{8.1.56ab}$$

where the wave speed U is to be determined and $u\,(\xi)$ represents the wave form with the property that it tends asymptotically to constant values u_1, as $\xi \to -\infty$, and u_2, as $\xi \to +\infty$. We assume that $u_1 > u_2$. Substituting (8.1.56ab) into the Burgers equation (8.1.49) gives the ordinary differential equation

$$-U\,u'(\xi) + u\,u'(\xi) - \nu\,u''(\xi) = 0.$$

Integrating this equation yields

$$-U\,u(\xi) + \frac{1}{2}u^2 - \nu\,u'(\xi) = A,$$

where A is a constant of integration. Or, equivalently,

$$u'(\xi) = \frac{1}{2\nu}\left(u^2 - 2Uu - 2A\right). \tag{8.1.57}$$

Clearly, this suggests that u_1 and u_2 are the roots of the quadratic equation

$$u^2 - 2Uu - 2A = 0. \tag{8.1.58}$$

Hence, U and A are determined from the sum and the product of the roots u_1 and u_2 of (8.1.58), and therefore,

$$U = \frac{1}{2}(u_1 + u_2), \qquad A = -\frac{1}{2}u_1 u_2. \tag{8.1.59ab}$$

Thus, the wave speed is the *average* of the two speeds at asymptotic states at infinity.

Equation (8.1.57) can now be written as

$$2\nu\,\frac{du}{d\xi} = u^2 - 2Uu - 2A = (u - u_1)(u - u_2).$$

Integrating this equation gives

$$\left(\frac{\xi}{2\nu}\right) = -\int \frac{du}{(u_1 - u)(u - u_2)} = \frac{du}{(u_1 - u_2)}\log\left(\frac{u_1 - u}{(u - u_2)}\right). \tag{8.1.60}$$

This leads to the solution for $u(\xi)$ in the form

$$u(\xi) = \frac{u_1 + u_2 \exp\left[\left(\frac{\xi}{2\nu}\right)(u_1 - u_2)\right]}{1 + \exp\left[\left(\frac{\xi}{2\nu}\right)(u_1 - u_2)\right]} \tag{8.1.61}$$

$$= u_2 + \frac{(u_1 - u_2)}{1 + \exp\left[\left(\frac{\xi}{2\nu}\right)(u_1 - u_2)\right]}. \tag{8.1.62}$$

Another useful expression for u can be written from (8.1.61) in the form

$$u(\xi) = \frac{1}{2}(u_1 + u_2) + \frac{u_1 + u_2 \exp\left[\left(\frac{\xi}{2\nu}\right)(u_1 - u_2)\right]}{1 + \exp\left[\left(\frac{\xi}{2\nu}\right)(u_1 - u_2)\right]} - \frac{1}{2}(u_1 - u_2)$$

$$= \frac{1}{2}(u_1 + u_2) - \frac{1}{2}(u_1 - u_2)\tanh\left[\left(\frac{\xi}{4\nu}\right)(u_1 - u_2)\right]. \tag{8.1.63}$$

As $u_1 > u_2$, the wave profile $u(\xi)$ decreases monotonically with ξ from the constant value u_1, as $\xi \to -\infty$, to the constant value u_2, as $\xi \to +\infty$, as shown in Figure 8.3. At $\xi = 0$, $u = \frac{1}{2}(u_1 + u_2)$. The shape of the waveform (8.1.63) is significantly affected by diffusion coefficient ν. This means that the presence of diffusion processes prevents the gradual distortion of the wave profile, and so, it does not break. On the other hand, if the diffusion term is absent ($\nu = 0$) in (8.1.49), the wave profile will suffer from gradual distortion and steepening, and hence, it would definitely break with development of a shock.

Shock waves are formed as a result of a balance between the steepening effect of the convective (nonlinear) term and the smoothing effect of the linear diffusive terms in the equation of motion. The tendency to steepening has been demonstrated for plane waves by many authors, including Bernhard Riemann (1858), who first introduced the most general approach.

The upshot of the above analysis is that the convection and diffusion terms in the Burgers equation exhibit opposite effects. The former introduces a sharp discontinuity in the solution profile, whereas the latter tends to diffuse (spread out) the discontinuity into a smooth profile. In view of this property, the solution is called the *diffusive wave*. In the context of fluid flow, ν represents the kinematic viscosity which measures the viscous dissipation.

Multiplying both numerator and denominator of (8.1.61) by

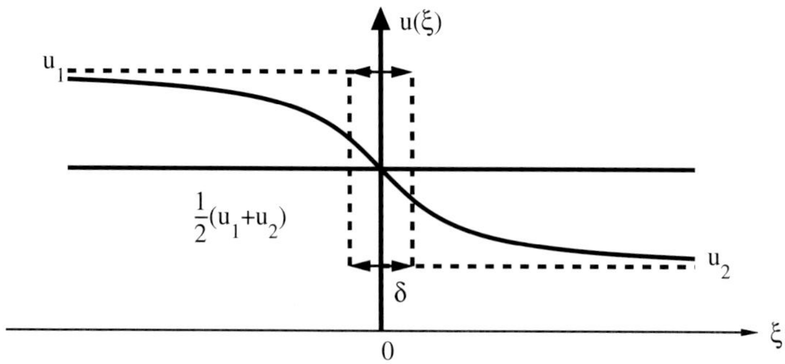

Fig. 8.3 Traveling wave solution $u(\xi)$, $\xi = x - Ut$.

$\exp\left[-\frac{\xi}{2\nu}(u_1-u_2)\right]$, we write the solution (8.1.61) in the form

$$u(\xi)=\frac{u_2+u_1\exp\left[-\left(\frac{\xi}{2\nu}\right)(u_1-u_2)\right]}{1+\exp\left[-\left(\frac{\xi}{2\nu}\right)(u_1-u_2)\right]}. \tag{8.1.64}$$

The exponential factor in this solution indicates the existence of a thin *transition layer* of thickness δ of the order $\nu/(u_1-u_2)$. This thickness δ can be referred to as the *shock thickness* , which tends to zero as $\nu\to 0$ for fixed u_1 and u_2. Also, δ increases as $u_1\to u_2$ for a fixed ν. If δ is small compared with other typical length scales of the problem, the rapid shock transition can satisfactorily be approximated by a discontinuity. Thus, in the limit as $\nu\to 0$, we might expect that solutions of (8.1.49) tend to solutions of the nonlinear equation

$$u_t+u\,u_x=0 \tag{8.1.65}$$

together with discontinuous shock waves satisfying the jump condition

$$\frac{1}{2}\left(u_1^2-u_2^2\right)=U\left(u_1-u_2\right), \tag{8.1.66}$$

where U is the shock speed given by

$$U=\frac{1}{2}\left(u_1+u_2\right). \tag{8.1.67}$$

This is in complete agreement with the above analysis.

The Exact Solution of the Burgers Equation

We solve the initial-value problem for the Burgers equation

$$u_t+u\,u_x=\nu\,u_{xx},\qquad x\in R,\quad t>0, \tag{8.1.68}$$

$$u(x,0)=F(x),\qquad x\in R. \tag{8.1.69}$$

Special attention will be given to small values of ν or to large values of Reynolds numbers.

Hopf (1950) and Cole (1951) independently discovered a transformation that reduces the Burgers equation to a linear diffusion equation. First, we write (8.1.68) in a form similar to a conservation law,

$$u_t+\left(\frac{1}{2}u^2-\nu\,u_x\right)_x=0. \tag{8.1.70}$$

This can be regarded as the compatibility condition for a function ψ to exist, such that

$$u=\psi_x \quad\text{and}\quad \nu\,u_x-\frac{1}{2}u^2=\psi_t. \tag{8.1.71ab}$$

We substitute the value of u from (8.1.71a) in (8.1.71b) to obtain

$$\nu\,\psi_{xx} - \frac{1}{2}\psi_x^2 = \psi_t. \tag{8.1.72}$$

Next, we introduce $\psi = -2\nu\log\phi$ so that

$$u = \psi_x = -2\,\nu\,\frac{\phi_x}{\phi}. \tag{8.1.73}$$

This is called the *Cole–Hopf transformation*, which, by differentiating, gives

$$\psi_{xx} = 2\,\nu\left(\frac{\phi_x}{\phi}\right)^2 - \frac{2\,\nu}{\phi}\phi_{xx}, \quad \text{and} \quad \psi_t = -2\,\nu\,\frac{\phi_t}{\phi}.$$

Consequently, (8.1.72) reduces to the linear diffusion equation

$$\phi_t = \nu\phi_{xx}. \tag{8.1.74}$$

Many solutions of this equation are well known in the literature. We substitute the given solution for ϕ to find solutions of the Burgers equation.

We now solve equation (8.1.74) subject to the initial condition

$$\phi\,(x,0) = \Phi\,(x) \qquad x \in R. \tag{8.1.75a}$$

This can be written in terms of the initial value $u\,(x,0) = F\,(x)$ by using (8.1.73):

$$F\,(x) = u\,(x,0) = -2\,\nu\,\frac{\phi_x\,(x,0)}{\phi\,(x,0)}. \tag{8.1.75b}$$

Integrating this result gives

$$\phi\,(x,0) = \Phi\,(x) = \exp\left\{-\frac{1}{2\nu}\int_0^x F\,(\alpha)\,d\alpha\right\}. \tag{8.1.76}$$

The Fourier transform method or the joint Fourier–Laplace transform technique can be used to solve the linear initial-value problem (8.1.74), (8.1.75a), and hence, the standard solution of this problem is

$$\phi\,(x,t) = \frac{1}{2\sqrt{\pi\nu t}}\int_{-\infty}^{\infty}\Phi\,(\zeta)\exp\left[-\frac{(x-\zeta)^2}{4\nu t}\right]d\zeta, \tag{8.1.77}$$

where $\Phi\,(\zeta)$ is given by (8.1.76). We then substitute the value of $\Phi\,(\zeta)$ to rewrite (8.1.77) in the convenient form

$$\phi\,(x,t) = \frac{1}{2\sqrt{\pi\nu t}}\int_{-\infty}^{\infty}\exp\left(-\frac{f}{2\nu}\right)d\zeta, \tag{8.1.78}$$

where

$$f(\zeta, x, t) = \int_0^\zeta F(\alpha)\, d\alpha + \frac{(x-\zeta)^2}{2t}. \tag{8.1.79}$$

Thus,

$$\phi_x(x,t) = -\frac{1}{4\nu\sqrt{\pi\nu t}} \int_{-\infty}^{\infty} \left(\frac{x-\zeta}{t}\right) \exp\left(-\frac{f}{2\nu}\right) d\zeta. \tag{8.1.80}$$

Therefore, the exact solution of the Burgers initial-value problem is obtained from (8.1.73) in the form

$$u(x,t) = \frac{\int_{-\infty}^{\infty} \left(\frac{x-\zeta}{t}\right) \exp\left(-\frac{f}{2\nu}\right) d\zeta}{\int_{-\infty}^{\infty} \exp\left(-\frac{f}{2\nu}\right) d\zeta}. \tag{8.1.81}$$

This is clearly single valued and continuous for all values of t. The physical interpretation of this exact solution can hardly be given unless a suitable simple form of $F(x)$ is specified. In many problems, an exact evaluation of the integrals involved in (8.1.81) is almost a formidable task. It is then necessary to resort to asymptotic methods. We next consider the following example to investigate the formation of discontinuities or shock waves.

Example 8.2. Find the solution of the Burgers initial-value problem with physical significance for the discontinuous data

$$F(x) = A\,\delta(x)\, H(x), \tag{8.1.82}$$

where A is a constant, $\delta(x)$ is the Dirac delta function, and $H(x)$ is the Heaviside unit step function.

To find the solution, first we calculate

$$f(\zeta, x, t) = A \int_{0+}^{\zeta} \delta(\alpha)\, d\alpha + \frac{(x-\zeta)^2}{2t} = \begin{bmatrix} \frac{(x-\zeta)^2}{2t} - A, & \zeta < 0 \\ \frac{(x-\zeta)^2}{2t}, & \zeta > 0 \end{bmatrix}.$$

Thus, the integral in the numerator of (8.1.81) is

$$\begin{aligned} \int_{-\infty}^{\infty} \left(\frac{x-\zeta}{t}\right) \exp\left(-\frac{f}{2\nu}\right) d\zeta &= \int_{-\infty}^{0} \left(\frac{x-\zeta}{t}\right) \exp\left\{\frac{A}{2\nu} - \frac{(x-\zeta)^2}{4\nu t}\right\} d\zeta \\ &\quad + \int_{0}^{\infty} \left(\frac{x-\zeta}{t}\right) \exp\left\{-\frac{(x-\zeta)^2}{4\nu t}\right\} d\zeta \\ &= 2\,\nu\left(e^R - 1\right) \exp\left(-\frac{x^2}{4\nu t}\right), \end{aligned}$$

which is obtained by substituting $(x-\zeta)/2\sqrt{\nu t}=\alpha$ and writing $\left(\frac{A}{2\nu}\right)=R$ which can be interpreted as the Reynolds number (Whitham 1974). Clearly, by small ν, we mean large $R\,(R\gg 1)$, and the large ν corresponds to small $R\,(R\ll 1)$.

Similarly, the integral in the denominator of (8.1.81) gives

$$\int_{-\infty}^{\infty}\left(-\frac{f}{2\nu}\right)d\zeta=2\sqrt{\nu t}\left[\sqrt{\pi}+\left(e^{R}-1\right)\operatorname{erfc}\left(\frac{x}{2\sqrt{\nu t}}\right)\right].$$

So the final solution $u(x,t)$ is obtained from (8.1.81) in the form

$$u(x,t)=\sqrt{\frac{\nu}{t}}\left[\frac{\left(e^{R}-1\right)\exp\left(-\frac{x^2}{4\nu t}\right)}{\sqrt{\pi}+\left(e^{R}-1\right)\frac{\sqrt{\pi}}{2}\operatorname{erfc}\left(\frac{x}{2\sqrt{\nu t}}\right)}\right]. \tag{8.1.83}$$

Thus, this solution takes the similarity form

$$u(x,t)=\sqrt{\frac{\nu}{t}}\quad g(\eta,R), \tag{8.1.84}$$

where $\eta=\left(x/2\sqrt{\nu t}\right)$ is the *similarity variable.* Two limiting cases (i) $R\to 0$ and (ii) $R\to\infty$ are of interest, and they are considered below.

Case (i) In the limit as $R\to 0\,(\nu\to\infty)$, the effect of diffusion would be more significant than that of nonlinearity. So, in this case, we can approximate $e^{R}\sim 1+R$ and $\operatorname{erfc}\left(\frac{x}{2\sqrt{\nu t}}\right)\to 0$ in (8.1.83) to obtain the final asymptotic solution

$$u(x,t)\sim\frac{A}{2\sqrt{\pi\nu t}}\exp\left(-\frac{x^2}{4\nu t}\right). \tag{8.1.85}$$

This is the *fundamental* (or *source*) solution of the linear diffusion equation. At any time t, the velocity field is Gaussian. The peak height of $u(x,t)$ decreases inversely with $\sqrt{\nu t}$, whereas the width of the peak $\left(x\sim\sqrt{\nu t}\right)$ increases with $\sqrt{\nu t}$. These are remarkable features of diffusion phenomena.

Case (ii) In this case, $R\to\infty\,(\nu\to 0)$. So, the nonlinear effect would dominate over diffusion, and hence, discontinuity as a shock is expected to develop. We introduce the similarity variable $\eta=x/\sqrt{2At}$ so that $x/\sqrt{4\nu t}=\eta\sqrt{R}$, and then, rewrite the solution (8.1.83) as

$$u(x,t)=\sqrt{\frac{\nu}{t}}\left[\frac{\left(e^{R}-1\right)\exp\left(-R\eta^2\right)}{\sqrt{\pi}+\left(e^{R}-1\right)\left(\frac{\sqrt{\pi}}{2}\right)\operatorname{erfc}\left(\eta\sqrt{R}\right)}\right], \tag{8.1.86}$$

which has the similarity form

$$u(x,t)=\left(\frac{A}{2Rt}\right)^{\frac{1}{2}}g(\eta,R). \tag{8.1.87}$$

Since $R \to \infty$, $e^R - 1$ can be replaced by e^R, hence, (8.1.86) gives

$$u(x,t) = \sqrt{\frac{\nu}{t}}\left[\frac{\exp\{R(1-\eta^2)\}}{\sqrt{\pi} + \left(\frac{\sqrt{\pi}}{2}\right)\exp(R)\,\mathrm{erfc}\left(\eta\sqrt{R}\right)}\right] \text{ for all } \eta, \qquad (8.1.88)$$

$$\sim 0 \text{ as } R \to \infty \text{ for } \eta < 0 \text{ and } \eta > 1.$$

When $R \to \infty$ and $0 < \eta < 1$, we use the asymptotic result

$$\mathrm{erfc}\left(\eta\sqrt{R}\right) \sim \frac{1}{\eta\sqrt{\pi R}}\exp\left(-\eta^2 R\right)$$

to obtain the solution from (8.1.88) as

$$u(x,t) = \left(\frac{2A}{t}\right)^{\frac{1}{2}}\left[\frac{\eta}{1 + 2\eta\sqrt{\pi R}\,\exp\{-R(1-\eta^2)\}}\right] \qquad (8.1.89)$$

$$\sim \eta\left(\frac{2A}{t}\right)^{\frac{1}{2}} = \frac{x}{t} \qquad \text{as } R \to \infty. \qquad (8.1.90)$$

It turns out that the final asymptotic solution, as $R \to \infty$, is

$$u(x,t) \sim \begin{bmatrix} \frac{x}{t}, & 0 < x < \sqrt{2At} \\ \\ 0, & \text{otherwise} \end{bmatrix}. \qquad (8.1.91)$$

This result represents a shock wave at $x = \sqrt{2At}$, and the shock speed is $U = \frac{dx}{dt} = \left(\frac{A}{2t}\right)^{\frac{1}{2}}$. The solution $u(x,t)$ has a jump from zero to $\frac{x}{t} = \left(\frac{2A}{t}\right)^{\frac{1}{2}}$, and hence, the shock condition is satisfied.

Whitham (1974) investigated the structure of the shock wave solution for large values of R. He found two transition layers—one at $\eta = 1$, and the other (weaker) layer at $\eta = 0$. For large but finite R, (8.1.89) indicates a rapid transition from exponentially small values in $\eta > 1$ to $u \sim \sqrt{\frac{2A}{t}}\,\eta$ in $\eta < 1$. In the transition layer, $\eta \approx 1$, solution (8.1.89) can be approximated by

$$u(x,t) \sim \sqrt{\frac{2A}{t}}\left[1 + 2\sqrt{\pi R}\,\exp\{2R(\eta - 1)\}\right]^{-1}. \qquad (8.1.92)$$

This shows that this transition layer has thickness $O\left(R^{-1}\right)$.

On the other hand, the second layer at $\eta = 0$ does smooth out the discontinuity in the derivative between $u \sim 0$ in $\eta < 0$ to $u \sim \frac{x}{t}$ in $0 < \eta < 1$. It follows from (8.1.88) that this weaker layer occurs for $\eta = O\left(\frac{1}{\sqrt{R}}\right)$, and hence, (8.1.89) assumes the approximate form

$$u(x,t) \sim 2\sqrt{\frac{\nu}{\pi t}} \cdot \frac{\exp\left(-R\eta^2\right)}{\mathrm{erfc}\left(\eta\sqrt{R}\right)}. \qquad (8.1.93)$$

This solution for large R is drawn in Figure 8.4, where $\sqrt{\frac{t}{2\nu}}\, u$ is plotted against $\eta = \frac{x}{\sqrt{2\nu t}}$. This represents a triangular wave solution of the Burgers equation. As $R \to \infty$, the shock layer at $x = 0$ becomes a discontinuity of u_x.

The Asymptotic Behavior of the Burgers Solution

We use the stationary phase approximation to examine the asymptotic nature of the Burgers solution (8.1.81). We consider any typical integral

$$u(x,t) = \int_a^b F(k) \exp\{it\theta(k)\}\, dk, \tag{8.1.94}$$

where $F(k)$ is a given function and $\theta(k) = \left(\frac{x}{t}\right) k - \omega(k)$. Integral (8.1.94) can be approximated asymptotically, as $t \to \infty$ for fixed x, by

$$u(x,t) \sim F(k_1) \left[\frac{2\pi}{t\,|\theta''(k_1)|}\right]^{\frac{1}{2}} \exp\left[i\left\{t\,\theta(k_1) + \frac{\pi}{4}\,\mathrm{sgn}\,\theta''(k_1)\right\}\right], \tag{8.1.95}$$

where k_1 is a stationary point determined by the solution of the equation

$$\theta'(k) = \frac{x}{t} - \omega'(k) = 0, \qquad a < k_1 < b \tag{8.1.96}$$

and $\theta''(k_1) \neq 0$.

According to this result, the significant contribution to integrals involved in (8.1.81) comes from stationary points for fixed x and t, that is, from the roots of the equation

$$\frac{\partial f}{\partial \zeta} = F(\zeta) - \left(\frac{x-\zeta}{t}\right) = 0. \tag{8.1.97}$$

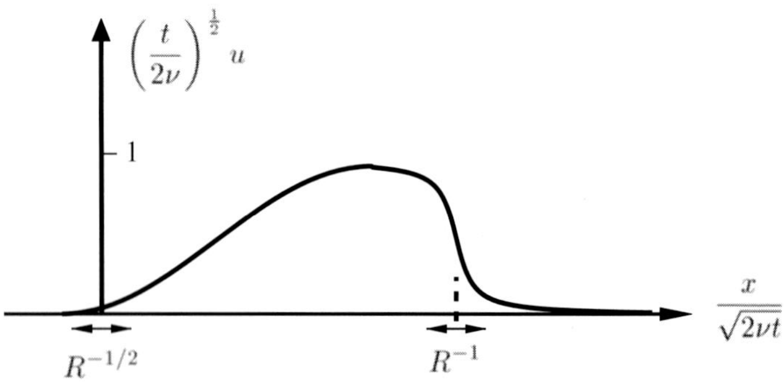

Fig. 8.4 Triangular wave solution of the Burgers equation.

Suppose that $\zeta = \xi(x,t)$ is a solution of (8.1.97) representing a stationary point. Invoking the stationary phase formula (8.1.95), the integrals in (8.1.81) have the asymptotic representation as $\nu \to 0$, in the form,

$$\int_{-\infty}^{\infty} \left(\frac{x-\zeta}{t}\right) \exp\left(-\frac{f}{2\nu}\right) d\zeta \sim \left(\frac{x-\zeta}{t}\right) \left[\frac{4\pi\nu}{|f''(\xi)|}\right]^{\frac{1}{2}} \exp\left\{-\frac{f(\xi)}{2\nu}\right\},$$

$$\int_{-\infty}^{\infty} \exp\left(-\frac{f}{2\nu}\right) d\zeta \sim \left[\frac{4\pi\nu}{|f''(\xi)|}\right]^{\frac{1}{2}} \exp\left\{-\frac{f(\xi)}{2\nu}\right\}.$$

Therefore, the final asymptotic solution is

$$u(x,t) \sim \frac{x-\xi}{t}, \tag{8.1.98}$$

where $\xi(x,t)$ satisfies equation (8.1.97).

In other words, the solution assumes the asymptotic form

$$\left.\begin{aligned} u &= F(\xi) \\ \xi &= x - t\,F(\xi) \end{aligned}\right\}. \tag{8.1.99ab}$$

This is identical with the solution of the Burgers equation without the diffusion term ($\nu = 0$). Here, the stationary point ξ corresponds to the characteristic variable in the context of the first-order, quasi-linear equation. It follows from the solution of nonlinear initial value problem (8.1.68)–(8.1.69) with $\nu = 0$ that the wave profile described by (8.1.99ab) suffers from gradual distortion and steepening, and leads to a multiple-valued solution after a sufficiently long time, and, eventually, it breaks with the development of discontinuity as a shock wave. When this state is reached, mathematically, there will be two stationary points of (8.1.97), and then, some modification is required to complete the asymptotic analysis. This leads to *Whitham's geometrical rule of equal area,* which is equivalent to the shock condition. We shall not pursue this analysis further, and refer to Whitham (1974) and Burgers (1974) for a detailed discussion.

The N-Wave Solution

To find an N-wave solution of the Burgers equation, we begin with the source solution of the linear diffusion equation (8.1.74) of the form

$$\phi(x,t) = 1 + \sqrt{\frac{\tau}{t}} \exp\left(-\frac{x^2}{4\nu t}\right), \tag{8.1.100}$$

where τ is a constant.

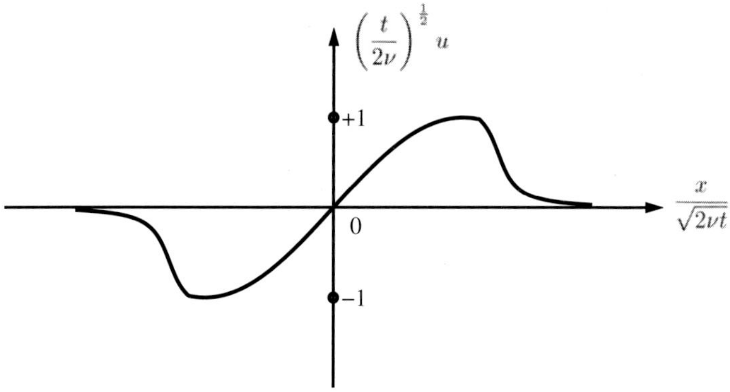

Fig. 8.5 N-wave solution of the Burgers equation.

Substituting (8.1.100) into (8.1.73) gives the solution of the Burgers equation as

$$u(x,t) = \left(\frac{x}{t}\right) \cdot \frac{\sqrt{\frac{\tau}{t}} \exp\left(-\frac{x^2}{4\nu t}\right)}{1 + \sqrt{\frac{\tau}{t}} \exp\left(-\frac{x^2}{4\nu t}\right)}, \tag{8.1.101}$$

$$= \frac{x}{t}\left[1 + \sqrt{\frac{t}{\tau}} \exp\left(\frac{x^2}{4\nu t}\right)\right]^{-1}. \tag{8.1.102}$$

For any time $t > 0$, this solution is shown in Figure 8.5 and has the N-shaped form. Because of this particular shape of the wave profile, it is known as the *N-wave solution* of the Burgers equation. At $t = t_0 > 0$, the profile may be taken as the initial profile. Such N waves are observed in many physical situations including problems governed by the cylindrical or spherical Burgers equation.

The area under the negative phase of the wave profile is the same as that over the positive phase of the profile. So, the area under the positive phase of the wave profile is given by

$$\int_0^\infty u(x,t)\,dx = -2\nu\left[\log \phi(x,t)\right]_0^\infty = 2\nu \log\left[1 + \sqrt{\frac{\tau}{t}}\right]. \tag{8.1.103}$$

We denote the right-hand side of (8.1.103) in the initial time $t = t_0$ by A and then introduce a Reynolds number R by

$$R = \frac{A}{2\nu} = \log\left[1 + \sqrt{\frac{\tau}{t_0}}\right], \tag{8.1.104}$$

so that $\sqrt{\frac{\tau}{t_0}} = \left(e^R - 1\right)$, and solution (8.1.102) reduces to

$$u(x,t) = \left(\frac{x}{t}\right)\left[1 + \sqrt{\frac{t}{t_0}}\left(e^R - 1\right)^{-1}\exp\left(\frac{x^2 R}{2At}\right)\right]^{-1}. \quad (8.1.105)$$

When $R \gg 1$, $e^R - 1 \sim e^R$, so that (8.1.105) reduces to the form

$$u(x,t) \sim \left(\frac{x}{t}\right)\left[1 + \sqrt{\frac{t}{t_0}}\exp\left\{-R\left(1 - \frac{x^2}{2At}\right)\right\}\right]^{-1}. \quad (8.1.106)$$

In the limit, as $R \to \infty$ with fixed t, (8.1.106) gives the shock-wave solution of the Burgers equation in the form

$$u(x,t) \sim \begin{bmatrix} \frac{x}{t}, & |x| < \sqrt{2At} \\ \\ 0, & |x| > \sqrt{2At} \end{bmatrix}. \quad (8.1.107)$$

In the limit as $t \to \infty$, for fixed ν and τ, solution (8.1.101) takes the form

$$u(x,t) \sim \left(\frac{x}{t}\right)\left(\frac{\tau}{t}\right)^{\frac{1}{2}}\exp\left(-\frac{x^2}{4\nu t}\right). \quad (8.1.108)$$

This corresponds to the dipole solution of the linear diffusion equation.

Finally, in the limit as $R \to 0$, $e^R - 1 \sim R$, and result (8.1.105) gives

$$u(x,t) \sim \left(\frac{x\,R}{t}\right)\sqrt{\frac{t_0}{t}}\exp\left(-\frac{x^2 R}{2At}\right). \quad (8.1.109)$$

This is identical with (8.1.108), as expected, because $R \to 0$ corresponds to $\nu \to \infty$.

Finally, Lighthill presented briefly fully-dispersed and partly-dispersed shock wave phenomena that deal with the interaction of convection and relaxation. His conclusion was that in many cases sound of frequencies up to and somewhat beyond the relaxation frequency is attenuated far more rapidly by relaxation than by diffusion. However, at higher frequencies, diffusion plays a significant role as the attenuation per wavelength due to relaxation falls off so much as to become negligible. This means that relaxation does not usually play the significant role in shock waves. In general, a very weak shock wave can derive sufficient energy dissipation from relaxation alone. Its effective frequency is then of the order of the relaxation frequency or less, and the effects of diffusion are negligible within it. Such shock waves may be called *fully-dispersed shock waves*. Relatively, strong shock waves can obtain sufficient energy dissipation only by including a far thinner region of dissipation by viscosity and heat conduction. Since this must still be followed by a region in which the internal energy does relax to its new equilibrium value, these strong shock waves may be called *partly-dispersed.*

There is an important analogy between the fully-dispersed shock waves, whose velocity relative to the fluid ahead of them lies between the wave velocities corresponding to small and large frequencies, and *weak bores* in rivers that satisfy the same condition. Thus, for both, stationary sinusoidal waves of intermediate frequency are possible features of the fluid flow. On the other hand, in a dispersive medium, the energy of waves is propagated at the group velocity which is less than the phase velocity for water waves. That is why the stationary wave train is observed behind the bore, energy liberated at the bore leaks back through the waves at a speed given by the difference between the group velocity and phase velocity. The similar situation may be expected for shock waves, except the group velocity exceeds the wave velocity, since the latter increases with frequency, so that the stationary wavetrain would lie ahead of the shock wave. Its absence in the solution for fully-dispersed waves can only be due to the heavy attenuation associated with the frequency range where dispersion is present, which causes the wave energy propagated upstream to be supercritically damped.

As a follow-up of his work in 1956 that was dedicated to Sir G.I. Taylor, Lighthill developed mathematical theory of gas dynamics interacting with gas physics. He then published two major papers "Dynamics of a Dissociating Gas, Part 1" and "Equilibrium Flow, Part 2, Quasi-equilibrium Transfer Theory" in the *Journal of Fluid Mechanics* in 1957. He dealt with predictions on the velocity field originated from acoustic noise and a generalized turbulence in a layer overlaying a convectively unstable atmospheric region.

8.2 Kinematic Waves

Classical wave motions are described by Newton's second law of motion together with some reasonable assumptions relating a stress to a displacement (as in gravity waves), to a strain (as in nondispersive longitudinal and transverse waves), or to a curvature (as in capillary waves and flexural waves). In contrast with the case of dynamic waves, a class of waves is called *kinematic waves* when an appropriate functional relation exists between the density and the flux of some physically observed quantity. Kinematic waves are not at all waves in the classical sense, and they are physically quite different from the classical wave motions involved in dynamical systems. They describe, approximately, many important real-world problems including traffic flows on long highways, flood waves in rivers, roll waves in an inclined channel, and chromatographic models in chemistry. Lighthill and Whitham (1955a,b) first gave a general and systematic treatment of kinematic waves and applications.

During his stay in Manchester, Lighthill and Whitham (1955a,b) first introduced the concept of *kinematic waves* to describe motions in which the dynamics is locally represented by a quasi-steady, one-dimensional, balance between the flux, $q(x,t)$ of a quantity per unit time, and a density, $\rho(x,t)$ of the quantity per unit length. They introduced a flow velocity $u\,(x,t)$ by $u = q/\rho$. Assuming that the material (or state) is conserved, they stipulate that the time rate of change of the total amount in any arbitrary interval $x_1 \le x \le x_2$ must be balanced by the net influx across x_1 and x_2. Physically, this states that the quantity in a small length segment changes at a rate equal to the difference between inflow and outflow. Mathematically, this can be formulated as

$$\frac{d}{dt}\int_{x_1}^{x_2} \rho\,(x,t)\,dx = q\,(x_1,t) - q\,(x_2,t) = -\int_{x_1}^{x_2}\left(\frac{\partial q}{\partial x}\right)dx, \tag{8.2.1}$$

or, equivalently,

$$\int_{x_1}^{x_2}\left(\frac{\partial \rho}{\partial t} + \frac{\partial q}{\partial x}\right)dx = 0. \tag{8.2.2}$$

Since this result holds for any arbitrary interval $x_1 \le x \le x_2$, the integrand must vanish identically, so that

$$\frac{\partial \rho}{\partial t} + \frac{\partial q}{\partial x} = 0, \tag{8.2.3}$$

provided ρ and q are sufficiently smooth functions. This is called the *conservation law, kinematic wave equation*, or the *equation of continuity*.

Based on theoretical or empirical grounds, we assume that there exists a relation between q and ρ, so that we can write

$$q = q\,(\rho)\,. \tag{8.2.4}$$

Thus, equations (8.2.3) and (8.2.4) form a closed system since there are two equations with two unknown functions. Substituting (8.2.4) in (8.2.3) gives

$$\frac{\partial \rho}{\partial t} + c\,(\rho)\,\frac{\partial \rho}{\partial x} = 0, \tag{8.2.5}$$

where the wave velocity $c(\rho)$ is defined by

$$c\,(\rho) = q'\,(\rho) = \frac{\partial q}{\partial \rho}. \tag{8.2.6}$$

Similarly, multiplying (8.2.3) by $c(\rho)$ leads to the equation

$$\frac{\partial q}{\partial t} + c(\rho) \frac{\partial q}{\partial x} = 0. \tag{8.2.7}$$

This means that q or ρ is constant in waves propagating with velocity $c(\rho)$ given by (8.2.6), and hence, $c(\rho)$ is called the *wave propagation velocity*. Mathematically, equation (8.2.7) has only one system of characteristics given by

$$\frac{dx}{dt} = c(\rho), \tag{8.2.8}$$

and along each of these characteristics the flow q or the density ρ is constant. The wave velocity c as given by (8.2.6) is the slope of the flow-density curve for fixed x. In terms of the mean flow velocity $u = q/\rho$, the wave propagation velocity is given by

$$c = q'(\rho) = \frac{d}{d\rho}(u\rho) = u + \rho \frac{du}{d\rho}. \tag{8.2.9}$$

Thus, $c < u$ when $\frac{du}{d\rho} > 0$, that is, the flow velocity increases with density as in flood waves in rivers, and $c < u$ when $\frac{du}{d\rho} < 0$, that is, it decreases with density as in traffic flows in highways.

Further, the following observations are in order. First, there is one important fundamental difference between kinematic waves and dynamic waves. A kinematic wave has only one wave velocity at each point, while dynamic waves possess at least two velocities (forwards and backwards relative to the medium). Second, kinematic waves are nondispersive, but they suffer from a change in form due to nonlinearity (dependence of the wave speed c on the flow q carried by the wave) exactly as do traveling sound waves of finite amplitude. Consequently, continuous wave forms may develop discontinuities because the faster waves overtake the slower ones. These discontinuities can be described as shock waves because their process of development is identically the same as that of shock waves in gas dynamics.

The law of motion of kinematic shock waves can be derived from the conservation laws, as was the law governing continuous kinematic waves. If the density and flow assume the values ρ_1 and q_1 on one side, and ρ_2 and q_2 on the other side of the shock wave which propagates with velocity U, then the quantity crossing it per unit time can be written either as $q_1 - U\rho_1$ or as $q_2 - U\rho_2$. This gives the velocity of the shock waves as

$$U = \frac{(q_2 - q_1)}{(\rho_2 - \rho_1)}. \tag{8.2.10}$$

This represents the slope of the chord joining the two points (ρ_1, q_1) and (ρ_2, q_2) on the density-flow curve which corresponds to the states behind and ahead of the shock wave when it reaches a given point x. In the limit, when the shock wave become a continuous wave, the slope of the chord becomes the slope of the tangent, and hence (8.2.10) reduces to (8.2.6). The development of kinematic shock waves was described by examples (see Lighthill and Whitham, 1955a,b).

Another important effect, known as *diffusion*, is confined entirely to the interior of kinematic shock waves, where its effect is crucial in smoothing the nonlinear steeping of the kinematic waves. The effect of diffusion is certainly small outside the shock regions, but inside the shock regions, the effect of diffusion is very important as a second-order approximation. Mathematically, diffusion corresponds to the inclusion in (8.2.7) of an additional term proportional to a second derivative of q. This will happen if the flow-density relation involves some dependence on a derivative of q or ρ in addition to q, ρ, and x. We assume that, for each x, q is a function of ρ and ρ_t; alternatively, ρ is a function of q and q_x because $\rho_t = -q_x$. Substituting for ρ in (8.2.3), we obtain

$$\frac{\partial \rho}{\partial q}\frac{\partial q}{\partial t} + \frac{\partial \rho}{\partial q_x}\frac{\partial^2 q}{\partial x \partial t} + \frac{\partial q}{\partial x} = 0. \tag{8.2.11}$$

Further, if the coefficients of the derivatives of q in (8.2.11) are approximated as functions of q and x alone, then equation (8.2.11) becomes

$$\frac{\partial q}{\partial t} + c(\rho)\frac{\partial q}{\partial x} + \nu \frac{\partial^2 q}{\partial x \partial t} = 0, \tag{8.2.12}$$

where $c = \frac{\partial q}{\partial \rho}$ is the kinematic wave velocity as before and $\nu = c\left(\frac{\partial \rho}{\partial q_x}\right)$. This is a typical nonlinear equation representing diffusive kinematic waves. Invoking the first approximation $q_t \sim -cq_x$, (8.2.12) may be rewritten as

$$\frac{\partial q}{\partial t} + c(\rho)\frac{\partial q}{\partial x} = \nu c \frac{\partial^2 q}{\partial x^2}. \tag{8.2.13}$$

Physically, there is no difference between the two equations (8.2.12) and (8.2.13).

Lighthill and Whitham, (1955a,b) provided a detailed treatment of several real-world problems as applications of kinematic waves, Riemann's simple waves, and Riemann's invariants and their extensions. In particular, they described traffic flow problems on highways and flood waves in rivers.

To investigate the nature of the discontinuous solution or shock waves, we assume a functional relation $q = Q(\rho)$ and allow a jump discontinuity for ρ and q. In many physical problems of interest, it would be a better approximation to assume that q is a function of the density gradient ρ_x as well as ρ. A simple model is to take

$$q = Q(\rho) - \nu\rho_x, \tag{8.2.14}$$

where ν is a positive constant. Substituting (8.2.14) into (8.2.13), we obtain the *nonlinear diffusion equation*

$$\rho_t + c(\rho)\rho_x = \nu\rho_{xx}, \tag{8.2.15}$$

where $c(\rho) = Q'(\rho)$.

We multiply (8.2.15) by $c'(\rho)$ to obtain

$$\begin{aligned} c_t + cc_x &= vc'(\rho)\rho_{xx} \\ &= \nu\left\{c_{xx} - c''(\rho)\rho_x^2\right\}. \end{aligned} \tag{8.2.16}$$

If $Q(\rho)$ is a quadratic function in ρ, then $c(\rho)$ is linear in ρ, and $c''(\rho) = 0$. Consequently, (8.2.16) becomes

$$c_t + cc_x = \nu c_{xx}. \tag{8.2.17}$$

As a simple model of turbulence, c is replaced by the fluid velocity field $u(x,t)$ to obtain the well-known *Burgers equation* as

$$u_t + uu_x = \nu u_{xx}, \tag{8.2.18}$$

where ν is the kinematic viscosity. Thus, the Burgers equation describes a balance between time evolution, nonlinearity, and diffusion. This is the simplest nonlinear model equation for diffusive waves in fluid dynamics. In 1948, Burgers first developed this equation primarily to shed light on the study of turbulence described by the interaction of the two opposite effects of convection and diffusion. However, turbulence is more complex in the sense that it is both three dimensional and statistically random in nature. Equation (8.2.18) arises in many physical problems including one-dimensional turbulence (where this equation had its origin), sound waves in a viscous medium, shock waves in a viscous medium, waves in fluid-filled viscous elastic tubes, and magnetohydrodynamic waves in a medium with finite electrical conductivity. We note that (8.2.18) is *parabolic*, whereas (8.2.18) with $\nu = 0$ is *hyperbolic*. More importantly, the properties of the solution of the parabolic equation are significantly different than those of the hyperbolic equation.

8.3 Linear Dispersive Waves

In 1960, Sir James Lighthill's deep and profound paper provided an elegant mathematical theory with physical interpretation of magnetohydrodynamic waves and other anisotropic wave motions. He formulated a fairly general linear magneto-hydrodynamic wave propagation problem in a compressible perfectly conductive fluid and other anistropic wave propagation problems. He formulated the wave propagation problem in term of partial differential equations in space variables, x, y, z and a time variable t which are linear with constant coefficients and represent wave motions that are dispersive or anisotropic or both. Using the theory of generalized functions combined with the multiple Fourier transforms, he developed a new method of asymptotic evaluation of multiple Fourier integrals for large values of their arguments. The major emphasis is to determine the asymptotic behavior of the unique solution of physical interest at large distances representing the waves generated by a source of disturbance of finite or infinitesimal spatial extent. Special attention is given to sources of disturbance of fixed frequency, and the solution which satisfies the Sommerfeld radiation condition at infinity. Included are mathematical results with physical interpretation and the velocity of energy propagation in a plane wave in an anisotropic and/or dispersive medium. He developed a new method for finding the asymptotic behavior at large distances (as $r \to \infty$) of the solution of an equation of the form

$$P\left(\frac{\partial^2}{\partial t^2}, \frac{\partial^2}{\partial x^2}, \frac{\partial^2}{\partial y^2}, \frac{\partial^2}{\partial z^2}\right) u = e^{i\omega t} f(\mathbf{r}), \tag{8.3.1}$$

where P is a polynomial in partial derivatives, $\mathbf{r} = (x, y, z)$ and $f(\mathbf{r})$ is a function with a compact support in $\mathbb{R}^3$.

Using the three-dimensional Fourier transform with respect to space variables x, y, and z, he solved equation (8.3.1) in the form

$$u = e^{i\omega t} \int_{-\infty}^{\infty}\int_{-\infty}^{\infty}\int_{-\infty}^{\infty} \exp\left[i(\boldsymbol{\kappa}\cdot\mathbf{r})\right] U(\boldsymbol{\kappa}) d\boldsymbol{\kappa}, \tag{8.3.2}$$

where $U(\boldsymbol{\kappa}) = F(\boldsymbol{\kappa})/D(\boldsymbol{\kappa})$, and

$$F(\boldsymbol{\kappa}) = \mathcal{F}\{f(\mathbf{r})\} = \int_{-\infty}^{\infty}\int_{-\infty}^{\infty}\int_{-\infty}^{\infty} e^{i(\boldsymbol{\kappa}\mathbf{r})} f(\mathbf{r}) d(\mathbf{r}), \tag{8.3.3}$$

$$D(\omega, \boldsymbol{\kappa}) = P\left(-\omega^2, -k^2, -l^2, -m^2\right), \tag{8.3.4}$$

and $\boldsymbol{\kappa} = (k, l, m)$. In general, the solution of the nonhomogeneous partial differential equation (8.3.1) is not unique because any complementary function (solution of the corresponding homogeneous equation) can be added to the particular solution of (8.3.1). However, in any physical situation, only one solution is of interest, that is, it satisfies the *Sommerfeld radiation condition*. This condition states that only waves originating from the given source of applied disturbance are present, no free waves cross the field from one side to the other, or simultaneously coming in from infinity in all directions. In other words, the waves exclusively behave like outgoing progressive waves propagating away from the source of applied disturbance. So, the radiation condition is, indeed, based on mathematical and physical reasons, absolutely essential to achieving a unique solution of the problem. He then obtained the unique solution of physical interest by different ways of applying the Sommerfeld radiation condition.

He developed a new and simple approach from a mathematical point of view is to replace ω by $\omega - i\varepsilon$, where $\varepsilon > 0$, and in the end to let $\varepsilon \to 0$. The physical idea behind it is that the disturbance is assumed as $\exp\left[i\left(\omega - i\varepsilon\right)t\right] f(\mathbf{r})$ which increases exponentially with time like $\exp\left(\varepsilon t\right)$, and so increase their amplitude in step with the strength of the source of applied disturbance. If the solution were contaminated by the incoming waves from the infinity, their amplitude during the travel time t_0 is of order $\exp(\varepsilon t_0)$, comparable with that of waves generated by the applied disturbance. Since the incoming waves are free waves with fixed amplitude still of order $\exp(\varepsilon t_0)$ so that this would be negligible compared with $\exp(\varepsilon t)$, the order of magnitude of waves generated at the source of disturbance, because of the very large time needed for the free waves to come in from infinity. Thus, if we seek only the solution of order $\exp(\varepsilon t)$, they must be absent. Consequently, with the source of disturbance $\exp\left[i(\omega - i\varepsilon)t\right]\ f(\mathbf{r})$, this solution proportional to $\exp\left[i(\omega - i\varepsilon)t\right]$ is the unique solution which satisfies the radiation condition and hence, its limit as $\varepsilon \to 0$ would represent the unique solution of physical interest satisfying the condition. This is an alternative way of applying the Sommerfeld radiation condition and may be referred to as Lighthill's formulation of the radiation condition. It is worth noting that Lighthill's method, an alternating way of applying the radiation condition, is fairly general and also useful for higher dimensional wave problem.

The most important and inherent difficulty associated with their problems is that the integrand of its integral solution has real singularities on the path of integration. He also expressed the asymptotic behavior of (8.3.2) as $r \to \infty$ in terms of the singularities of the integrand.

The solution of (8.3.2), which satisfies the radiation condition, has the asymptotic representation, as $r \to \infty$,

$$u(r,t) \approx \left(\frac{4\pi^2}{r}\right) \sum \frac{C\,F(\boldsymbol{\kappa})}{|\nabla D|\sqrt{K}} \exp\left[i(\omega t + \boldsymbol{\kappa} \cdot \mathbf{r})\right] + O\left(\frac{1}{r^2}\right), \qquad (8.3.5)$$

along any fixed radius vector L and the summation is taken over all points (k, l, m) of the surface $D = 0$, where the normal to D is parallel to L and $(\mathbf{r} \cdot \nabla D)/(\partial D/\partial \omega) > 0$ provided that the surface $D = 0$ has non-zero Gaussian curvature K at each of these points $\boldsymbol{\kappa}$, that C is (a) $\pm i$ when $K < 0$ and ∇D is in the direction of $\pm r$, (b) ± 1 when $K > 0$ and the surface is convex to the direction of $\pm \nabla D$, that

$$F(\boldsymbol{\kappa}) = \frac{1}{(2\pi)^3} \int_{-\infty}^{\infty} \int_{-\infty}^{\infty} \int_{-\infty}^{\infty} \exp\left[i(\boldsymbol{\kappa} \cdot \mathbf{r})\right] f(\mathbf{r}) d\mathbf{r}. \qquad (8.3.6)$$

In particular, for any plane wave

$$u(\mathbf{r}, t) = a \exp\left[i(\boldsymbol{\kappa} \cdot \mathbf{r} + \omega t)\right], \qquad (8.3.7)$$

with the associated dispersion relation

$$D(\omega, \boldsymbol{\kappa}) = 0 \qquad (8.3.8)$$

the velocity of energy propagation (or the group velocity) can be given by

$$U = \frac{\nabla D}{(\partial D/\partial \omega)}, \qquad (8.3.9)$$

provided

$$(\mathbf{r} \cdot \nabla D)/(\partial D/\partial \omega) > 0. \qquad (8.3.10)$$

Therefore, this inequality means that $\mathbf{r} \cdot \nabla D > 0$ implying that only waves are present for which the group velocity has a positive (outward) component along a radius vector. In other words, all waves travel outwards and none come in from infinity. This satisfies the radiation condition.

The nondispersive waves are also governed by the basic equation (8.3.1) where the polynomial P is homogeneous in its four variables. Obviously, the phase velocity (ω/k) is independent of wavelength for plane waves propagating in a fixed direction. If P is a homogeneous function of degree n, then D is a homogeneous function of degree $2n$, hence, by Euler's equation for homogeneous functions

$$\omega \frac{\partial D}{\partial \omega} + \boldsymbol{\kappa} \cdot \nabla D = 2nD = 0 \qquad (8.3.11)$$

throughout the wavenumber surface $D = 0$. Consequently, inequality (8.3.10) requires that $\mathbf{r} \cdot \nabla D$ and $\boldsymbol{\kappa} \cdot \nabla D$ are of opposite signs. Since $\mathbf{r}$

and ∇D are parallel, thus, for nondispersive waves, the restriction involved in (8.3.5) leads to $\boldsymbol{\kappa}\cdot\mathbf{r}<0$. In this case, only waves whose phase velocity has an outward radial component are present. In view of (8.3.11), the group velocity can be written as

$$U=-\frac{\omega\nabla D}{\boldsymbol{\kappa}\cdot\nabla D}, \tag{8.3.12}$$

while the phase velocity c can be written as a vector

$$\mathbf{c}=-\frac{\omega\boldsymbol{\kappa}}{\kappa^2}. \tag{8.3.13}$$

It follows from (8.3.12) and (8.3.13) that $\mathbf{U}\cdot\boldsymbol{\kappa}=\mathbf{c}\cdot\boldsymbol{\kappa}$, and hence the resultant of $\mathbf{U}$ in the direction of the phase velocity $\mathbf{c}$ is $\mathbf{c}$. In other words, the group velocity normal to the wave fronts is the phase velocity.

Finally, he solved a more general initial value problem of a more general equation of the form

$$P\left(\frac{\partial^2}{\partial t^2},\frac{\partial^2}{\partial x^2},\frac{\partial^2}{\partial y^2},\frac{\partial^2}{\partial z^2}\right)u=f(\mathbf{r},t). \tag{8.3.14}$$

Application of the four-dimensional Fourier transform with respect to three space variables and time t gives the solution in the form

$$u(\mathbf{r},t)=\int_{-\infty}^{\infty}\int_{-\infty}^{\infty}\int_{-\infty}^{\infty}\int_{-\infty}^{\infty}\exp\left[i(\boldsymbol{\kappa}\cdot\mathbf{r}+\omega t)\right]\frac{F(\boldsymbol{\kappa},\omega)}{D(\boldsymbol{\kappa},\omega)}\,d\boldsymbol{\kappa}\,d\omega, \tag{8.3.15}$$

where $\mathfrak{F}\{f(\mathbf{r},t)\}=F(\boldsymbol{\kappa},\omega)$ is the four-dimensional Fourier transform and

$$F(\boldsymbol{\kappa},\omega)=\int_{-\infty}^{\infty}\int_{-\infty}^{\infty}\int_{-\infty}^{\infty}\int_{-\infty}^{\infty}\exp\left[-i(\boldsymbol{\kappa}\cdot\mathbf{r}+\omega t)\right]f(\mathbf{r},t)\,d\mathbf{r}\,dt. \tag{8.3.16}$$

Thus, he obtained a solution for u which is zero for $t<0$, and there is no complex roots of ω for which $D=0$ with real k, l, and m. Consequently, the resulting solution becomes

$$u(\mathbf{r},t)=\int_{-\infty}^{\infty}\int_{-\infty}^{\infty}\int_{-\infty}^{\infty}\int_{-i\varepsilon-\infty}^{-i\varepsilon+\infty}\exp\left[i(\boldsymbol{\kappa}\cdot\mathbf{r}+\omega t)\right]\frac{F(\boldsymbol{\kappa},\omega)}{D}\,d\boldsymbol{\kappa}\,d\omega. \tag{8.3.17}$$

Then the asymptotic representation of (8.3.17) is obtained as $r\to\infty$ and $t\to\infty$ simultaneously.

In his 1964 inaugural lecture as the first President of the Institute of Mathematics and Its Applications, Sir James discussed a wide range of different aspects of the theory of group velocity. In this paper, he obtained a new formula of the theory of group velocity in three dimensions. He also developed the method of stationary phase in three dimensions for a problem of forced wave motion due to a steady source of fixed frequency ω_0 with a compact support. His inaugural lecture on "Group Velocity" published in the *Journal of the Institute of Mathematics and Its Applications* in **1** (1965) 1-28 is concerned with an elaborate review of the linear theory of dispersive wave phenomena based on either the method of superposition of plane wave solutions or the kinematics of surfaces of constant phase.

He provided an excellent review of the theory of group velocity for three-dimensional isotropic and anisotropic, homogeneous and non homogeneous, conservative and dissipative, linear and nonlinear, classical and relativistic wave propagation under free and forced-motion conditions. He formulated a three-dimensional problem of forced motion generated by a steady source $\exp(-2\pi i\omega_0 t)f(\mathbf{x})$ of fixed frequency ω_0 so that $f(\mathbf{x})$ has a compact support in $\mathbb{R}^3$. So the problem of a small disturbance ϕ is governed by a non-homogeneous linear partial differential equation with constant coefficients in the form

$$P\left(\frac{1}{2\pi i}\frac{\partial}{\partial x_1}, \frac{1}{2\pi i}\frac{\partial}{\partial x_2}, \frac{1}{2\pi i}\frac{\partial}{\partial x_3}, -\frac{1}{2\pi i}\frac{\partial}{\partial t}\right)\phi = \exp(-2\pi i\omega_0 t)f(\mathbf{x}), \tag{8.3.18}$$

where $\mathbf{x} = (x_1, x_2, x_3)$, and P is a polynomial in partial derivatives.

Assuming that the corresponding homogeneous equation $P\phi = 0$ admits solutions in the form of waves, the problem is to find the solution of the forced motion at large distances along each direction out of the source region. In general, such problem does not have a unique solution of physical interest.

It is necessary to impose a Sommerfeld radiation condition at infinity or an equivalent devise in order to determine a unique solution of physical interest. The Lighthill equivalent devise is used to solve this problem, that is, the right hand side of (8.3.18) is replaced by $\exp[2\pi(\varepsilon - i\omega_0)t]\, f(\mathbf{x})$ where ε is a small positive quantity, so that the source term has grown to its present strength from zero during all the time from $t \to -\infty$. Thus, the solution for ϕ proportional to $\exp[2\pi(\varepsilon - i\omega_0)t]$ is to be determined. This is equivalent to allowing the frequency ω to have a small positive imaginary part ε which is later allowed to tend to zero. In order to obtain the solution which tends to zero at large distances from the source region, we take the three-dimensional Fourier transforms of $f(\mathbf{x})$ and $\phi(\mathbf{x})$ so that

$$f(\mathbf{x}) = \int_{-\infty}^{\infty}\int_{-\infty}^{\infty}\int_{-\infty}^{\infty} F(\boldsymbol{\kappa})\exp[2\pi i(\boldsymbol{\kappa}\cdot\mathbf{x})]\, d\boldsymbol{\kappa}, \tag{8.3.19}$$

$$\phi(\mathbf{x}, t) = \exp[2\pi(\varepsilon - i\omega_0)t]\, \phi_\varepsilon(\mathbf{x}), \tag{8.3.20}$$

where

$$\phi_\varepsilon(\mathbf{x}) = \int_{-\infty}^{\infty}\int_{-\infty}^{\infty}\int_{-\infty}^{\infty} \Phi_\varepsilon(\boldsymbol{\kappa}) \exp\left[2\pi i(\boldsymbol{\kappa}\cdot\mathbf{x})\right] d\boldsymbol{\kappa}, \tag{8.3.21}$$

where $\boldsymbol{\kappa} = (k_1, k_2, k_3)$ is the wavenumber vector and $\Phi_\varepsilon(\boldsymbol{\kappa})$ is the Fourier transform of $\phi_\varepsilon(\mathbf{x})$.

Evidently, equation (8.3.18) leads to the following result

$$P\left(k_1, k_2, k_3, \omega_0 + i\varepsilon\right) \Phi_\varepsilon(\boldsymbol{\kappa}) = F(\boldsymbol{\kappa}). \tag{8.3.22}$$

Thus, the problem is to find

$$\phi_0(\mathbf{x}) = \lim_{\varepsilon\to 0} \phi_\varepsilon(\mathbf{x}), \tag{8.3.23}$$

at large distances from the source region along any straight line L stretching away from it.

Based on the Lighthill method with a rotation of axes, it turns out that

$$\phi_\varepsilon = \int_{-\infty}^{\infty}\int_{-\infty}^{\infty} dk_2 dk_3 \int_{-\infty}^{\infty} \frac{F(\boldsymbol{\kappa})\exp(2\pi i k_1 x_1) dk_1}{P(\boldsymbol{\kappa}, \omega_0 + i\varepsilon)}. \tag{8.3.24}$$

Using the Lighthill method, it turns out that the asymptotic form of ϕ_0 takes the form

$$\phi_0 \sim 2\pi i \iint_S \frac{F(\boldsymbol{\kappa})\exp(2\pi i k_1 x_1) dk_2 dk_3}{\partial P(k_1, k_2, k_3, \omega_0)/\partial k_1}, \tag{8.3.25}$$

where S represents the wavenumber surface with $\omega = \omega_0$ so that $(\partial\omega/\partial k_1) > 0$ on S.

Finally, the method of stationary phase is employed to determine the solution along L in the form

$$\phi = \phi_0 \exp\left(-2\pi i\omega_0 t\right) \sim \exp\left(-2\pi i\omega_0 t\right) \sum \phi_r, \tag{8.3.26}$$

where the summation is taken over points $k_\alpha^{(r)}$ of the surface $\omega = \omega_0$ where the normal to the surface is parallel to L and ω increases in the direction of L. The amplitude of the solution (8.3.26) is proportional to R^{-1} where $R = \left(x_1^2 + x_2^2 + x_3^2\right)^{\frac{1}{2}}$.

He then extended the linear wave theory in another deep and profound paper entitled "On Waves Generated in Dispersive Systems to Travelling Forcing Effects with Applications to the Dynamics of Rotating Fluids" published in *JFM* in 1967. His general problem of the propagation of dispersive waves generated by travelling forcing effects, that may be steady, oscillatory or transient in nature, is governed by a linear partial differential equation

$$P\left(i\frac{\partial}{\partial t}, -i\frac{\partial}{\partial x}, -i\frac{\partial}{\partial y}, -i\frac{\partial}{\partial z}\right)\phi = e^{-i\omega_0 t} f(\mathbf{r} - \mathbf{U}t), \tag{8.3.27}$$

where P is a polynomial, $\phi(x, y, z, t)$ is the disturbance, and the forcing term f vanishes outside a finite region around the origin. Using the three dimensional Fourier transform with respect to x, y, and z, he obtained the general formal solution

$$\phi(\mathbf{r}, t) = \int_{-\infty}^{\infty}\int_{-\infty}^{\infty}\int_{-\infty}^{\infty} \frac{F(\boldsymbol{\kappa})\exp\left[i\left\{-\omega_0 t + \boldsymbol{\kappa}\cdot(\mathbf{r} - \mathbf{U}t)\right\}\right]}{P(\omega_0 + \mathbf{U}\cdot\boldsymbol{\kappa},\ \boldsymbol{\kappa})} d\boldsymbol{\kappa}, \quad (8.3.28)$$

where $\boldsymbol{\kappa} = (k, l, m)$ is the wavenumber vector, and $F(\boldsymbol{\kappa})$ is the Fourier transform of $f(\mathbf{r})$. He then used his own asymptotic method (1960) of evaluation of integrals such as (8.3.28) at distances from the forcing region large compared with the dimension. The asymptotic value of (8.3.28), as defined by replacing ω_0 by $\omega_0 + i\varepsilon$ in (8.3.28) and making $\varepsilon \to 0$ is determined. In the wavenumber space $\boldsymbol{\kappa} = (k, l, m)$, at each point of the surface

$$P(\omega_0 + \mathbf{U}\cdot\boldsymbol{\kappa},\ \boldsymbol{\kappa}) = 0 \quad (8.3.29)$$

on which the denominator of (8.3.28) vanishes, an arrow is drawn normal to the surface, selecting from the two normal directions the one pointing towards the surface $P(\omega_0 + \mathbf{U}\cdot\boldsymbol{\kappa} + \delta,\ \boldsymbol{\kappa}) = 0$, where δ is small and positive. Then, waves are determined in any particular direction stretching out from the forcing region with their amplitude given asymptotically by the formula

$$\frac{4\pi^2}{|K|^{\frac{1}{2}}R}\cdot\frac{F(\boldsymbol{\kappa})}{|\nabla_{\boldsymbol{\kappa}} P(\omega_0 + \mathbf{U}\cdot\boldsymbol{\kappa},\ \boldsymbol{\kappa})|}, \quad (8.3.30)$$

where $R = |\mathbf{r} - \mathbf{U}t|$ represents distance from the forcing region, and K is the Gaussian curvature (product of principal curvatures) of the wavenumber surface (8.3.29). The waves whose frequency is ω_0 relative to a forcing region moving at velocity $\mathbf{U}$ must have absolute frequency $\omega_0 + \mathbf{U}\cdot\boldsymbol{\kappa}$ (the Doppler effect), and hence their wavenumber vector must lie on the surface (8.3.29). Moreover, the group velocity for waves satisfying the dispersion relation $P(\omega, \boldsymbol{\kappa}) = 0$ is given by

$$\left(\frac{\partial\omega}{\partial k}, \frac{\partial\omega}{\partial l}, \frac{\partial\omega}{\partial m}\right) = -\frac{\nabla P}{\partial P/\partial\omega}. \quad (8.3.31)$$

When $\omega_0 = 0$, the equation of the surface (8.3.29) becomes $P(\mathbf{U}\cdot\boldsymbol{\kappa}, \boldsymbol{\kappa}) = 0$ which may be interpreted as a statement that a steady forcing effect can generate only waves whose crests are stationary relative to the velocity of the forcing region.

His mathematical theory as well as physical significance of this paper is so general that he employed his theory to study the following problems in great detail:

(i) Rossby waves in a β-plane ocean due to travelling steady or travelling transient disturbances. For a steady wind stress moving westward with velocity $(-U, 0)$, *semi-circular waves* of wavenumber $\sqrt{\beta/U}$ trail behind it. Lighthill obtained wavenumber curves for Rossby waves generated on a β-plane ocean by an oscillating forcing effect moving with constant velocity U in a direction making an angle 30° with the east direction.

(ii) The classical Kelvin ship wave problem ($\omega_0 = 0$) deals with waves generated in deep water by an oscillating disturbance moving with a uniform velocity. The *Kelvin ship waves* are found outside the Kelvin wedge only for frequencies ω_0 satisfying $(U\omega_0/g) < 1.63$. Surface waves are also found in front of the obstacle only if $(U\omega_0/g) < 0.27$.

(iii) Internal gravity waves in a stratified fluid due to a vertically moving steady disturbance. The wavenumber surface for *internal gravity waves* generated by steady vertical motion of an obstacle with velocity $(0, 0, U)$ through a uniformly stratified medium with the Brunt-Väisäla frequency N. The wave shapes with cusps at finite distance behind the obstacle are in excellent agreement with experimental findings of Mowbray (1966).

(iv) Motion of an obstacle along the axis of uniformly rotating homogeneous fluid which leads to the well-known formation of the *Taylor column* ahead of the obstacle at low Rossby numbers. In this problem, the wavenumber surface for *inertial waves* generated by steady axial motion of an obstacle with velocity $(0, 0, U)$ through fluid rotating at angular velocity $(0, 0, \Omega)$ consists of a sphere and two coincident planes. The more general case of waves generated by a periodic forcing effect with frequency ω_0 moving along the axis of rotation in unbounded fluid was solved by Nigam and Nigam (1962). When $(\omega_0/2\Omega) < 1$, waves are found ahead of the obstacle, but a cone of semi-angle $\sin^{-1}(\omega_0/2\Omega)$ is empty of such waves. When $(\omega_0/2\Omega) > 1$, there exist trailing waves confined within a cone.

(v) Fully three-dimensional wave problem when an obstacle moves at right angle to the axis of rotating fluid. An obstacle of dimension a with Rossby number $(U/2\Omega a)$ generates waves with wavenumber κ satisfying $0 \leq (U\kappa/2\Omega) \leq \varepsilon$, where ε is the maximum Rossby number of the waves generated. In this case, the disturbance takes the form of the trailing cones (behind the obstacle) containing waves where crests have cusps on the boundaries of the cones.

In summary, his 1967 remarkable paper led to a unified approach to many types of complex dispersive waves that can be excited in a non-rotating, rotating and/or stratified fluid including the atmosphere and the oceans.

Chapter 9

Geophysical Fluid Dynamics

"Nature, unaware of its splendour and glory,
Closed to its past, without desire or memory,
Alien to her creator,
Is nothing more than a pendulum that oscillates,
And strikes in the infinity of time and space,
The rhythm of weight."

Friedrich von Schiller

"Yet not every solution of the equations of motion, even if it is exact, can actually occur in Nature. The flows that occur in Nature must not only obey the equations of fluid dynamics, but also be stable."

L. D. Landau and E. M. Lifshitz

9.1 Introduction

At Imperial College, Lighthill worked extensively on geophysical fluid dynamics and biofluid dynamics. He delivered an expository research and survey lecture in the IUTAM Symposium on Rotating Fluid Systems at La Jolla, California on March 28, 1966. This lecture on Dynamics of Rotating Fluids: A Survey was published in *J. Fluid Mech.* **26** (1966) 411-431. In 1969, he published his work on "Dynamic Response of the Indian Ocean to Onset of the Southwest Monsoon" in the *Philosophical Transactions of the Royal Society*, London, **A265** (1969) 45-92. Remarkably, his linearized theory of unsteady wind-driven currents in a horizontally stratified ocean with applications to the northern part of the Indian Ocean was fairly general so that it can be applied to other oceanic models. He also discussed barotropic response, baroclinic wave propagation, and baroclinic response to monsoon onset in an equatorial ocean in some detail. Subsequently, he wrote a few articles on monsoon dynamics and made several presentations on hurricane dynamics and tropical cyclones. From his extensive study and research in flood waves, monsoon dynamics, and hurricane dynamics, Lighthill became seriously interested in all kinds of natural hazards that include floods, hurricanes, cyclones, tornadoes, earthquakes, thunderstorms, underwater explosions, landslides, and volcanic eruptions. These hazards are the most common and frequently observed to have catastrophic effects on life, economic, coastal and marine resources. During the last three decades, Sir James showed a tremendous increase in interest in the dynamics of oceans and atmosphere with special reference to waves and currents caused by hurricanes, tropical cyclones, underwater explosions and earthquakes. He became more and more involved in developing protective measures to save life, economic, coastal and marine resources from natural hazards. During the last several years of his life, he served as Chairman of the Special Committee on the International Decade for Natural Disaster Reduction of the International Council of Scientific Unions, and traveled and lectures world-wide. His deep concern and active interest in reduction of all natural hazards and of wasteful use of energy reveal the unique character of this man.

9.2 Dynamics of Rotating Fluids

In his 1966 research expository and survey article, Lighthill described dynamics of rotating fluids in a coordinate frame fixed to the earth rotating with angular velocity $\mathbf{\Omega}$. With $\mathbf{r}$ representing the position vector of an arbitrary fluid element, the velocity $\mathbf{v}$ in the non-rotating (inertial) frame is equal to the velocity $\mathbf{u}$ observed in a rotating frame plus the velocity imparted to the fluid by the rigid-body rotation $\mathbf{\Omega} \times \mathbf{r}$ so that $\mathbf{v} = \mathbf{u} + \mathbf{\Omega} \times \mathbf{r}$. Taking curl of this result gives

$$\boldsymbol{\zeta} = \boldsymbol{\omega} + 2\mathbf{\Omega}, \tag{9.2.1}$$

where $\boldsymbol{\zeta} = \text{curl}\,\mathbf{v}$ is the absolute vorticity of the fluid (the vorticity relative to the fixed frame of reference), $\boldsymbol{\omega} = \text{curl}\,\mathbf{u}$ is the relative vorticity in the rotating system, and $2\mathbf{\Omega} = \text{curl}\,(\mathbf{\Omega} \times \mathbf{r})$ is the planetary vorticity. The interaction of the latter two vorticities play a significant role in geophysical fluid dynamics.

In all cases when the horizontal lines of fluid normal to the flow remain nearly horizontal, the vertical distance H between two such lines can change as in the flow over some topographical feature. The absolute vorticity vector will then change its vertical component in direct proportion to H. This vertical component is $(\zeta + f)$ where $f = 2\Omega \sin\theta$ is the Coriolis parameter (vertical component of planetary vorticity) and θ is the latitude. Thus, the rate of change of absolute vorticity is conserved, that is,

$$\frac{D}{Dt}\left(\frac{\zeta + f}{H}\right) = 0. \tag{9.2.2}$$

According to Lighthill, this property of vorticity is generally very useful for an understanding of the dynamics of a rotating fluid.

Lighthill has also investigated the existence of the Ekman layer on an infinite rigid disk at $z = 0$ in a semi-infinite expanse of an incompressible uniformly rotating viscous fluid bounded by the disk. Both the fluid and the disk are in a state of rigid body rotation with constant angular velocity $\mathbf{\Omega}$ about an axis normal to the disk. We also assume that there is a uniform flow along the x-axis with velocity U past the disk. The equations of motion for the horizontal velocity field $\mathbf{u} = (u, v)$, $q = u + iv$, have to be solved under the boundary and initial conditions

$$q = 0 \text{ on } z = 0; \;\; q = U \text{ as } z \to \infty \text{ for } t > 0, \tag{9.2.3}$$

$$q = U, \qquad \text{at } t = 0 \text{ for all } z > 0. \tag{9.2.4}$$

It follows from the boundary conditions that, far away from the disk, the pressure gradient $\frac{\partial p}{\partial y}$ must balance the Coriolis force so that

$$\frac{\partial p}{\partial x} = 0 \quad \text{and} \quad -\frac{1}{\rho}\frac{\partial p}{\partial y} = Uf. \tag{9.2.5ab}$$

Near the disk the Coriolis and viscous forces are of the same order of magnitude so that the resulting equations of motion for the horizontal velocity field (u, v) are

$$\frac{\partial u}{\partial t} - fv = \nu\frac{\partial^2 u}{\partial z^2}, \qquad \frac{\partial v}{\partial t} + f(u - U) = \nu\frac{\partial^2 v}{\partial z^2}. \tag{9.2.6}$$

With $\widetilde{q} = q - U$, these equations are combined to obtain

$$\frac{\partial \widetilde{q}}{\partial t} + if\,\widetilde{q} = \nu\,\frac{\partial^2 \widetilde{q}}{\partial z^2}. \tag{9.2.7}$$

Using the Laplace transform, the equation (9.2.7) is solved with the transformed boundary conditions and the initial condition (9.2.4). The solution of the unsteady flow field is given by

$$\widetilde{q}\,(z,t) = -\frac{1}{2}U\left[\exp\left(-\eta z\right) erfc\left\{\zeta - \sqrt{ift}\right\} + \exp\left(\eta z\right) erfc\left\{\zeta + \sqrt{ift}\right\}\right], \tag{9.2.8}$$

where

$$\eta = \left(\frac{f}{2\nu}\right)^{\frac{1}{2}} \quad \text{and} \quad \zeta = \frac{z}{2\sqrt{\nu t}}. \tag{9.2.9}$$

In the limit as $t \to \infty$, the ultimate steady-state flow is given by

$$\widetilde{q}(z) \sim -U \exp\{-\eta(1+i)z\}. \tag{9.2.10}$$

Consequently, the velocity components are given by

$$u(z,t) = U\left[1 - \exp(-\eta z)\cos(\eta z)\right], \tag{9.2.11}$$

$$v(z,t) = U\exp(-\eta z)\sin(\eta z). \tag{9.2.12}$$

These represent well defined Ekman's boundary layer solutions near the disk and this boundary layer is known as the *Ekman boundary layer* of thickness of order $\sqrt{2\nu/f} = \sqrt{\nu/\Omega}$ which is small compared with the dimension l of the system. Thus, the above solutions exist provided $\sqrt{\nu/\Omega} \ll l$, that is, the Taylor number $T = \left(\Omega l^2/\nu\right)$ is large. The Ekman layer has a thickness of penetration of vorticity generated by the solid disk of order $(\nu/\Omega)^{\frac{1}{2}}$ which depends on the Coriolis parameter f and viscosity ν. It vanishes with viscosity and has the structure similar to that of the Stokes boundary layer in a non-rotating fluid system. The plot of u and v represents the Ekman spiral distribution traced out by the endpoints of the velocity vector (u, v) within the Ekman layer as shown in Figure 9.1. The velocity near the disk is linear and makes an angle $\left(\frac{\pi}{4}\right)$ with the positive x-axis in the clockwise sense.

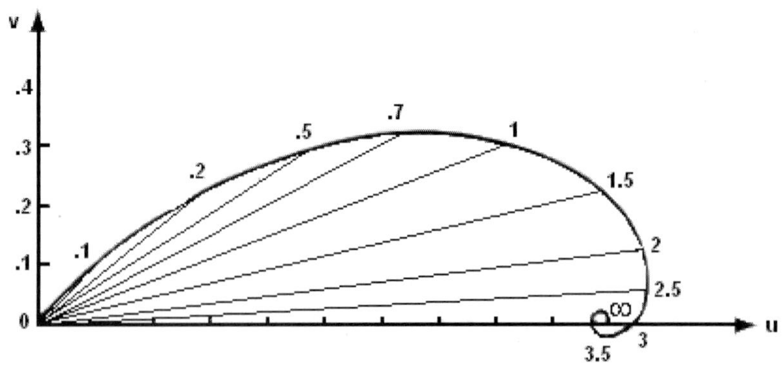

Fig. 9.1 The Ekman Spiral.

The total mass flux in the Ekman layer along the direction normal to the uniform stream outside the layer is given by

$$M_y = \rho \int_0^\infty v\,dz = \rho U \left(\frac{\nu}{2f}\right)^{\frac{1}{2}} = \rho U \left(\frac{\nu}{\Omega}\right)^{\frac{1}{2}}. \tag{9.2.13}$$

We next consider the classical solution of the Sverdrup problem where the steady solution is dominated by the variation of Coriolis parameter with latitude so that the rate of change of planetary vorticity

$$\frac{Df}{Dt} = \nu \frac{df}{dy} = \beta v, \tag{9.2.14}$$

where v is the northward component of current. In order to describe wave motions in systems with variable Coriolis parameter with purely horizontal zero divergence, the law of the Rossby wave propagation is used and this law states that the vertical component of relative vorticity changes simply due to north-south convection of planetary vorticity as

$$\frac{\partial \zeta}{\partial t} = -\beta v. \tag{9.2.15}$$

This can be written in terms of the stream function ψ $\left(u = -\psi_y,\ v = \psi_x,\ \zeta = \nabla^2\psi\right)$ which satisfies the *Rossby wave equation*

$$\frac{\partial}{\partial t}\left(\nabla^2 \psi\right) + \beta \frac{\partial \psi}{\partial x} = 0. \tag{9.2.16}$$

This admits plane wave solutions in the form

$$\psi = A \exp\left[i\left(\boldsymbol{\kappa}\cdot\mathbf{r} - \omega t\right)\right], \tag{9.2.17}$$

where A is a constant and $\boldsymbol{\kappa} = (k, l)$ is the wavenumber vector and ω is the frequency provided the dispersion relation

$$\omega\left(k^2 + l^2\right) + \beta k = 0 \tag{9.2.18}$$

is satisfied.

The phase velocity of these waves is

$$c_p = \frac{\omega}{k} = -\frac{\beta}{(k^2 + l^2)} = -\frac{\beta}{\kappa^2} = -\frac{\beta\lambda^2}{4\pi^2}. \tag{9.2.19}$$

The whole wave pattern propagates towards the west with the phase velocity of magnitude $\left(\beta/\kappa^2\right)$ which depends only on the wavenumber κ (or wavelength λ), but independent of the orientation α of the waves where $\boldsymbol{\kappa} = (k, l) = (\kappa\cos\alpha,\ \kappa\sin\alpha)$. These waves are called the *Rossby waves.* The group velocity of the Rossby waves are given by

$$c_g = \left(\frac{\partial\omega}{\partial k}, \frac{\partial\omega}{\partial l}\right) = \frac{\beta}{\kappa^2}\left(\cos 2\alpha,\ \sin 2\alpha\right). \tag{9.2.20}$$

Thus, the group velocity has the same magnitude of c_p, but it propagates in a different direction and makes an angle with the east which is twice the angle that the phase velocity makes (or the wavenumber vector $\boldsymbol{\kappa} = (k, l)$).

For the study of really large-scale phenomena in rotating fluids, it is necessary to use the true spherical geometry. We consider equations of motion of a thin layer of fluid of uniform depth h and uniform density ρ on the surface of a rotating sphere of radius R. It is convenient to introduce the spherical coordinates (r, θ, ϕ) which rotates with the sphere with the origin at the center of the sphere and $r = R$ at the spherical outer boundary of the layer. We take $\theta = 0$ at the north pole so that θ is the colatitude and $\left(\frac{\pi}{2} - \theta\right)$ is the latitude, and ϕ is the longitude with the east directed the direction of ϕ increasing. The corresponding velocity components are $(u_r, u_\theta, u_\phi) = (u, v, w)$.

We investigate the small amplitude wave motions within the scope of the linearized theory under the assumptions that the depth h of the fluid is small compared with the radius R of the sphere $(h \ll R)$ and the pressure satisfies the hydrostatic law. Under these assumptions the horizontal equations of motion in the linearized form with $r = R$ are given by

$$\frac{\partial u}{\partial t} - fv = -\frac{g}{R}\frac{\partial \zeta}{\partial \theta}, \tag{9.2.21}$$

$$\frac{\partial v}{\partial t} + fu = -\frac{g}{R\sin\theta}\frac{\partial \zeta}{\partial \phi}, \tag{9.2.22}$$

where ζ is the vertical surface elevation.

The associated equation of continuity is given by

$$\frac{h}{R\sin\theta} = \left[\frac{\partial}{\partial \theta}(u\sin\theta) + \frac{\partial v}{\partial \phi}\right] = -\frac{\partial \zeta}{\partial t}. \tag{9.2.23}$$

With small vertical surface elevation ζ, the continuity equation implies the existence of a stream function ψ so that the velocities are

$$u = \frac{1}{R\sin\theta}\frac{\partial \psi}{\partial \theta}, \qquad v = -\frac{1}{R}\frac{\partial \psi}{\partial \theta}. \tag{9.2.24ab}$$

Substituting (9.2.24ab) into (9.2.21)–(9.2.21) and eliminating ζ by cross-differentiation gives

$$\frac{\partial}{\partial t}(\nabla^2 \psi) + 2\Omega\frac{\partial \psi}{\partial \theta} = 0, \tag{9.2.25}$$

where ∇^2 is the two-dimensional Laplacian given by

$$\nabla^2 \equiv \frac{1}{\sin\theta}\frac{\partial}{\partial \theta}\left(\sin\theta\frac{\partial}{\partial \theta}\right) + \frac{1}{\sin^2\theta}\frac{\partial^2}{\partial \phi^2}. \tag{9.2.26}$$

The steady state solution of (9.2.25) is very simple. In fact, $\frac{\partial}{\partial \theta} = 0$ and hence $\frac{\partial \psi}{\partial \phi} = 0$ which implies that the streamlines are everywhere circles of longitude ϕ.

Evidently, ocean waves can again be generated by disturbances moving westward. But, the sphere is a limited region, there are certain velocities at which the disturbance can propagate for which these waves get intensified by resonance. The normal mode solution of (9.1.25) on the sphere in the form

$$\psi = S_n (\theta, \phi - \phi t) \tag{9.2.27}$$

exists, where S_n is a spherical harmonic which satisfies the equation

$$\nabla^2 S_n + n(n+1) S_n = 0, \tag{9.2.28}$$

provided $-(2\Omega/\phi) = n(n+1)$, or

$$\phi = -\frac{2\Omega}{n(n+1)}. \tag{9.2.29}$$

This implies that the streamlines and hence, the whole current system rotate westward around the axis of rotation with the angular velocity $\{2\Omega/n(n+1)\}$. These westward speeds of the wave pattern have been observed by the spherical-shell experiments.

For $n = 1$, the angular velocity of the current system is just Ω so that the system is stationary relative to the fixed axes. However, if $n > 1$ the angular velocity is always less than Ω in numerical value so that the system approaches to be carried round with the rotating globe. Longuet Higgins (1964, 1965) developed an elaborate theory of the nature of spherical harmonic solutions in a rotating globe. He pointed out that some of the resonant solutions involve wave motions only at the lower latitudes; for example, the spherical harmonic P_n^s for large n does so only below the latitude $\cos^{-1}(s/n)$, where the wave crests reflected back at a locus of cusps. Such resonant modes have also been observed.

Finally, the dispersion relation (9.2.18) shows that a small unsteadiness, with a frequency of order βL, where L is a characteristic length scale of a fluid bounded above and below by surfaces $z = H(x, y)$ and $z = h(x, y)$ is equally significant in wind driven ocean currents. But also the tidal term contribution, $(gH)^{-1}(f^2 - \omega^2)$ which might be small if the phase velocity (9.2.19) is small compared with $\sqrt{gh}$, may not in fact, small and needs to be added to (9.1.18) so that it becomes

$$\omega \left[\left(k^2 + l^2\right) + \left(f^2 - \omega^2\right) \,/ gH \right] + \beta k = 0. \tag{9.2.30}$$

If ω is of order βL, the additional term $\left(\omega^2/\,gH\right)$ may be neglected, but *not* the term $\left(f^2/\,gH\right)$. Consequently, the modified dispersion relation becomes

$$\omega\left[\left(k^2+l^2\right)+\left(f^2/\,gH\right)\right]+\beta k=0. \tag{9.2.31}$$

Or, equivalently

$$c_p=-\frac{\omega}{k}=\frac{\beta}{k^2+l^2+\left(f^2/\,gH\right)}, \tag{9.2.32}$$

giving

$$k^2+l^2=\kappa^2=\left(\frac{\beta}{c_p}-\frac{f^2}{gH}\right). \tag{9.2.33}$$

In real situations, therefore, the tidal term makes a small, but usually not negligible, contribution to the Rossby wave propagation.

9.3 The Navier-Stokes Equations of Motion in a Rotating Frame of Reference

We consider large scale oceanic or atmospheric motions that are really only small deviation from the rigid body rotation imposed upon the fluid due to the earth rotation. In this rotating frame of reference, the motion of a fluid is governed by the Navier–Stokes equations (see Debnath 1976a)

$$\rho\left[\frac{D\mathbf{u}}{Dt}+2\boldsymbol{\omega}\times\mathbf{u}+\boldsymbol{\Omega}\times(\boldsymbol{\Omega}\times\mathbf{r})\right]=-\nabla P+\mathbf{g}+\mu\,\nabla^2\,\mathbf{u}+\frac{1}{3}\,\mu\,\nabla\left(\nabla\cdot\mathbf{u}\right), \tag{9.3.1}$$

where the operator $\frac{D}{Dt}$ is the total derivative defined by

$$\frac{D}{Dt}=\frac{\partial}{\partial t}+\mathbf{u}\cdot\nabla, \tag{9.3.2}$$

ρ is the fluid density, $\mathbf{u}=(u,v,w)$ is the fluid velocity vector, $\boldsymbol{\Omega}$ is the angular velocity of the earth, $\mathbf{r}=(x,y,z)$, P is the pressure, $\mathbf{g}$ is the acceleration due to gravity and μ is the dynamic viscosity,

Or, equivalently, (9.3.1) can be written as

$$\frac{D\mathbf{u}}{Dt}+2\boldsymbol{\Omega}\times\mathbf{u}=-\frac{1}{\rho}\,\nabla p+\nu\,\nabla^2\mathbf{u}+\frac{\nu}{3}\,\nabla\left(\nabla\cdot\mathbf{u}\right), \tag{9.3.3}$$

where $p=P+\rho V-\frac{1}{2}\rho\Omega^2r^2$ is called the **modified** (or **reduced**) pressure, $\nu=(\mu/\rho)$ is the kinematic viscosity, $\mathbf{g}=-\nabla V$ and $\boldsymbol{\Omega}\times(\boldsymbol{\Omega}\times\mathbf{r})=\nabla\left(\frac{1}{2}\Omega^2r^2\right)$.

The equation of continuity is

$$\frac{D\rho}{Dt} + \rho\,\nabla \cdot \mathbf{u} = 0 \tag{9.3.4}$$

which is, by (9.3.2),

$$\frac{\partial \rho}{\partial t} + \nabla \cdot (\rho \mathbf{u}) = 0. \tag{9.3.5}$$

For an incompressible fluid, the density is constant so that (9.3.5) becomes

$$div\ \mathbf{u} = \nabla \cdot \mathbf{u} = 0. \tag{9.3.6}$$

The Navier-Stokes equations of motion becomes

$$\frac{D\mathbf{u}}{Dt} + 2\boldsymbol{\Omega} \times \mathbf{u} = -\frac{1}{\rho}\nabla p + \nu \nabla^2 \mathbf{u}. \tag{9.3.7}$$

In terms of some representative length scale L, time scale Ω^{-1} and velocity scale U, it is convenient to introduce the nondimensional flow variables

$$\mathbf{x}^* = \frac{\mathbf{x}}{L},\ t^* = \Omega t,\ \mathbf{u}^* = \frac{\mathbf{u}}{U} \ \text{ and } \ p^* = \rho \Omega L U P. \tag{9.3.8}$$

In terms of these nondimensional flow variables, the continuity equation and the Navier–Stokes equations for an incompressible fluid can be written as, dropping the asterisks,

$$\frac{\partial \mathbf{u}}{\partial t} + R_0\,(\mathbf{u} \cdot \nabla)\,\mathbf{u} + 2\,\widehat{\mathbf{k}} \times \mathbf{u} = -\nabla p + E\,\nabla^2 \mathbf{u}, \tag{9.3.9}$$

$$\nabla \cdot \mathbf{u} = 0, \tag{9.3.10}$$

where $\boldsymbol{\Omega} = \widehat{\mathbf{k}}\,\Omega$, $\widehat{\mathbf{k}}$ is the unit vector, $R_0 = \frac{U}{\Omega L}$ is the *Rossby number* and $E = \frac{\nu}{\Omega L^2}$ is the *Ekman number*.

Equation (9.3.9) shows that the Rossby number and the Ekman number play an important role in the dynamics of oceans.

9.4 Linear Rossby Waves in a Non-Uniformly Rotating Ocean

The atmosphere and oceans move ceaselessly in complex patterns and on diverse scales. Geophysical motions that are naturally arise on the surface of the earth, range in small from small ripples to the majestic current systems of the oceans (such as the Gulf Stream) and the atmosphere (such as the Jet Stream). It is almost impossible to discuss the dynamics of this enormous spectrum of motions. So we focus our attention to the so called *large scale motions* that are significantly influenced by the effects of the rotation of the earth. There are several important parameters that measure the importance of earth's rotation. One of the most important such parameters is known as the *Rossby number* defined by $R_0 = U/(\Omega L)$, where L is the typical horizontal length scale of the motion, U is the typical velocity scale relative to the earth's surface and $\Omega\left(= 7.3 \times 10^{-5}\,\text{sec}^{-1}\right)$ is the frequency of the earth rotation. A necessary condition for rotation to be dominant is $R_0 \ll 1$, that is $O\left(10^{-3}\right)$. When the Rossby number is small, the Coriolis force is a dominant factor in the balance of forces acting on a fluid element. In addition, the dynamics of large scale motions have two other features. First, the motions occur in a thin spherical shell, that is, the horizontal scale of the motion is very large compared to the vertical scale. In general, the oceanic circulation with a horizontal scale of thousands of kilometers has a vertical scale of only about 2 kilometers. Thus, the oceanic flow deals with very shallow water motions. Second, both the atmosphere and oceans have the significant density variation which is almost always stable in the sense that heavy fluid underlies lighter fluid. Indeed, oceanic motion is caused by both mechanical and thermal processes. The former is essentially due to action of wind stress at the ocean surface, whereas the latter is associated with the density variation produced by heat transfer at the ocean surface due to solar heating. So the dynamics of oceans deals with the study of physical phenomena concerning waves and currents in the ocean on the rotating earth.

Historically, Rossby (1939) first introduced the equatorial β-plane approximation where the surface of the spherical globe is replaced by a plane and the Coriolis parameter is allowed to vary linearly with latitude so that its northward derivative $\frac{df}{dy} = \beta$ remains constant. It is shown that the solution of the β-plane linear Rossby wave equation represents long periodic oscillations with periods of several days or longer. Subsequently, several authors including Pedlosky (1971), Stewartson (1978), Haynes (1985), Killworth and McIntyre (1985) studied the dynamics of large scale oceanic and atmospheric motions that give rise to general features of Rossby waves.

To study large scale oceanic motions, Rossby's β-plane approximation is used. This β-plane approximation simply states that the variation of the Coriolis parameter $f = 2\Omega \sin\theta$, where θ is the latitude, and Ω is the frequency of Earth's rotation, must be included but the effects of earth's curvature may be neglected. So we can write the Coriolis parameter as

$$f = f_0 + \beta y, \tag{9.4.1}$$

where

$$f_0 = 2\Omega \sin\theta_0 \quad \text{and} \quad \beta = \left(\frac{df}{dy}\right)_{\theta=\theta_0}, \tag{9.4.2}$$

θ_0 is a reference latitude.

With large horizontal length scale L and small vertical scale $h\,(L \gg h)$ representing depth of ocean and with β-plane approximation, the Navier–Stokes equations and the continuity equation reduce to

$$\frac{\partial \mathbf{u}}{\partial t} + (\mathbf{u} \cdot \nabla)\,\mathbf{u} + \left[-\left(f_0 + \beta y\right), \left(f_0 + \beta y\right),\ 0\right]\mathbf{u} = -\frac{1}{\rho}\nabla p + \nu \nabla^2 \mathbf{u}, \quad (9.4.3)$$

$$div\ \mathbf{u} = 0, \quad (9.4.4)$$

Remarks.

(1) For uniformly rotating liquid $f = 2\Omega \sin\theta_0 =$ constant, the equations deal with uniformly rotating fluid.
(2) For non-uniformly rotating liquid (f varies with latitude, $\beta = \frac{df}{dy}$), equations deal with real dynamics of oceans. So, the dynamics of oceans essentially deals with non-uniformly rotating oceans.

With turbulent momentum transfer neglected, the horizontal equations of motion for an inviscid homogeneous ocean of depth h due to wind stress $\boldsymbol{\tau} = (\tau^x, \tau^y)$ at the ocean surface are given by

$$\frac{\partial \mathbf{u}}{\partial t} + (\mathbf{u} \cdot \nabla \mathbf{u}) + (-fv, fu) = -\frac{1}{\rho}\nabla p + \frac{1}{\rho h}\boldsymbol{\tau}, \quad (9.4.5)$$

$$\frac{\partial u}{\partial x} + \frac{\partial v}{\partial y} = -\frac{1}{h}\frac{\partial \zeta}{\partial t}, \quad (9.4.6)$$

where $\mathbf{u} = (u, v)$, and $\zeta = \zeta\,(x, y, t)$ is free surface elevation of the ocean. Neglecting the nonlinear terms, eliminating the pressure, and the free surface elevation gives

$$\left[\left(\nabla^2 - \frac{f^2}{c^2}\right)\frac{\partial}{\partial t} + \beta\frac{\partial}{\partial x}\right] v = \frac{1}{\rho h}\ \mathrm{curl}_z\ \tau, \quad (9.4.7)$$

where $c^2 = gh$ is the square of the shallow water speed and $\nabla^2 = \frac{\partial^2}{\partial x^2} + \frac{\partial^2}{\partial y^2}$ is the horizontal Laplacian.

With small free surface elevation, there exists a stream function $\psi\,(x, y, t)$ that satisfies

$$u = -\psi_y \quad \text{and} \quad v = \psi_x, \quad (9.4.8)$$

so that ζ can be eliminated from (9.4.5)–(9.4.6) to obtain

$$\nabla^2 \psi_t + \beta \psi_x = \mathrm{curl}_z \boldsymbol{\tau}. \quad (9.4.9)$$

This is the classical celebrated linear *forced Rossby wave equation* which relates the rate of change of vertical vorticity to the curl of the wind stress.

With no wind stresses at the ocean surface, the plane wave solution exists in the form

$$\psi = \Psi \exp\left[i\left(\boldsymbol{\kappa} \cdot \mathbf{r} - \omega t\right)\right] \tag{9.4.10}$$

provided the following dispersion relation is satisfied

$$\omega = -\frac{\beta k}{\left(\kappa^2 + a^2\right)}, \tag{9.4.11}$$

$\boldsymbol{\kappa} = (k, l)$ and $\mathbf{r} = (x, y)$ and $a^2 = \left(f^2/c^2\right)$. The phase velocity of the free Rossby waves is

$$c_p = \frac{\omega}{k} = -\frac{\beta}{\kappa^2 + a^2}, \tag{9.4.12}$$

where $R = \frac{1}{a} = (c/|f|) = \left(\sqrt{gh}/|f|\right)$ is a length scale of fundamental importance for the description of rotating oceans. This parameter R is called the *Rossby deformation radius*. In the limit as $f \to 0$, $R \to \infty$ indicating that for length scale large or comparable with R, the effect of rotation is dominant, whereas for length scale small compared with R, the effect of rotation is small.

Formula (9.4.12) reveals that the whole wave pattern propagates towards the west with velocity $\beta/\left(\kappa^2 + a^2\right)$ which depends only on the wavenumber κ but independent of the orientation α where $\boldsymbol{\kappa} = (k, l) = (\kappa \cos\alpha, \ \kappa \sin\alpha)$. At a given latitude, frequency ω is maximum when $(k, l) = (a, 0)$ and the maximum value of the frequency is

$$\omega_{\max} = -\frac{\beta}{2a}. \tag{9.4.13}$$

This can be interpreted as the *cut-off* frequency and the corresponding waves are called the *cut-off Rossby* waves.

With $\gamma = \frac{\beta}{2a}$, the dispersion relation can be written as

$$(k + \gamma)^2 + l^2 = \gamma^2 - a^2. \tag{9.4.14}$$

This shows that the locus of the wavenumber $\boldsymbol{\kappa} = (k, l)$ is a circle in the wavenumber plane with center $(-\gamma, 0)$ and radius $\sqrt{\gamma^2 - a^2}$.

The group velocity of the Rossby waves is

$$\mathbf{c}_g = \hat{k}\,\frac{\partial \omega}{\partial k} + \hat{l}\,\frac{\partial \omega}{\partial l}, \tag{9.4.15}$$

$$= \frac{\beta\,\kappa^2}{\left(\kappa^2 + a^2\right)^2}\left[\hat{k}\left\{\cos 2\alpha - \left(a^2/\kappa^2\right)\right\} + \hat{l}\sin 2\alpha\right]. \tag{9.4.16}$$

9.5 General Class of Traveling Wave Solutions of the Rossby Wave Equation

With no wind stresses, the plane wave solution (9.4.10) is one of the general class of solutions of the free Rossby wave equation (9.4.9). Any expression of the form

$$\psi = F\,(x - ct,\ y)\,, \tag{9.5.1}$$

is a solution of (9.4.9) with zero right hand side provided F satisfies

$$\left(\nabla^2 + \kappa^2\right) F = 0, \tag{9.5.2}$$

where

$$\kappa^2 + a^2 = -\frac{\beta}{c}. \tag{9.5.3}$$

This solution represents motion drifting westward with velocity

$$c = -\frac{\beta}{\kappa^2 + a^2}. \tag{9.5.4}$$

Even more general solution representing a current system with a moving center $(ct, 0)$ and decaying with distance from the center like $\frac{1}{\sqrt{r'}}$ at infinity is possible in the form

$$\psi = J_n\,(\kappa r')\exp\,(\pm in\theta)\,, \tag{9.5.5}$$

where

$$r' = \sqrt{\left\{(x - ct)^2 + y^2\right\}} \quad \text{and} \quad \theta = \tan^{-1}\left(\frac{y}{x - ct}\right). \tag{9.5.6}$$

For large values of (kr'), the asymptotic solution is obtained from (9.5.5) in the form

$$\psi \sim \left(\frac{2}{\pi\kappa r'}\right)^{\frac{1}{2}} \exp\left[i\left(\kappa r' - \frac{n\pi}{2} - \frac{\pi}{4}\right)\right] \exp\,(\pm in\theta)\,. \tag{9.5.7}$$

Even more general solution of the Rossby wave equation incorporating the effects of viscosity, bottom friction and bottom topography has been found. For details, see Longuet Higgins (1964, 1965a) and Debnath (1976b, 1995).

9.6 Small Amplitude Waves in a Rotating Ocean

We consider the motion of an inviscid incompressible homogenous shallow rotating ocean of depth h due to an arbitrary wind stress distribution acting on the free surface of the ocean. For small amplitude wave motions, we make the following assumptions to derive the basic equations.

(i) The depth $h(x, y, t)$ of the ocean can be written as

$$h(x, y, t) = h_0(x, y) + \zeta(x, y, t), \tag{9.6.1}$$

where $h_0(x, y)$ is the mean depth of the ocean at rest and $\zeta(x, y, t)$ is the vertical surface elevation about the mean depth.

(ii) Small amplitude motions imply that $\zeta \ll h_0$. Further, we assume that horizontal velocity field $\mathbf{u} = (u, v)$ is small enough so that

$$\frac{\partial \mathbf{u}}{\partial t} \gg \mathbf{u} \cdot \nabla \mathbf{u}. \tag{9.6.2}$$

(ii) The vertical acceleration of the fluid is assumed to be small so that the equation of motion in the vertical z-direction reduces to the hydrostatic equation

$$p = p_0 + g\rho(\zeta - z), \tag{9.6.3}$$

where p is the pressure, z is the depth below the mean free surface, and p_0 is the constant atmospheric pressure at the free surface.

(iv) The Coriolis parameter $f = 2\Omega \sin\theta$ is taken as constant, where Ω is the angular velocity of the earth and θ is the north latitude. The effects of curvature of the earth on the motion is neglected.

The linearized non-dissipating horizontal equations of motion and the continuity equation in the rotating Cartesian coordinate system are

$$\frac{\partial \mathbf{u}}{\partial t} + f\,\widehat{\mathbf{k}} \times \mathbf{u} = -\frac{1}{\rho}\nabla p + \frac{1}{\rho h_0}\boldsymbol{\tau}, \tag{9.6.4}$$

$$\frac{\partial \zeta}{\partial t} + \nabla \cdot (h_0 \mathbf{u}) = 0, \tag{9.6.5}$$

where $\widehat{\mathbf{k}}$ is the unit vector normal to the horizontal (x, y)-plane and $\boldsymbol{\tau} = (\tau^x, \tau^y)$ is the wind stress vector.

Invoking the assumptions (i)–(iv) and equation (9.6.3), the above equations (9.6.4)–(9.6.5) reduce to the form

$$\frac{\partial u}{\partial t} - fv = -g\frac{\partial \zeta}{\partial x} + \frac{\tau^x}{\rho h_0}, \tag{9.6.6}$$

$$\frac{\partial v}{\partial t} + fu = -g\frac{\partial \zeta}{\partial y} + \frac{\tau^y}{\rho h_0}, \tag{9.6.7}$$

$$\frac{\partial \zeta}{\partial t} + \frac{\partial}{\partial x}(uh_0) + \frac{\partial}{\partial y}(vh_0) = 0. \tag{9.6.8}$$

In terms of the mass flux vector, $\mathbf{U} = \mathbf{i}U + \mathbf{j}V$ where $U = h_0 u$ and $V = h_0 v$, equations (9.6.6)–(9.6.8) become

$$\frac{\partial U}{\partial t} - fV = -gh_0 \frac{\partial \zeta}{\partial x} + \frac{1}{\rho}\tau^x, \tag{9.6.9}$$

$$\frac{\partial V}{\partial t} + fU = -gh_0 \frac{\partial \zeta}{\partial y} + \frac{1}{\rho}\tau^y, \tag{9.6.10}$$

$$\frac{\partial \zeta}{\partial t} + \nabla \cdot \mathbf{U} = 0. \tag{9.6.11}$$

Simple manipulation of (9.6.9) and (9.6.10) gives

$$\frac{\partial}{\partial t}(\nabla \cdot \mathbf{U}) - f \operatorname{curl} \mathbf{U} = -g\nabla \cdot (h_0 \nabla \zeta) + \frac{1}{\rho}\nabla \cdot \tau, \tag{9.6.12}$$

$$\frac{\partial}{\partial t}(\operatorname{curl} \mathbf{U}) + f(\nabla \cdot \mathbf{U}) = -gJ(h_0, \zeta) + \frac{1}{\rho}\operatorname{curl} \tau, \tag{9.6.13}$$

where the Jacobian $J(h_0, \zeta)$ is defined by

$$J(h_0, \zeta) = \left(\frac{\partial h_0}{\partial x}\frac{\partial \zeta}{\partial y} - \frac{\partial h_0}{\partial y}\frac{\partial \zeta}{\partial x}\right). \tag{9.6.14}$$

It follows from (9.6.12) and (9.6.13) that

$$D(\nabla \cdot \mathbf{U}) = -\frac{\partial}{\partial t}\left[\nabla \cdot (c^2 \nabla \zeta)\right] - gfJ(h_0, \zeta) + E(x, y, t), \tag{9.6.15}$$

where $c^2 = gh_0$, and the differential operator D is given by

$$D = \frac{\partial^2}{\partial t^2} + f^2, \tag{9.6.16}$$

and the forcing function $E(x, y, t)$ has the form

$$E(x, y, t) = \frac{1}{\rho}\left[\frac{\partial}{\partial t}(\nabla \cdot \boldsymbol{\tau}) + f \operatorname{curl} \boldsymbol{\tau}\right]. \tag{9.6.17}$$

Finally, elimination of $\mathbf{U}$ from (9.6.11) and (9.6.15) gives the differential equation for ζ in the form

$$\left[D - \nabla \cdot (c^2 \nabla)\right]\zeta_t = gfJ(h_0, \zeta) - E(x, y, t). \tag{9.6.18}$$

It follows from (9.6.6)–(9.6.8) that u and v can be found in terms of ζ as

$$Du = -g\left(\frac{\partial^2}{\partial x \partial t} + f\frac{\partial}{\partial y}\right)\zeta + \frac{1}{\rho h_0}\left(\frac{\partial \tau^x}{\partial t} + f\tau^y\right), \tag{9.6.19}$$

$$Dv = -g\left(\frac{\partial^2}{\partial y \partial t} - f\frac{\partial}{\partial x}\right)\zeta + \frac{1}{\rho h_0}\left(\frac{\partial \tau^y}{\partial t} - f\tau^x\right). \tag{9.6.20}$$

For each case of constant mean depth, h_0 is constant, equation (9.6.18) further simplifies to

$$\left(c^2 \nabla^2 - D\right)\zeta_t = E(x, y, t). \tag{9.6.21}$$

9.7 Poincaré, Kelvin and Rossby Waves in a Rotating Ocean

We consider free harmonic ocean waves in a shallow rotating channel of constant depth h_0. In the absence of the wind stress distribution, we seek solutions of the homogeneous equation (9.6.21) of the form

$$\zeta(x, y, t) = A \exp\left[i\left(\boldsymbol{\kappa} \cdot \mathbf{r} - \omega t\right)\right], \tag{9.7.1}$$

provided the following dispersion relation is satisfied with $\boldsymbol{\kappa} = (k, l)$,

$$\omega^2 = f^2 + \left(k^2 + l^2\right) c^2. \tag{9.7.2}$$

Waves associated with the dispersion relation (9.7.2) are called the *Poincaré waves* which are dispersive in a rotating ocean. In terms of the Rossby deformation radius, $R = (c/f)$, relation (9.7.2) reads as

$$\left(\omega^2/f^2\right) = 1 + \kappa^2 R^2. \tag{9.7.3}$$

Two limiting cases are of special interest:

(i) Waves $\left(\kappa R \ll 1 \text{ or } \kappa^{-1} \gg R\right)$ are long compared with the Rossby deformation radius R so that the dispersion relation (9.7.3) becomes

$$\omega \sim f = 2\Omega. \tag{9.7.4}$$

Thus, the frequency is approximately equal to f or twice the rotation of the fluid. In this limit, gravity has no effect, so fluid particles move under their own inertia. In other words, long waves are dominated by the rotation effects and have frequency close to the inertial frequency $f = 2\Omega$.

(ii) Waves $\left(\kappa R \gg 1, \text{ or } \kappa^{-1} \ll R\right)$ are short compared with the Rossby deformation radius R so that the dispersion relation (9.7.3) becomes

$$\omega \sim \kappa c. \tag{9.7.5}$$

This corresponds to very much like shallow water gravity waves in a nonrotating ocean. In other words, for large wavenumber, short waves are hardly affected by rotation.

The group velocity $\mathbf{c}_g$ of Poincaré waves is equal to the slope of the dispersion curve (9.7.2) or (9.7.3) as shown in Figure 9.2. It follows that

$$\mathbf{c}_g = \frac{c^2}{\omega}\boldsymbol{\kappa} \sim \left(\frac{c}{\kappa}\right)\boldsymbol{\kappa} \qquad \text{for large } \kappa, \tag{9.7.6}$$

$$= \frac{c^2}{\omega}\boldsymbol{\kappa} \sim \left(\frac{c^2}{f}\right)\boldsymbol{\kappa} \qquad \text{for small } \kappa. \tag{9.7.7}$$

With $\omega^* = (\omega/f)$ and $\kappa^* = \kappa R$ the dispersion curve for the Poincaré waves are drawn in Figure 9.2.

These results show that the group velocity tends to zero as $\kappa \to \infty$ (infinitely long waves), and increases monotonically in magnitude with κ to a maximum value of c for very short waves. Thus, Poincaré waves can be used as the most suitable example of how ocean waves disperse when the group velocity is a minimum (zero) for long waves and a maximum for short waves. Finally, Gill (1976) obtained the solution of the transient problem by a suitable Fourier linear superposition of wave solutions (9.7.1).

We next consider free harmonic waves in a shallow rotating channel of variable depth $h_0(y)$ and width $d(0 \le y \le d)$ bounded by the two rigid walls at $y = 0$ and $y = d$. We assume the channel is parallel to the x-axis and the depth h_0 varies slightly in the y direction, that is,

$$h_0 = d_0 \left(1 - \frac{\gamma\, y}{d}\right), \tag{9.7.8}$$

where d_0 is a constant and the slope $\gamma\,(\gamma \ll 1)$ of the bottom is assumed to be small.

The lines of constant h_0 are parallel to the x-axis, and pure geostrophic motion is possible only if the fluid velocity in the y direction vanishes. Motion across the line of constant depth h_0 of liquid columns will cause them to stretch or contract which in combination with f will generate

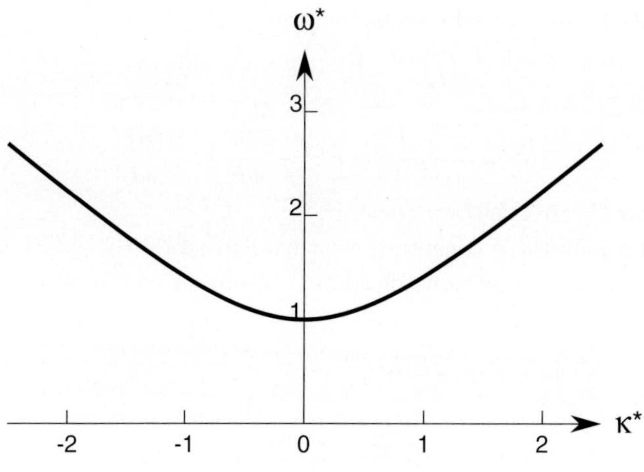

Fig. 9.2 The dispersion curve for the Poincaré waves.

relative vorticity. This indicates the strong possibility of new waves whose existence depends on the joint effects of rotation and bottom slope.

In the absence of the wind stress distribution, the governing equation (9.6.21) for ζ admits periodic wave solutions in the form

$$\zeta(x,y,t) = \text{Re } A(y)\, e^{i(kx-\omega t)}, \tag{9.7.9}$$

where $A(y)$ is the complex wave amplitude, k is the wavenumber, ω is the frequency and Re stands for the real part.

In view of the fact that the fluid velocity in the y direction must vanish, we set the boundary conditions (9.6.20) with no wind stress in the form

$$\frac{\partial^2 A}{\partial y \partial t} - f\frac{\partial A}{\partial x} = 0 \quad \text{on} \quad y = 0,\ d. \tag{9.7.10ab}$$

We next substitute (9.7.9) into equation (9.6.18) and the boundary conditions (9.7.10ab) to formulate the linear eigenvalue problem for $A(y)$:

$$\left(1 - \frac{\gamma y}{d}\right)\frac{d^2 A}{dy^2} - \frac{\gamma}{d}\frac{dA}{dy} + \left[\frac{\omega^2 - f^2}{g d_0} - k^2\left(1 - \frac{\gamma y}{d}\right) - \frac{f\gamma k}{\omega d}\right] A = 0, \tag{9.7.11}$$

$$\frac{dA}{dy} + \frac{fk}{\omega}A = 0 \qquad \text{on} \quad y = 0,\ d. \tag{9.7.12ab}$$

In view of the fact that the bottom slope is small ($\gamma \ll 1$) and (y/d) is always less than one, the approximate form of equation (9.7.11) is

$$\frac{d^2 A}{dy^2} - \frac{\gamma}{d}\frac{dA}{dy} + \left[\frac{\omega^2 - f^2}{c^2} - k^2 - \frac{f\gamma k}{\omega d}\right] A = 0. \tag{9.7.13}$$

This equation admits solutions in the form

$$A(y) = \exp\left(\frac{\gamma y}{2d}\right)(C_1 \cos \alpha y + C_2 \sin \alpha y), \tag{9.7.14}$$

where

$$\alpha^2 = \frac{\omega^2 - f^2}{c^2} - \left(k^2 + \frac{\gamma^2}{4d^2}\right) - \frac{\gamma k f}{\omega d}, \tag{9.7.15}$$

and C_1 and C_2 are arbitrary constants.

To determine these constants, we then substitute the solution (9.7.15) into the boundary conditions (9.7.12ab) to obtain two linear homogenous equations for C_1 and C_2:

$$\left(\frac{\gamma}{2d} + \frac{fk}{\omega}\right) C_1 + \alpha C_2 = 0, \tag{9.7.16}$$

$$\left[\left(\frac{\gamma}{2d} + \frac{fk}{\omega}\right) \cos \alpha d - \alpha \sin \alpha d\right] C_1 + \left[\left(\frac{\gamma}{2d} + \frac{fk}{\omega}\right) \sin \alpha d + \alpha \cos \alpha d\right] C_2 = 0. \tag{9.7.17}$$

The nontrivial solutions for C_1 and C_2 exist provided the coefficient determinant of (9.7.16) and (9.7.17) vanishes. After some simple algebraic calculation, the coefficient determinant yields the eigenvalue equation

$$\left(\omega^2 - f^2\right)\left(\omega^2 - k^2c^2\right)\sin\,\alpha d = 0. \tag{9.7.18}$$

This leads to three possibilities: either $\sin\alpha d = 0$, or $\omega^2 - k^2c^2 = 0$, or $\omega^2 - f^2 = 0$. The first equation $\sin\,\alpha d = 0$ can be satisfied provided

$$\alpha = \left(\frac{n\pi}{d}\right), \qquad n = 1,\ 2,\ 3,\ \ldots, \tag{9.7.19}$$

so that the dispersion relation (9.7.15) assumes the form

$$\omega^2 = \frac{f\gamma kc^2}{\omega d} + c^2\left(k^2 + \frac{f^2}{c^2} + \frac{\gamma^2}{4d^2}\right) + \omega_n^2, \qquad n = 1,\ 2,\ 3, \ldots, \tag{9.7.20}$$

which, for the case $\gamma \ll 1$, becomes

$$\omega^2 = c^2\left[\frac{f\gamma k}{\omega d} + \left(k^2 + \frac{f^2}{c^2}\right)\right] + \omega_n^2, \qquad n = 1,\ 2,\ 3, \ldots, \tag{9.7.21}$$

where

$$\omega_n = \frac{n\pi c}{d}. \tag{9.7.22}$$

It should be noted $\alpha = n = 0$ does not correspond to a possible solution. In fact, such a solution represents a plane wave with no y-variation in the wave field. Although such a solution exists in a non-rotating fluid system, result (9.6.20) with no wind stresses and $\frac{\partial\zeta}{\partial y} = 0$ implies that, for the velocity field v,

$$v = \frac{gf}{(f^2 - \omega^2)}\frac{\partial\zeta}{\partial x} \neq 0. \tag{9.7.23}$$

This means that v is not only non-zero, but also independent of y. Consequently, it fails to satisfy the boundary condition on $y = 0$ and $y = d$.

The dispersion relation (9.7.21) has several remarkable properties. First, there is an infinite spectrum of solutions corresponding to $n = 1,\ 2,\ 3, \ldots$. Second, it provides the existence of two distinctly separate classes of solutions. The first class corresponds to the case $\gamma = 0$ which gives an infinite number of solutions in the channel with flat bottom ($\gamma = 0$). And the associated dispersion equation is obtained from (9.7.21) with $\gamma = 0$ in the form

$$\omega^2 = \left(c^2k^2 + f^2 + \omega_n^2\right) + O(\gamma) \tag{9.7.24}$$

or,

$$\omega = \pm\left(c^2k^2 + f^2 + \omega_n^2\right)^{\frac{1}{2}}, \tag{9.7.25}$$

where $n = 1,\ 2,\ 3, \ldots$.

These wave modes are called the *Poincaré waves*. The frequency of these waves is not greater than both f and ω_n, but the y component of the wave vector is quantized as an integral multiple of $\frac{\pi}{d}$, since $\omega_n = \frac{n\pi c}{d}$. Also, the two solutions (9.7.25) for ω of the same magnitude but opposite sign imply that the Poincaré modes propagate in both positive and negative x-directions. The phase velocity, c_p, and the group velocity, c_g of the modes are given by

$$c_p = \frac{\omega}{k} = \left[c^2 + \frac{1}{k^2}\left(f^2 + \omega_n^2\right)\right]^{\frac{1}{2}}, \qquad c_g = \frac{\partial \omega}{\partial k} = \frac{c^2}{c_p^2}. \tag{9.7.26ab}$$

It is evident from these results that $c_g < c < c_p$. It follows from (9.7.26a) that, in addition to introducing dispersion, the effect of rotation is to increase the phase velocity of these waves. Moreover, (9.7.21) implies that the high frequency Poincaré modes are essentially independent of the small bottom slope. Finally, it is clear that Poincaré modes are dynamically similar to the dispersive waves in a shallow rotating ocean of constant depth. The free surface elevation ζ can be found from (9.7.9) and (9.7.14) with the dispersion relation (9.7.24) in the form

$$\zeta\,(x, y, t) = \operatorname{Re} C_1\left[\cos\left(\frac{n\pi y}{d}\right) - \left(\frac{fkd}{n\pi\omega}\right)\sin\left(\frac{n\pi y}{d}\right)\right]e^{i(kx-\omega t)}. \tag{9.7.27}$$

Similarly, the explicit representation of the velocity field can be obtained from the basic equations of motion.

The second important class of solutions corresponds to the case of small but non-zero γ, and has a frequency $\omega = O\,(\gamma)$. The associated dispersion relation is obtained from (9.7.20) in the form

$$\omega^2 - c^2\left(\frac{kf\gamma}{\omega d}\right) - \left(c^2k^2 + f^2\right) - \omega_n^2 = 0. \tag{9.7.28}$$

Since $\omega = O\,(\gamma)$, the first term of this equation is small and hence, negligible, while the second term is $O\,(1)$. Consequently, (9.7.28) reduces to

$$\omega = -\gamma\left(\frac{f}{d}\right)\frac{c^2k}{\left(c^2k^2 + f^2 + \omega_n^2\right)}, \tag{9.7.29}$$

where $n = 1,\ 2,\ 3, \ldots$.

These new waves with the dispersion relation (9.7.29) are called the *topographic Rossby waves* which exist only when both rotation f and slope γ are non-zero. These waves have several remarkable properties. First, in contrast to the Poincaré waves, the frequency of the Rossby waves decreases with increasing wavenumbers. Second, the Rossby wave frequency attains its maximum value, $\omega_{\max}$, at $k = k_n$ given by

$$\omega = \omega_{\max} = -\left(\frac{\gamma c}{2d}\right)\frac{f}{\left(f^2+\omega_n^2\right)^{\frac{1}{2}}}, \qquad k = k_n. \qquad (9.7.30\text{ab})$$

Third, for small γ, the frequency of the Rossby waves is always less than f. Thus, these waves have very low frequencies, that is, they have a very large period compared to a rotation period. Another striking feature of the Rossby waves is that their phase velocity in the x-direction is always negative and has the value

$$c_p = \frac{\omega}{k} = -\left(\frac{\gamma f}{d}\right)\frac{1}{\left(c^2k^2+f^2+\omega_n^2\right)}. \qquad (9.7.31)$$

The group velocity of the waves is

$$c_g = \frac{\partial\omega}{\partial k} = \left(\frac{\gamma f c^2}{d}\right)\left[\frac{f^2+\omega_n^2-c^2k^2}{\left(c^2k^2+f^2+\omega_n^2\right)^2}\right]. \qquad (9.7.32)$$

Neglecting all small terms of $O(\gamma)$, the free surface elevation is obtained in the form

$$\zeta(x,y,t) = \text{Re}\; C_2 \;\sin\, \alpha y\; e^{i(kx-\omega t)}. \qquad (9.7.33)$$

Similarly, the velocity fields for these waves can be obtained.

The second solution of the eigenvalue equation (9.7.18) is the dispersion relation $\omega^2 = c^2k^2$ which describes the plane waves whose crests are parallel to the y-axis in a non-rotating ocean. The wave modes with the frequency $\omega = \pm ck$ are called the *Kelvin waves* which propagate in both positive and negative x-directions. One of the remarkable features of the *Kelvin waves* is that the dispersion relation is independent of rotation, and can be derived from the dispersion equation for the Poincaré waves with $n = 0$ and $f = 0$.

We next obtain the dynamical fields for the Kelvin waves traveling in both the positive and the negative x-directions with $\omega = \pm ck$. With $\gamma = 0$, (9.7.15), yields $\alpha^2 = -f^2/c^2$, that is $\alpha = if/c$. Thus, the dynamical fields for the Kelvin waves traveling in the positive x-direction are given by

$$\zeta = \zeta_0 \;\exp\left(-\frac{y}{R}\right)\;\exp\;[ik(x-ct)], \qquad (9.7.34)$$

$$u = \frac{g\zeta_0}{c}\;\exp\left(-\frac{y}{R}\right)\;\exp\left[ik(x-ct)\right], \qquad (9.7.35\text{a})$$

$$= -\frac{g}{f}\frac{\partial\zeta}{\partial y}, \qquad (9.7.35\text{b})$$

$$v = 0, \quad \text{by (9.6.7)}, \tag{9.7.36}$$

where the Rossby deformation radius R represents the decay scale.

There are several striking features of these solutions. First, the wave amplitude decays exponentially from a maximum value at the wall $y = d$ as y decreases to zero. Thus, the existence of the Kelvin waves requires at least one boundary. Second, the flow in the x-direction is in precise geostrophic balance, and the equations of motion assume the form

$$\frac{\partial u}{\partial t} = -g\,\frac{\partial \zeta}{\partial x}, \tag{9.7.37}$$

$$\frac{\partial \zeta}{\partial t} = -h_0\,\frac{\partial u}{\partial x}, \tag{9.7.38a}$$

$$= -\frac{g}{f}\frac{\partial \zeta}{\partial y}. \tag{9.7.38b}$$

These lead to the classical wave equation

$$\frac{\partial^2 \zeta}{\partial t^2} = c^2\,\frac{\partial^2 \zeta}{\partial x^2} \tag{9.7.39}$$

which also gives the dispersion relation $\omega^2 = c^2 k^2$.

Third, the cross-channel velocity component is identically zero. Fourth, the low frequency ($\omega \ll f$), Kelvin waves imply the result $kR = kc/f = \omega/f \ll 1$. This means that the wavelength in the x-direction is much greater than the decay scale R. This anisotropy is responsible for the non-geotrophic balance (9.7.37). On the other hand, the geostrophic balance (9.7.38b) produces a free-surface slope to balance the Coriolis acceleration due to u. Finally, it is noted that $R \to \infty$ as $f \to 0$. In this limit, the Kelvin wave becomes the $n = 0$ mode of the Poincaré waves with the dispersion relation (9.7.25).

The third solution of (9.7.18) corresponds to the dispersion relation $\omega^2 = f^2$. This case apparently represents oscillations with the inertial frequency $\omega = \pm f$. At these frequencies, the differential operator D involved in (9.6.19)–(9.6.20) is identically zero and so it cannot be inverted to determine u and v in the terms of ζ. In order to resolve this difficulty, and to pursue this case further, we recall the original equations (9.6.6)–(9.6.7) with $\boldsymbol{\tau} = 0$ and $\omega = f$, and seek solutions in the form

$$(u, v, \zeta) = \text{Re}\left[\overline{u}\,(y)\,,\ \overline{v}\,(y)\,,\ \overline{\zeta}\,(y)\right]\ \exp\left[i\,(kx - \omega t)\right], \tag{9.7.40abc}$$

where $\overline{u}$, $\overline{v}$ and $\overline{\zeta}$ are the complex amplitudes of u,v and ζ respectively.

Consequently, equation (9.6.6) and (9.6.7) become

$$f\,(\overline{u} - i\,\overline{v}) = gk\,\overline{\zeta}, \tag{9.7.41}$$

$$f\,(\overline{u} - i\,\overline{v}) = -g\,\frac{d\overline{\zeta}}{dy}. \tag{9.7.42}$$

Obviously, these equations cannot be solved to find $\overline{u}$ and $\overline{v}$ in terms of ζ. However, we subtract them to obtain a single equation for ζ

$$\frac{d\overline{\zeta}}{dy} + k\,\overline{\zeta} = 0, \tag{9.7.43}$$

which gives the solution

$$\overline{\zeta} = \zeta_0\, e^{-ky}, \tag{9.7.44}$$

where ζ_0 is constant.

Using (9.7.41)–(9.7.42) with $\omega = f$, and constant h_0, equation (9.6.8) can be transformed into the form

$$\frac{d\overline{v}}{dy} - k\,\overline{v} = \frac{if\zeta_0}{h_0}\left(1 - k^2R^2\right)e^{-ky}. \tag{9.7.45}$$

The solution of this equation with the boundary condition $\overline{v} = 0$ is given by

$$\overline{v}(y) = \left(\frac{if\zeta_0}{kh_0}\right)\left(1 - k^2R^2\right)\,\sinh\,ky. \tag{9.7.46}$$

This solution does not satisfy the other boundary condition $\overline{v} = 0$ on $y = d$ unless $kR = 1$. Thus when $k = R^{-1}$, $\overline{v}$ is identically zero and $\omega = ck = f$. This means that the wave mode cannot be distinguished from the Kelvin waves at the wavenumber $k = R^{-1}$. So there exists a new wave mode at the inertial frequency $\omega = f$.

Thus, the above analysis reveals that the complete spectrum of the solutions consists of the Poincaré waves, the Kelvin waves and the Rossby waves, including the $\omega = 0$ mode which corresponds to the geostrophic flow. The dispersion curves of these waves is shown in Figure 9.3, which is a graph of the nondimensional frequency $\omega^* = (\omega/f)$ as a function of the nondimensional wavenumber $\kappa^* = kR$.

Finally, it is possible to obtain dispersive plane wave solution of (9.6.21) in the form

$$\zeta = \zeta_0\,\exp\left[i\,(kx + \ell y - \omega t)\right], \tag{9.7.47}$$

where ζ_0 is a constant amplitude, and then put it into (9.6.21).

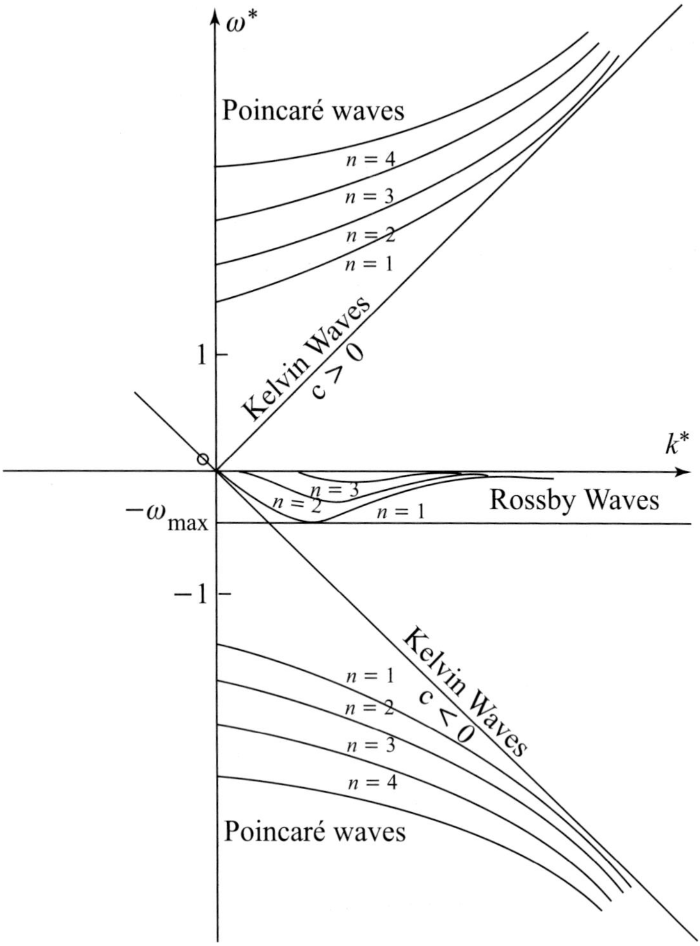

Fig. 9.3 Dispersion diagram for Poincaré, Kelvin and Rossby waves.

This gives the famous dispersion relation (9.7.2) for Poincaré waves

$$\omega^2 = f^2 + c^2\kappa^2, \qquad \kappa^2 = \left(k^2 + \ell^2\right). \tag{9.7.48}$$

The phase velocity of the wave is

$$c_p = \frac{\omega}{\kappa} = \pm\left(c^2 + \frac{f^2}{\kappa^2}\right)^{\frac{1}{2}}. \tag{9.7.49}$$

In the absence of rotation ($f = 0$), the free surface elevation satisfies the well known wave equation. The solution of this equation represents the non-dispersive shallow water waves which propagate with the constant phase velocity c. However, the presence of rotation is fully responsible for the dispersion. And the dispersive waves propagate with the phase velocity c_p which is greater than c.

9.8 Dynamic Response of the Indian Ocean to the Southwest Monsoon

The regular fluctuation in surface currents between summer and winter in the northern part of the Indian Ocean was observed by mariners for over one thousand years and is usually shown in all good atlases. Normally in June, soon after the onset of the Southwest Monsoon, when winds north of the Equator go into reverse and an extensive region of strong southwest wind comes into existence, there is an appropriate reversal of most current directions in this region. This phenomenon is usually regarded as related to the changes in prevailing winds. Important information of the corresponding distributions of current in depth was included during the International Indian Ocean Expedition. Thus, the strongest seasonal variation in the prevailing winds in the northern Indian Ocean is very suitable for studies of the dynamic response of the ocean current patterns to changes in wind stresses associated with the Monsoon onset.

(i) Barotropic Response

With the notation $(G^x, G^y) = \frac{1}{\rho h_0}(\tau^x, \tau^y)$ and h_0 is constant, representing the force per unit mass, we eliminate ζ from (9.6.6) and (9.6.7) by means of (9.6.8) to obtain

$$u_{tt} - f v_t = c^2(u_{xx} + v_{xy}) + G_t^x, \tag{9.8.1}$$

$$v_{tt} + f u_t = c^2(u_{xy} + v_{yy}) + G_t^y, \tag{9.8.2}$$

where $c^2 = \sqrt{gh_0}$ is the velocity of shallow water waves.

From $\frac{\partial^2}{\partial t^2}$ (9.8.2) $-f\frac{\partial}{\partial t}$ (9.8.1), it follows that

$$v_{tttt} + f^2 v_{tt} = c^2\left(u_{xytt} + v_{yytt} - f u_{xxt} - f v_{xyt}\right) + \left(G_{ttt}^y - f G_{tt}^x\right). \tag{9.8.3}$$

Differentiating (9.8.1) with respect to x and y gives

$$\beta v_{tx} = u_{ttyx} - f v_{txy} - c^2\left(u_{xxxy} + v_{xxyy}\right) - G_{txy}^x. \tag{9.8.4}$$

Also, from the expression $\frac{\partial^2}{\partial x^2}$ (9.8.2) $+\frac{\partial^2}{\partial y^2}$ (9.8.2), we obtain

$$\begin{aligned}\nabla^2 v_{tt} = c^2\left(u_{xxxy} + v_{xxyy} + u_{xyyy} + v_{yyyy}\right) - f u_{txx}\\ -(f u_t)_{yy} + G_{xxt}^y + G_{yyt}^y.\end{aligned} \tag{9.8.5}$$

A simple combination of (9.8.3) and (9.8.5)

$$\nabla^2 v_{tt} - \frac{1}{c^2}\left(\frac{\partial^2}{\partial t^2} + f^2\right) v_{tt} + \beta v_{tx} = \frac{1}{c^2}\left(fG^x_{tt} - G^y_{ttt}\right) + \left(G^y_x - G^x_y\right)_{xt}, \tag{9.8.6}$$

where $\nabla^2 = \left(\frac{\partial^2}{\partial x^2} + \frac{\partial^2}{\partial y^2}\right)$ is the horizontal Laplacian.

The linearized barotropic response (u, v) of a homogeneous ocean to an external wind force (G^x, G^y) per unit mass is effectively uniform with respect to depth h_0 and is governed by equations (9.8.1)–(9.8.2). The real meaning of the term *barotropic* is that the pressure is constant on the surfaces of constant density. The above equations for a homogeneous ocean describe the unrealistic feature that currents are distributed uniformly with respect to depth.

Elimination of u from equations (9.8.1)–(9.8.2) or integration of (9.8.6) once with respect to t gives

$$\left[\frac{\partial^2}{\partial t^2} + f^2 - c^2\nabla^2\right] v_t - \beta c^2 v_x = G^y_{tt} - fG^x_t + c^2\left(G^x_{xy} - G^y_{xx}\right). \tag{9.8.7}$$

In the absence of rotation ($f = 0$), this equation describes the well known shallow water waves due to atmospheric disturbances given by the right hand side of (9.8.7). In a uniformly rotating system, ($\beta = 0$), (9.8.7) reduces to the forced long wave equation representing isotropic dispersive waves with a low frequency cut-off below the Coriolis frequency f. On the other hand, in a rotating system with β-effect ($\beta > 0$), there exist waves of far lower frequency. These are dispersive anisotropic Rossby waves.

In order to describe the barotropic response of the homogeneous ocean of constant depth to a wind stress representing the Southwest Monsoon, it is convenient to formulate the problem in terms of the stream function ψ so that $u = -\psi_y$ and $v = \psi_x$ which follow from the approximate continuity equation $u_x + v_y = 0$. Expressing equation (9.8.7) in terms of ψ and then integrating the terms involving the factor c^2 in (9.8.7) with respect to x gives the standard forced Rossby wave equation

$$\nabla^2 \psi_t + \beta\psi_x = G^y_x - G^x_y. \tag{9.8.8}$$

The corresponding homogeneous equation (9.8.8) admits the Rossby wave solutions proportional to $\exp\left[i\left(\boldsymbol{\kappa}\cdot\mathbf{x} - \omega t\right)\right]$, where $\boldsymbol{\kappa} = (k, l)$ with the dispersion relation $\omega\left(k^2 + l^2\right) + \beta k = 0$. Thus, the group velocity traveling westward has magnitude $\beta/\left(k^2 + l^2\right)$ and makes an angle with the eastward direction twice that which makes the wavenumber vector $\boldsymbol{\kappa} = (k, l)$

makes. More generally, when the wind stress curl forcing the Rossby waves is distributed approximately zonally, the energy of the Rossby waves propagates nearly westward. This gives a clear understanding of the westward intensification of the barotropic components of ocean currents. The onset of the Southwest Monsoon in the northern part of the Indian Ocean produces a region of strong negative wind stress curl in a region centered approximately on the Equator. Across the region, the eastward component of wind force G^x changes from its negative value in the south of the Equator to a significantly large positive value in the Monsoon itself in the north of the Equator. Consequently, the barotropic response of the northern part of the Indian Ocean (with no effect of boundaries) to Monsoon consists of waves with k much smaller than l ($|k| \ll |l|$), whose energy moves westward with velocity β/l^2. These waves leave behind a southward motion determined by the steady form of the Sverdrup equation balanced by an opposite northward flow which the waves carry progressively farther towards west direction.

We denote the solution of equation (9.8.8) in an unbounded ocean by ψ_U, which is very useful for determining the solution ψ_B for a bounded ocean so that $\psi = \psi_U + \psi_B$. Since both ψ and ψ_U satisfy the forced Rossby wave equation (9.8.8), it follows by subtraction that ψ_B satisfies the free Rossby wave equation. In fact, the solution ψ_B represents the reflection of the wave group from the coast and satisfies the boundary condition

$$\psi_B = -\psi_U \text{ on the boundary.} \tag{9.8.9}$$

This means that there is no barotropic flow across the boundary.

In order to determine the oceanic response to a pattern of wind stress curl $\left(G_x^y - G_y^x\right)$, the Fourier analysis in terms of the y component of the wavenumber, l is very useful because the barotropic response consists mainly of waves with the x component of the wavenumber k much smaller than l. Consequently, the waves generated are Fourier analyzed only with respect to y by introducing the Fourier transforms

$$(\Psi, \Phi) = \int_{-\infty}^{\infty} e^{-ily} \left(\psi_U, \psi_B\right) dy. \tag{9.8.10}$$

Since our main interest is in the response to the changes in wind stress curl associated with the Monsoon onset, we denote $W(x, l; t)$ for the Fourier transform of the difference between the value of $G_x^y - G_y^x$ at some time t and its values at a time $t = 0$ just before the Monsoon onset. In view of the Fourier transform, equation (9.4.9) reduces to the form

$$\frac{\partial}{\partial t}\left(\Psi_{xx} - l^2\Psi\right) + \beta\Psi_x = W(x, l; t). \tag{9.8.11}$$

With the forcing term W vanishing for $t < 0$, we seek the solution that vanishes for $t < 0$ and represents the changes in ocean currents due to changes in wind stress pattern. The Laplace transform with respect to time t is the natural choice for solving the transient problem. Application of the Laplace transform to (9.8.11) gives

$$\overline{\Psi}_{xx} + \left(\frac{\beta}{s}\right)\overline{\Psi}_x - l^2\,\overline{\Psi} = \frac{W}{s}, \tag{9.8.12}$$

where $\overline{\Psi}(x, l; s)$ is the Laplace transform of $\psi(x, l; t)$ and W is independent of s, representing step function onset of the Monsoon.

The complementary function of equation (9.8.12) is

$$\overline{\Psi} = Ae^{\lambda_1 x} + Be^{\lambda_2 x}, \tag{9.8.13}$$

where λ_1 and λ_2 are given by

$$\lambda_1 = -\frac{\beta}{2s} + \left(l^2 + \frac{\beta^2}{4s^2}\right)^{\frac{1}{2}}, \qquad \lambda_2 = -\frac{\beta}{2s} - \left(l^2 + \frac{\beta^2}{4s^2}\right)^{\frac{1}{2}}. \tag{9.8.14ab}$$

The complete solution of (9.8.12) in an unbounded ocean is

$$\overline{\Psi} = \frac{1}{s\,(\lambda_2 - \lambda_1)}\left[e^{\lambda_1 x}\int_x^{\infty} W(z)e^{-\lambda_1 z}dz + e^{\lambda_2 x}\int_{-\infty}^{x} W(z)e^{-\lambda_2 z}dz\right]. \tag{9.8.15}$$

For determining the time scale of the barotropic response, an asymptotic analysis of the solution for small s is needed. In the limit $s \to 0$, λ_1, λ_2 are approximately equal to

$$\lambda_1 \sim s\left(\frac{l^2}{\beta}\right), \qquad \lambda_2 \sim -\frac{\beta}{s} - \left(\frac{l^2}{\beta}\right)s \approx -\frac{\beta}{s}. \tag{9.8.16ab}$$

Also the limiting form of the solution for $\overline{\Psi}$ as $s \to 0$ is

$$\overline{\Psi} = -\frac{1}{\beta}\left[\int_x^{\infty} \exp\{\lambda_1(x - z)\}\,W(z)\,dz + \int_{-\infty}^{x} \exp\{\lambda_2(x - z)\}\,W(z)\,dz\right], \tag{9.8.17}$$

where λ_1 and λ_2 are given by (9.8.16ab).

The second term of this solution seems likely to be gradually insignificant as $s \to 0$ and in fact, it tends to zero as $s \to 0$. Consequently, the first term in (9.8.17) constitutes the solution whose inverse Laplace transform gives

$$\Psi = -\frac{1}{\beta}\int_x^{\infty} W(z)H\left[t - \left(\frac{l^2}{\beta}\right)(z - x)\right]dz = -\frac{1}{\beta}\int_x^{x+\frac{\beta t}{l^2}} W(z)dz, \tag{9.8.18}$$

where $H(t)$ is the Heaviside unit step function.

Thus, Ψ_x, the Fourier component of the northward velocity v which has north-south wave number l, is given by

$$\Psi_x = \frac{W(x)}{\beta} - \frac{1}{\beta} W\left(x + \frac{\beta t}{l^2}\right). \tag{9.8.19}$$

The first term of this solution represents a steady local response related to the wind stress curl by classical Sverdrup's law. The second term describes an equal and opposite response propagated to the west with velocity (β/l^2) which is the group velocity of the wave groups with east-west wavenumber k small compared with north-south wavenumber l.

It is noted that the oceanic response to the step-function onset of Monsoon is obtained. However, the dynamic response to wind stresses varying more generally with time can be determined by means of the convolution theorem for the Laplace transform.

For further investigation of the solution, it is important to determine ψ_B satisfying the free Rossby wave equation (9.8.8) and the boundary condition (9.8.9). The condition $\psi_B = -\psi_U$ will specify non-zero values for ψ_B only on the western boundary. As an additional simplification, it is convenient to assume that the western boundary is a meridian whose equation without loss of generality may be taken as $x = 0$. Thus, the Fourier-Laplace transform $\overline{\Phi}$ of ψ_B satisfies the equation (9.8.11) with zero on the right hand side and the boundary condition

$$\overline{\Phi} = -\overline{\Psi} \quad \text{on } x = 0. \tag{9.8.20}$$

With a step-function change in $(\Psi)_{x=0}$, the bounded solution for $\overline{\Phi}$ is given by

$$\overline{\Phi} = -\frac{1}{s}(\Psi)_{x=0} \exp(\lambda_2 x), \qquad \lambda_2 = -\beta s^{-1}. \tag{9.8.21}$$

It is noted that the asymptotic form of the inverse Laplace transform of $s^{-1}\exp(\lambda_2 x)$ is $J_0\left(2\sqrt{\beta x t}\right)$ for large t. This falls from 1 at $x = 0$ to zero at $x = (1.4)(\beta t)^{-1}$ in a progressively thinning boundary layer outside which it oscillates about zero and its gradients are much less. Also the time taken for the thickness of the boundary current to be reduced to a typical value 100 km is about $\frac{1.4}{\beta x} = 6 \times 10^5$ sec $= 1$ week. It is inferred from the above that the solution must be independent of l so that ψ_B and ψ_U possess a relationship similar to their Fourier transforms Φ and Ψ respectively.

It follows from the inverse Laplace and Fourier transformations that

$$\psi_B = -\psi_0 J_0\left[2\left\{\beta x(t - t_0)\right\}^{1/2}\right], \tag{9.8.22}$$

where ψ_0 represents a change in $(\psi_U)_{x=0}$ at time $t = t_0$.

Within the thining boundary current (once its thickness is small compared with any east-west distances in which ψ_U changes significantly), ψ_0 must be added to this solution to obtain the total contribution to the stream function $\psi = \psi_U + \psi_B$ so that

$$\psi \sim \psi_0 \left\{1 - J_0 \left[2 \left\{\beta x(t - t_0)\right\}^{1/2}\right]\right\}. \qquad (9.8.23a)$$

Thus, all the flux which are propagated to the western boundary piles up there in a continually thinning boundary current. In fact, the entire flux is confined to a current width $(1.4)[\beta(t - t_0)]^{-1}$ when $t - t_0$ is very large. There is a time lag of about $t - t_0 = \frac{1.4}{\beta x} = 1$ week before this width is reduced to a typical value 100 km characteristic of the Somali Current or other western boundary currents. This time lag must be added to the propagation time lag required for the main northward flow signal to arrive at the western boundary.

To the next approximation $\lambda_2 \sim -\frac{\beta}{s} - \frac{sl^2}{\beta}$ which introduces an additional exponential factor $\exp\left(-\frac{sl^2 x}{\beta}\right)$ into (9.8.17). In the physical space, this factor induces a change in the argument of the Bessel function to give

$$\psi \sim \psi_0 \left\{1 - J_0 \left[2 \left\{\beta x \left(t - t_0 - \frac{xl^2}{\beta}\right)\right\}^{1/2}\right]\right\}, \qquad t \to \infty. \qquad (9.8.24b)$$

It has already been shown that to the first approximation, the time taken for the thickness of the boundary current to be reduced to 100 km is $\frac{1.4}{\beta x} = 6 \times 10^5$ sec $= 1$ week. To the next approximation, the corresponding time scale is $\frac{1.4}{\beta x} + \frac{xl^2}{\beta}$ which shows a small change by an amount $\frac{xl^2}{\beta} = 1.1 \times 10^5$ sec for the median wave number $l = 1/(200$ km$)$. This confirms that the variation in the boundary response with l is only small after a week. Once the boundary current has become thin, the second approximation produces a small correction.

(ii) Baroclinic Response

Lighthill presented his study of the baroclinic modes of propagation in the ocean with special attention to the response in the first baroclinic mode. The word 'baroclinic' means that the pressure is not constant on surfaces of constant density. He formulated the linear problem without any particular baroclinic mode(s) and hence, no subscripts are placed on u, v, h, etc., to signify which mode is being considered. Most of the theory to be discussed can be applied equally to higher baroclinic modes. Like the barotropic response, the study is based on equation (9.8.7) with an important fact that h is much smaller in the baroclinic case so that the term $f^2 v_t$ in (9.8.7) plays a far more significant role. In fact, this term is almost indispensable for the study of the baroclinic modes of propagation near the equator. With f^2 and β as constants, such wave motions are governed by the equation (9.8.7) with zero right hand side. The plane wave solution of this equation, proportional to $\exp\left[i\left(\boldsymbol{\kappa}\cdot\mathbf{r}-\omega t\right)\right]$, $\boldsymbol{\kappa}=(k,l)$ satisfies the dispersion equation

$$-\omega^3+\omega\left(c^2\kappa^2+f^2\right)+\beta k c^2=0. \tag{9.8.25}$$

Two cases, $\omega>f$ and $\omega<f$ are involved in the solution of (9.8.25). In the former case, the last term in (9.8.25) is completely negligible and hence, the solution is given by

$$\omega^2=f^2+c^2\kappa^2. \tag{9.8.26}$$

This shows only a moderate correction to the inertial frequency except at high wavenumbers. However, this case is not of interest in the dynamics of ocean waves and currents.

In the latter case, as the frequency ω decreases below the inertial frequency f, there is an extensive gap without any solution until $\omega<\frac{c\beta}{2f}$. When this condition is satisfied, the first term in (9.8.25) can be neglected to obtain

$$\omega\left(\kappa^2c^2+f^2\right)+\beta k c^2=0, \tag{9.8.27}$$

or

$$\left(k+\frac{\beta}{2\omega}\right)^2+l^2=\left(\frac{\beta}{2\omega}\right)^2-\frac{f^2}{c^2}. \tag{9.8.28}$$

This represents a circle in the (k,l) plane with center $\left(-\frac{\beta}{2\omega},0\right)$ and radius $\left\{\left(\frac{\beta}{2\omega}\right)^2-\frac{f^2}{c^2}\right\}^{\frac{1}{2}}$.

Wind stresses of nearly zonal distribution preferentially generate waves with $|k|\ll|l|$ which is satisfied on the circle only if its radius tends to $\frac{\beta}{2\omega}$.

It also follows from (9.8.27) that

$$\omega=-\frac{\beta k}{\kappa^2+a^2}, \qquad a^2=\frac{f^2}{c^2}, \tag{9.8.29}$$

or

$$c_p=\frac{\omega}{k}=-\frac{\beta}{\kappa^2+a^2}. \tag{9.8.30}$$

Thus, the wave pattern moves towards west with velocity c_p. The group velocity is given by

$$\mathbf{c}_g = \frac{\beta}{(\kappa^2 + a^2)^2} \left\{ \hat{k} \left(k^2 - l^2 - a^2 \right) + 2\hat{l}\, kl \right\}, \tag{9.8.31}$$

$$c_g \sim -\frac{\beta}{(l^2 + a^2)} \qquad \text{provided } |k| \ll |l|. \tag{9.8.32}$$

This means that the group velocity is approximately westward with magnitude given by

$$c_g = \frac{\beta}{l^2 + a^2} < \frac{\beta}{a^2}. \tag{9.8.33}$$

Evidently, the baroclinic waves generated by wind stresses with a predominantly zonal distribution have group velocity directed westward with magnitude c_g given by (9.8.32) which, as l decreases, makes a transition from the barotropic value $\left(\beta/\, l^2\right)$ to the constant maximum value $\left(\beta/\, a^2\right)$.

In a classic paper, Veronis and Stommel (1956) studied baroclinic response in regions away from the Equator that is expected to be far slower. This is possibly because the effective depth for baroclinic wave modes is at most about 1 meter which means that the second and the fourth terms in equation (9.8.7) are more important compared with the first and the second terms so that equation (9.8.7) takes the approximate first order form

$$f^2 v_t - \beta c^2 v_x = \left(G_{tt}^y - f G_t^x\right) + c^2 \left(G_{xy}^x - G_{xx}^y\right). \tag{9.8.34}$$

All solutions of this equation propagate westward with a uniform velocity $\left(\beta/\, a^2\right)$, and are called the *Veronis and Stommel non-dispersive baroclinic Rossby waves*.

Lighthill made a more satisfactory study of baroclinic waves near the Equator based on the variation of f with β constant so that f is replaced by βy in (9.8.7). Consequently, (9.8.7) becomes

$$\left(\frac{\partial^2}{\partial t^2} + y^2\beta^2 - c^2\nabla^2\right) v_t - \beta c^2 v_x = \left(G_{tt}^x - \beta y G_t^x\right) + c^2 \left(G_{xy}^x - G_{xx}^y\right). \tag{9.8.35}$$

This equation was solved for a wind-stress distribution representing the Southwest Monsoon after finding the wave modes proportional to $\exp\left(ikx - i\omega t\right)$ of the corresponding homogeneous equation (9.8.35). These waves modes satisfy the equation

$$v_{yy} + \left(\frac{\omega^2}{c^2} - k^2 - \frac{\beta k}{\omega} - \frac{\beta^2 y^2}{c^2}\right) v = 0. \tag{9.8.36}$$

A solution of (9.8.36) is confined to near the Equator (rather than exponentially large solutions as y tends to either $+\infty$ or $-\infty$ or both) only if

$$\frac{\omega^2}{c^2} - k^2 - \frac{\beta k}{\omega} = (2M+1)\,\frac{\beta}{c}, \tag{9.8.37}$$

where $M = 0, 1, 2, 3, \cdots$. The associated solutions of (9.8.36) represent the parabolic cylinder functions. In fact, equations (9.8.36) and (9.8.37) were obtained by Blandford (1966) in his study of baroclinic modes of wave propagation. He noted a peculiarity of the mode $M = 0$ which propagate westward with velocity c, but did not observe the fact that this mode is not really possible. This is because, although a solution for v that is confined near the Equator exists for $M = 0$, equations (9.8.1) and (9.8.2) reveal that no such solution for u exists. For $M = 0$ mode case, the solution for u must be exponentially large away from the Equator.

Thus, the dispersion relation (9.8.37) admits only true wave modes confined near the equator for $M = 1, 2, 3, \cdots$. Each mode has a westward group velocity that is maximum when east-west wavenumber k is small compared with $\sqrt{\beta/c} \geq (1/370\ km)$ as is for the case for forcing by the Monsoon type disturbance. In such a case, there exists a non-dispersive westward wave mode moving with a velocity $(\omega/k) = -c/(2M+1)$. This is possible when the first two terms in (9.8.35) are neglected (that is, the corresponding terms v_{ttt} and v_{txx} in (9.8.35) are ignored). Therefore, a first wave mode from any disturbance will arrive at a speed $\frac{1}{3}c$ corresponding to $M = 1$ which with the estimated depth, 0.75 m obtained for the first baroclinic mode $n = 1$ in the equatorial Indian Ocean is 0.9 m/s. This speed is in agreement with that of the barotropic mode. The second mode reaches at a velocity $\frac{1}{5}c \approx 0.55\ m/s$ and so on.

Especially, in an equatorial ocean, the above description needs modification in the dispersion relation (9.8.29) and the Veronis and Stommel equation (9.8.34) with zero right hand side reduces to

$$f^2 v_t - c^2 \beta v_x = 0 \tag{9.8.38}$$

which becomes complicated due to local rates of vorticity (mainly, of its u_y component) in the mid-latitude ocean, these are important only when the wavenumber l is at least of order $(f/c) = 1/(30\ km)$. Near the Equator, this critical wavenumber (f/c) assumes much smaller values, and the u_y component of vorticity must be taken into account in all cases. Moreover, the variation of f makes large changes in the group velocity (9.8.32) over a distance even as small as l^{-1} so that the variability of the coefficient of

f^2 in (9.8.7) needs to be included in the analysis. This discussion leads to the representation of (9.8.7) near the Equator by (9.8.35), which, when propagation in regions without local forcing effects, takes the form

$$\left(\frac{\partial^2}{\partial t^2} + y^2\beta^2 - c^2\nabla^2\right) v_t - \beta c^2 v_x = 0. \tag{9.8.39}$$

It follows from a simple dimensional analysis of (9.8.39) that baroclinic propagation near the Equator takes place on fundamental length scale, $\sqrt{c/2\beta}$ and time scale $(2c\beta)^{-\frac{1}{2}}$. Using these scales, we make the equation nondimensional so that (9.8.39) becomes

$$\left(\frac{\partial^2}{\partial t^2} + \frac{1}{4}y^2 - \nabla^2\right) v_t - \frac{1}{2}v_x = 0. \tag{9.8.40}$$

This equation admits wave-like solutions that are concentrated near the Equator in the form

$$v = \exp\left[i\left(kx - \omega t\right)\right] D_M(y), \tag{9.8.41}$$

where the dispersion relation is

$$\left(\omega^2 - k^2 - \frac{k}{2\omega}\right) = M + \frac{1}{2}, \tag{9.8.42}$$

$M = 0,\ 1,\ 2,\ 3,\ \cdots$, and $D_M(y)$ is the parabolic cylinder function given by

$$D_M(y) = \left[\exp\left(\frac{y^2}{4}\right)\right]\left(-\frac{d}{dy}\right)^M \left[\exp\left(-\frac{y^2}{2}\right)\right]. \tag{9.8.43}$$

This function $D_M(y)$ is the only solution of the second order ordinary differential equation

$$w'' + \left(M + \frac{1}{2} - \frac{1}{4}y^2\right) w(y) = 0, \tag{9.8.44}$$

which tends to zero as $|y| \to \infty$.

The dispersion relation (9.8.42) can be plotted in the (k, ω) plane for different values of $M = 0,\ 1,\ 2,\ \cdots$. In particular, when $M = 0$, the dispersion relation gives two separate solution curves

$$\omega + k = 0 \qquad \text{and} \qquad k = \omega - \frac{1}{2\omega}. \tag{9.8.45}$$

The former has the group velocity $\frac{d\omega}{dk} = -1 < 0$ (westward) leading to the propagation of energy of baroclinic waves to produce western boundary currents. On the other hand, the latter gives exclusively positive group velocity $\frac{d\omega}{dk} = \frac{\omega^2}{1+2\omega} > 0$ (eastward).

On the other hand, for the higher baroclinic modes $M = 1, 2, 3, \cdots$, there exist waves with both k and ω small, represented by points near the origin ($\omega = 0,\ k = 0$) where the curves are nearly straight lines with slope

$$\frac{\omega}{k} = -\frac{1}{2M+1}. \tag{9.8.46}$$

This shows that non-disersive waves propagate westward with non-dimensional phase velocity $1/(2M + 1)$, or with dimensional velocity $c/(2M+1)$.

In the first approximation, k^2 and ω^2 are both small compared with 1 (or in practice, both $|k|$ and $|\omega|$ are less than 0.4). This implies that k^2 and ω^2 can be neglected in (9.8.42) which in turn means neglect of terms v_{ttt} and v_{txx}, but not v_{tyy} in (9.8.39) so that it becomes

$$-c^2 v_{tyy} + \beta^2 y^2 v_t - \beta c^2 v_x = 0, \tag{9.8.47}$$

where the second and third terms can be interpreted in terms of the absolute vorticity conservation law described before for the case f constant, while the first term represents the effects of shear, that is, of additional vorticity dues to changes in the gradient of eastward velocity u in the north-south direction. Equation (8.9.47) admits the same fundamental solutions (9.8.41) as does (8.9.39), the associated dispersion relation is simplified to (8.9.46). There is now no solution for $M = 0$, which has $\omega + k = 0$ as only dispersion relation in this approximation, leading to are exponentially unbounded solution for u as $|y| \to \infty$

In order to determine the velocity component u from equation (9.8.41) for v, it is necessary to use (9.8.1) and (9.8.2) which, for propagation in regions without forcing terms, in non-dimensional form (using the fundamental length and time scales and f replaced by βy as before),

$$u_{tt} - u_{xx} = v_{xy} + \frac{1}{2}\, y\, v_t, \tag{9.8.48}$$

$$u_{xy} - \frac{1}{2}\, y\, u_t = v_{tt} - v_{yy}. \tag{9.8.49}$$

If $\omega^2 \neq k^2$, (8.9.48) leads us to derive that, corresponding to the mode (9.8.41) for v, there is a solution for u in the form

$$u = \left(k^2 - \omega^2\right)^{-1} \exp\left[i\,(kx - \omega t)\right] \left[ikD'_M(y) - \frac{1}{2}\, i\, \omega\, y\, D_M(y)\right]. \tag{9.8.50}$$

It turns out from (8.9.49) that solution for u with $u \to \infty$ as $y \to -\infty$, is given by

$$u = \left(\frac{i}{k}\right) \exp\left[ik(x+t) + \frac{1}{4}y^2\right] \int_{-\infty}^{y} \exp\left(-\frac{1}{4}y^2\right) \left[D''_0(y) + k^2 D_0(y)\right] dy \tag{9.8.51}$$

where $D_0(y) = \exp\left(-\frac{1}{4}y^2\right)$.

We return to the non-homogeneous equation (9.8.35) for wave motion by expanding the forcing term as a function of y in a series of normal mode solutions of (9.8.36), that is, the solutions with the dispersion relation (9.8.37) for $M = 0, 1, 2, \cdots$. Similarly, expanding the velocity component v in a series leads to the solution for $M = 0$ which does not produce any westward propagation mode. However, each solution for $M = 1, 2, 3, \cdots$ produce a westward propagating wave mode.

Finally, the first two wave modes are obtained by substituting for $D_1(y)$ and $D_2(y)$ in (8.9.51) and (8.9.41) in the form

$$M = 1: \ (u, v) = \left[\left(\frac{3i}{8k}\right)\left(3 - y^2\right), y\right] \exp\left[ik\left(x + \frac{1}{3}t\right) - \frac{1}{4}y^2\right], \tag{9.8.52}$$

$$M = 2: \ (u, v) = \left[\left(\frac{5i}{12k}\right)\left(6y - y^3\right), y^2 - 1\right] \exp\left[ik\left(x + \frac{1}{5}t\right) - \frac{1}{4}y^2\right]. \tag{9.8.53}$$

Further studies reveal that the initial baroclinic response of the western equatorial region of the Indian Ocean to monsoon onset is particularly influenced by these two baroclinic modes of lowest order (8.9.52) and (8.9.53). Some observations shows that the equatorial baroclinic modes not only are propagated fast enough to account for the rate of response of the Indian Ocean, but also can be sufficiently generated by monsoon winds to account for much of the amplitude of response near the western boundary. Lighthill also presented a simplified account of the linear theory of long ocean waves in a horizontally stratified ocean of uniform depth, with turbulent momentum transfer neglected. He then formulated a more general Sturm-Liouville problem for the nth baroclinic mode of variation of (u, v) with depth z, namely $c_0(z)$. It is shown that there exist both barotropic $c_0(z)(n = 0)$, and baroclinic modes $c_0(z)(n = 1, 2, 3, \cdots)$ which change sign just n times between $z = 0$ and $z = H^\circ$ (total depth). He then applied his general theory to describe baroclinic response in the Indian Ocean near the Equator.

In general, real oceans possess stratification of both density and current velocity. A linearized model of a real ocean is one based on small perturbations to a stationary ocean with horizontal density stratification, that is, the density depends only on the vertical coordinate z. The distribution of the horizontal velocity (u, v) with respect to depth can be expanded as a sum of normal modes, each with its own characteristic distribution proportional to $c_n(z)$, where $n = 0, 1, 2, 3, \cdots$. The coefficients of each of these functions $c_n(z)$ in (u, v) satisfy (9.8.1) and (9.8.2) provided that the depth h_0 is replaced by an effective depth h_{0n} characteristic of the mode. The value $n = 0$ corresponds to the barotropic mode with $c_0(z) = 1$, that is, currents are the same at all depths and h_0 is the actual depth for the homogeneous ocean. The values $n = 1, 2, 3, \cdots$ correspond to the *baroclinic modes*. However, the first baroclinic mode $n = 1$ is the most significant baroclinic response and the effective depth in this mode is usually of the order 1 meter. The surface stress can effectively force surface currents through the first baroclinic mode about 14 times as strongly as through the barotropic mode $n = 0$. For higher baroclinic modes, the corresponding coefficient is much smaller than even the barotropic value 1 and the effective depths h_{0n} are much smaller than for the first mode $n = 1$. In fact, the response of the northern Indian Ocean in the higher baroclinic modes is less than 0.3 times as rapid as the first mode. So, the response is quantitatively irrelevant to any description of surface-current response. In summary, the wind-stress pattern associated with the onset of the Southwest Monsoon generates both the barotropic and the baroclinic modes of dynamic response in the Indian Ocean. However, the barotropic mode $n = 0$ and the first baroclinic mode seem to be more important compared with higher order baroclinic modes $n = 2, 3, \cdots$.

Although linearized theory of Lighthill and others provided new results and information about the barotropic and baroclinic waves and propagation of current patterns in oceans, it is relevant to mention possible nonlinear effects on the dynamics of oceans. Since governing equations and/or free surface conditions are nonlinear, nonlinear effects should play significant role in the dynamics of oceans, in general and in the Western boundary current, in particular. However, nonlinear effects may be less important because typical current velocities are usually not very large compared with the speeds of propagation of baroclinic response, which are found to be much higher near the Equator than elsewhere.

9.9 Concluding Remarks

In recent years, considerable attention has been given to linear and nonlinear Rossby waves by Benney (1979), Gill (1974, 1976), Debnath (1976a,b; 1984, 2000), Debnath and Kulchar (1982), Lighthill (1969), Derzho and Grimshaw (2005). It has been shown that Rossby waves represent a fundamental component of planetary scale oceanic and atmospheric flows. It has been shown by Benney (1979) that the finite amplitude long steady Rossby waves riding on a background flow with a weak shear are described by a generalized Korteweg-de Vries (KdV) equation. This equation contains linear dispersive term, but the nonlinear term depends on the specific shear profile. Benney's analysis also included finite amplitude short waves corresponding to a particular mode. In this case, a slowly varying wavetrain of such waves satisfies the nonlinear Schrödinger equation. The quantity which governs stability was also obtained. In this case finite amplitude envelope solitons are expected for unstable waves. On the other hand, Warn (1983) studied weak transient effects into Benney's analysis and showed that the temporal term becomes nonlocal in this case. These studies are based on the assumption that there is no stagnation point or closed streamlines in the flow, that is, there is an implicit limitation on the amplitude of the waves. On the other hand, Derzho and Grimshaw (2005) generalized Benney's work to the case of Rossby waves containing stagnation points in the interior.

Chapter 10

Nonlinear Dispersive Waves

"True Laws of Nature cannot be linear."

Albert Einstein

"The progress of physics will to a large extent depend on the progress of nonlinear mathematics, of methods to solve nonlinear equations ... and therefore we can learn by comparing different nonlinear problems."

Werner Heisenberg

"It seems doubtful whether we can expect to understand fully the instability of fluid flow without obtaining a mathematical representation of the motion of a fluid in some particular case in which instability can actually be observed, so that a detailed comparison can be made between the results of analysis and those of experiment."

G. I. Taylor

10.1 Introduction

Sir James published many deep and comprehensive papers on linear and nonlinear dispersive waves. In particular, he also wrote a deep and comprehensive paper entitled, "Some special cases treated by the Whitham theory" published in the Special Volume A299 (1967) of the *Proceedings of the Royal Society*. This paper is based on Whitham's (1965a,b) general nonlinear theory of one, two and more dimensional dispersive wave propagation in a general conservative system. Lighthill gave an elegant analysis of the implications of Whitham's theory in a restricted class of homogeneous conservative wave systems. Special attention is given not only to hyperbolic, elliptic or parabolic nature of the Whitham fundamental dispersion equation for the averaged Lagrangian $\mathcal{L}$, but also its method of solution in three different cases. When the equation is hyperbolic, the nonlinear group velocity splits into two different velocities. In this case, there is one form of this problem that is particularly appropriate for Whitham's conjecture regarding the possibility of discontinuous solutions. In the elliptic case, there is a pronounced effect of nonlinearity to change the nature of the wave amplitude as well as the wavenumber of a group of waves produced by a wave generating mechanisms. The total energy of the group gradually becomes concentrated within a narrower distance. At the same time, the amplitude at the center of the group, that is, the peak amplitude increases rapidly. As time t reaches $t_{\max}$, the rate of increase of peak amplitude becomes infinite. In other words, the peak of the wave group has become so enhanced that is, actually a *cusp* is formed, so that the gradient, $(\partial k/\partial x)$ of wavenumber has become infinite at the center. However, in the limiting case of infinitesimal amplitude, the Whitham dispersion equation is parabolic and there is only one group velocity in this case.

In his two famous papers, Lighthill (1965, 1967) developed a general theory of nonlinear dispersive waves in homogeneous conservative systems under conditions when the proportional change in amplitude or wavenumber over a distance of one wavelength is very small. In his pioneering work Whitham (1965) developed a new approach to linear and nonlinear dispersive waves in two or more dimensions in a general conservative system. Using Hamilton's principle (that the time integral of the Lagrangian is stationary) in an averaged form, he first derived the equations governing wave dispersion in term of the Whitham averaged Lagrangian $\mathcal{L}$ per unit volume for periodic waves of given frequency and wavenumber vector. In certain wave problems, $\mathcal{L}$ becomes a function of other variables (besides the frequency and wavenumber vector) which are called *pseudo-frequencies* by Whitham. Lighthill described the implications of Whitham's general theory in a restricted class of homogeneous conservative systems. These are one-dimensional wave systems without pseudo-frequencies, where waveform of these systems is defined uniquely by the wavelength and amplitude for a periodic wave.

10.2 Lighthill's Theory of Nonlinear Dispersive Waves

Based on Whitham's (1965a, 1965b) general theory of the effects of nonlinearity on one, two and more dimensional dispersive wave propagation in a general conservative system, Lighthill (1967) gave an elegant analysis of the implications of Whitham's theory in a restricted class of homogeneous conservative wave systems. Under the assumptions that the proportional changes in wavenumber or wave amplitude over a distance of one wavelength is very small, his analysis reveals certain important and interesting features of Whitham's general theory in a restricted class of homogeneous conservative systems.

For any dynamical system in this restricted class, a periodic waveform $u = \phi(\theta)$, $(\theta = \omega t - kx)$ of given wavenumber k and amplitude a, has a given frequency $\omega = \omega(k, a)$. When k and a are known, the average Lagrangian $\mathcal{L}(\omega, k, a)$ per unit length can be determined. It is then possible to eliminate a from ω and $\mathcal{L}$ and treat $\mathcal{L}$ as a function of ω and k so that

$$\mathcal{L} \equiv \mathcal{L}(\omega, k) = \frac{1}{2\pi} \int_0^{2\pi} L(\phi, \phi_t, \phi_x) d\theta, \tag{10.2.1}$$

where L is the Lagrangian of the system.

According to Whitham's theory of dispersive waves, the dispersion of a fairly general slowly varying wavetrain can be described by the phase function $\theta(x, t)$ such that the local frequency and the local wavenumber are defined by

$$\omega = -\theta_t, \qquad k = \theta_x. \tag{10.2.2ab}$$

With the specified phase function θ, the Whitham's averaged variational principle

$$\delta \iint \mathcal{L}(-\theta_t, \theta_x)\, dt\, dx = 0, \tag{10.2.3}$$

gives Euler's equation

$$\frac{\partial}{\partial t}\mathcal{L}_\omega - \frac{\partial}{\partial x}\mathcal{L}_k = 0. \tag{10.2.4}$$

This is the *Whitham fundamental equation* for one-dimensional nonlinear dispersion. This equation represents the fundamental *wave-action conservation law* for a gradually varying wave system on a scale of wavelengths.

In the theory of nonlinear dispersion, the *wave-action density* $\mathcal{A}$ is of profound significance and defined as the rate of change of average Lagrangian density with frequency, keeping wavenumber constant, so that

$$\mathcal{A} = \frac{\partial \mathcal{L}}{\partial \omega}. \tag{10.2.5}$$

If $\mathcal{A}$ is calculated using the expression (10.2.1) for $\mathcal{L}$, it is not necessary to allow for the small change in the waveform $\phi(\theta)$ that accompanies a small change in ω, because $\mathcal{L}$ is stationary with respect to any small changes in $\phi(\theta)$. We use (10.2.1) to evaluate $\mathcal{A}$ for fixed waveform $\phi(\theta)$, so that

$$\mathcal{A} = \frac{\partial \mathcal{L}}{\partial \omega} = \frac{1}{2\pi} \int_0^{2\pi} \phi'(\theta) \left(\frac{\partial L}{\partial \dot{\phi}} \right) d\theta. \tag{10.2.6}$$

This shows that the quantity $\omega \mathcal{L}_\omega$ is the average value of $\dot{\phi}(\partial L / \partial \dot{\phi})$, which is twice the average kinetic energy T per unit distance. So the wave-action density is

$$\mathcal{A} = \frac{\partial \mathcal{L}}{\partial \omega} = \frac{2T}{\omega}. \tag{10.2.7}$$

If this result is applied to long-crested water waves, all the energies are taken per unit length of the wave-crest so that these energies per unit distance represent energy densities per unit horizontal area. It then follows from (10.2.7) that, for waves of a given wavelength, the difference of the kinetic and potential energies ($\mathcal{L} = T - V$, where T and V are not necessarily equal in nonlinear wave problems) increases as ω increases above its linear-theory value $\sqrt{gk}$, so that

$$\mathcal{L} = T - V = \int_{\sqrt{gk}}^{\omega} \mathcal{A}\, d\omega = \int_{\sqrt{gk}}^{\omega} \left(\frac{2T}{\omega} \right) d\omega. \tag{10.2.8}$$

This is a fairly general result for $\mathcal{L}$ obtained from the Whitham variational formulation. It also follows from $\omega \mathcal{L}_\omega = 2T$ that

$$\frac{(\omega \mathcal{L}_\omega - 2\mathcal{L})}{\omega^3} = \frac{2V}{\omega^3}, \tag{10.2.9}$$

or

$$\frac{\partial}{\partial \omega} \left(\frac{\mathcal{L}}{\omega^2} \right) = \frac{2V}{\omega^3}, \tag{10.2.10}$$

so that $\mathcal{L}$ can be obtained from this result in the form

$$\mathcal{L} = 2\omega^2 \int_{\sqrt{gk}}^{\omega} \left(\frac{V}{\omega^3} \right) d\omega. \tag{10.2.11}$$

Equation (10.2.4) can now be interpreted most directly as a *wave-action conservation law*

$$\frac{\partial \mathcal{A}}{\partial t} + \frac{\partial \mathcal{B}}{\partial x} = 0, \tag{10.2.12}$$

where $\mathcal{B}$ is referred to as the *wave-action flux* defined by

$$\mathcal{B} = -\frac{\partial \mathcal{L}}{\partial k}. \tag{10.2.13}$$

The preceding results can be used to find the energy conservation law. From the total wave-energy density $\mathcal{E} = 2T - \mathcal{L} = \omega\mathcal{L}_\omega - \mathcal{L}$, it follows that

$$\frac{\partial \mathcal{E}}{\partial t} = \left(\omega\frac{\partial}{\partial t}\mathcal{L}_\omega + \omega_t\,\mathcal{L}_\omega\right) - \left(\omega_t\,\mathcal{L}_\omega + k_t\,\mathcal{L}_k\right),$$

which is, by (10.2.2ab) and (10.2.4),

$$\frac{\partial \mathcal{E}}{\partial t} = \omega\,\frac{\partial}{\partial x}\mathcal{L}_k + \omega_x\,\mathcal{L}_k = \frac{\partial}{\partial x}\left(\omega\,\mathcal{L}_k\right). \tag{10.2.14}$$

It is natural to define the *energy flux* $\mathcal{F}$ by

$$\mathcal{F} = -\omega\,\mathcal{L}_k, \tag{10.2.15}$$

so that the *energy conservation law* assumes the form

$$\frac{\partial \mathcal{E}}{\partial t} + \frac{\partial \mathcal{F}}{\partial x} = 0. \tag{10.2.16}$$

It follows from the Whitham theory that the averaged Lagrangian, $\mathcal{L}$ for deep water waves is given by

$$\mathcal{L} = \frac{1}{2}E\left(\frac{\omega^2}{gk} - 1\right) - \frac{1}{2}\frac{k^2E^2}{g\rho} + \cdots, \tag{10.2.17}$$

where $E = \frac{1}{2}g\rho a^2$.

Evidently, $\mathcal{L}$ depends only on three fundamental wave parameters ω, k and a^2. The dispersion relation for deep water waves is obtained from $\mathcal{L}_E = 0$ so that

$$\omega^2 = gk\left(1 + \frac{2Ek^2}{g\rho}\right) + \cdots = gk\left(1 + a^2k^2 + \cdots\right). \tag{10.2.18}$$

This is the famous *Stokes expansion* of the square of the frequency in deep water waves.

Using (10.2.17), a straightforward calculation shows that

$$k\,\mathcal{L}_k + \frac{1}{2}\,\omega\,\mathcal{L}_\omega = -2\mathcal{L}. \tag{10.2.19}$$

This implies that the energy flux $\mathcal{F}$ can be written as

$$\mathcal{F} = c\,(2\mathcal{L} + T), \qquad c = \frac{\omega}{k}. \tag{10.2.20}$$

One of the most remarkable features of the present theory is that the nonlinear effects on wave dispersion come from the wave-action conservation law (10.2.4). We expand the derivatives in this law and use (10.2.2ab) to derive a second-order quasi-linear partial differential equation for the phase function $\theta(x,t)$:

$$\mathcal{L}_{\omega\omega}\,\theta_{tt} - 2\mathcal{L}_{\omega k}\,\theta_{tx} + \mathcal{L}_{kk}\,\theta_{xx} = 0. \tag{10.2.21}$$

We now transform this equation into a linear equation by a Legendre transformation that uses the first derivatives (10.2.2ab) as new independent variables. Thus, the new dependent variable and its first derivatives are

$$\phi(\omega, k) = kx - \omega t - \theta, \qquad \frac{\partial \phi}{\partial k} = x, \qquad \frac{\partial \phi}{\partial \omega} = -t, \tag{10.2.22}$$

so that (10.2.21) reduces to the linear equation

$$\mathcal{L}_{\omega\omega}\,\phi_{kk} - 2\mathcal{L}_{\omega k}\,\phi_{\omega k} + \mathcal{L}_{kk}\,\phi_{\omega\omega} = 0. \tag{10.2.23}$$

It can be shown that all solutions of (10.2.21) are solutions of (10.2.23) except the simple solutions which can easily be found by other ways, where ω is a given function of k. The characteristics of (10.2.23), which is a fixed system of curves, satisfy the equation

$$\mathcal{L}_{\omega\omega}\,d\omega^2 + 2\mathcal{L}_{\omega k}\,d\omega\,dk + \mathcal{L}_{kk}\,dk^2 = 0. \tag{10.2.24}$$

At every point (ω, k) this equation defines two characteristic directions $(d\omega/dk)$, which may be real, imaginary or coincident. These directions at any point are those in which the tangent plane to the surface $\mathcal{L} \equiv \mathcal{L}(\omega, k)$ at that point intersects the surface. If they are real, the equation is the *hyperbolic* type, where the surface has principal curvatures of opposite sign. On the other hand, if the characteristic directions are imaginary, the equation is *elliptic*, where the surface has the same principal curvatures. What we have stated earlier is that the dispersion equation is elliptic or hyperbolic at a point (ω, k) where the surface $\mathcal{L} = \mathcal{L}(\omega, k)$ is *synclastic* or *anticlastic*, respectively. A line separating a synclastic from an anticlastic part of the surface is a transition curve on which the equation is parabolic.

We next use the Lagrangian for arbitrary amplitude to investigate the development of weak and slow modulations to a deep water wave train during the period while they remain weak. We consider waves produced by a wavemaker oscillating with fixed frequency ω_0 at the point $x = 0$ so that the amplitude of oscillations deviates from a large constant value by a small quantity that varies periodically at a frequency $\sigma \ll \omega_0$. If the unmodulated values of ω and k are ω_0 and k_0, respectively, the small deviation of the phase θ from the undisturbed value $\theta_0 = (k_0 x - \omega_0 t)$ satisfies (10.2.21) with its coefficients replaced by constants $(\mathcal{L}_{\omega\omega})_0$, $(\mathcal{L}_{\omega k})_0$ and $(\mathcal{L}_{kk})_0$ equal to the values of the second derivatives of $\mathcal{L}$ at $\omega = \omega_0$ and $k = k_0$. The appropriate conditions at the wavemaker $x = 0$ are

$$\theta = -\,\omega_0 t, \qquad \frac{\partial \theta}{\partial x} = k_0 + \varepsilon \exp(i\sigma t), \tag{10.2.25}$$

so that the frequency $\theta_t = \omega_0$ remains constant, the amplitude and also the wavenumber $(\partial\theta/\partial x)$ are modulated at the frequency σ.

We seek solution of (10.2.21) as a linear superposition of solutions proportional to $\exp[i(\sigma t - \kappa x)]$, where κ is either of the roots of the quadratic equation

$$(\mathcal{L}_{kk})_0\, \kappa^2 + 2(\mathcal{L}_{\omega k})_0\, \kappa\sigma + (\mathcal{L}_{\omega\omega})_0\, \sigma^2 = 0. \tag{10.2.26}$$

Two roots of this equation may be real or complex. For deep water waves, a simple calculation based upon the preceding condition shows that these roots are real or complex according to whether the finite amplitude $a > 0.054\lambda$ or $a < 0.054\lambda$. For the case of complex roots, the solution of (10.2.21) subject to the preceding conditions is

$$\theta = \theta_0 + \varepsilon \left(\frac{\sinh \gamma x}{\gamma}\right) \exp\left[i\sigma\left(t - \frac{x}{C}\right)\right], \tag{10.2.27}$$

where γ is the rate of exponential growth of modulation with distance given by

$$\gamma = \sigma \left[(\mathcal{L}_{\omega\omega})_0\, (\mathcal{L}_{kk})_0 - (\mathcal{L}_{\omega k}^2)_0\right]^{\frac{1}{2}} \div (\mathcal{L}_{kk})_0, \tag{10.2.28}$$

and the velocity of propagation of the amplitude modulation is

$$C = -\frac{(\mathcal{L}_{kk})_0}{(\mathcal{L}_{\omega k})_0}. \tag{10.2.29}$$

When the roots are complex, then γ is real. The solution containing the hyperbolic sine in (10.2.27) suggests that the pure gravity waves of finite amplitude $a < 0.054\lambda$ on deep water become *unstable* in the sense that the slow modulation grows as the waves propagate in the far field, as $x \to \infty$.

With the increase in depth of the modulation, the coefficients in (10.2.21) change substantially, and the preceding predictions show a modified shape of modulation. The crest of largest amplitude propagates forward with fastest speed, which reduces the wavelength in front of the amplitude peak and increases that behind; then, the energy behind moves forward rapidly and the energy in front slowly, leading to a highly localized development of the amplitude peak. After a finite time the amplitude becomes a cusped shape, and then the assumption that the amplitude varying slowly breaks down. Benjamin and Feir (1967) first demonstrated experimentally the growth of a small, slow modulation to waves of moderate amplitude. Several authors have theoretically predicted weak instability of waves of moderate amplitude in deep water. In a series of papers during 1975-1985, Longuet-Higgins (see Debnath, 1994) developed breaking phenomena of water waves including the implications for wave breaking of the important discovery that the maximum of wave energy is found at an amplitude below its maximum value.

On the other hand, when roots of (10.2.26) are real, γ is purely imaginary. The solution exhibits no exponential growth as $x \to \infty$, but the slow modulation splits into two modulations propagating with two separate group velocities C_1 and C_2 given by

$$C_1 = \frac{\sigma}{\left(\frac{\sigma}{C} - |\gamma|\right)}, \qquad C_2 = \frac{\sigma}{\left(\frac{\sigma}{C} + |\gamma|\right)}. \tag{10.2.30}$$

This prediction is in perfect agreement with that of Whitham (1965a), who first discovered such a splitting of group velocity in a nonlinear dispersive wave system. It is noted that one modulation propagates with the group velocity C_1 close to the linear-theory value, while the other travels ahead with the significantly greater group velocity C_2.

We follow Lighthill's (1967) elegant analysis to discuss approximate versions of the dispersion equation (10.2.23) for moderately small amplitude and to determine whether, under certain conditions, very gradual variations in the properties of a wave train are governed by the elliptic or hyperbolic equations, and also to ensure instability associated with the elliptic case. According to his analysis, it is possible to retain terms of the order $\tau \equiv \omega - f(k)$ in the coefficients of (10.2.23) and to ignore terms of order τ^2. The line $\tau = 0$ corresponds to infinitesimal amplitude, and the characteristic directions both coincide with it thereon. To derive the approximate forms of equation (10.2.23), the term of order τ^3 is included in the expansion of $\mathcal{L}$ so that

$$\mathcal{L} = \frac{1}{2}\, g(k)\tau^2 + \frac{1}{3}\, h(k)\tau^3. \tag{10.2.31}$$

As we shall see that the second term in (10.2.31) produces terms of order τ in the coefficients of (10.2.23), but they cancel out so that there will be no contribution.

If k and τ are treated as new independent variables, terms $\partial/\partial\omega$ and $\partial/\partial k$ in (10.2.23) can be replaced by $\partial/\partial\tau$ and $\partial/\partial k - f'(k)\partial/\partial\tau$ respectively, so that equation (10.2.23) becomes

$$\mathcal{L}_{\tau\tau}\phi_{kk} - 2\mathcal{L}_{\tau k}\phi_{\tau k} + \mathcal{L}_{kk}\phi_{\tau\tau} - (\mathcal{L}_{\tau\tau}\phi_\tau + \mathcal{L}_\tau\phi_{\tau\tau})f''(k) = 0. \tag{10.2.32}$$

Substituting (10.2.31) into (10.2.32) and neglecting terms $O(\tau^2)$ yields

$$\phi_{kk} - 2\tau\left[\frac{g'(k)}{g(k)}\right]\phi_{\tau k} - f''(k)(\tau\phi_{\tau\tau} + \phi_\tau) = 0. \tag{10.2.33}$$

This equation in independent of $h(k)$ to this order of approximation. Comparing the two terms involving τ, we see that the term $\tau f''(k)\phi_{\tau\tau}$ has more significant effects than the term $2\tau\left(g'(k)/g(k)\right)\phi_{\tau k}$. The charactersitics of equation (10.2.33) are given by

$$\frac{d\tau}{dk} = -\tau\frac{g'(k)}{g(k)} \pm \left[\tau f''(k) + \tau^2\frac{g'^2(k)}{g^2(k)}\right]^{\frac{1}{2}} \tag{10.2.34}$$

which becomes, for small τ,

$$\frac{d\tau}{dk} = \pm\left[\tau f''(k)\right]^{\frac{1}{2}}. \tag{10.2.35}$$

Thus, equation (10.2.33) without the second term is

$$\phi_{kk} - f''(k)\left(\tau\phi_{\tau\tau} + \phi_\tau\right) = 0. \tag{10.2.36}$$

This equation is *elliptic* or *hyperbolic* according to whether $[\tau f''(k)] < 0$ or > 0. So the characteristics of (10.2.32) are imaginary or real in the former or latter inequality, respectively. Further, (10.2.36) is similar to the familiar Tricomi equation in the hodograph equations for a plane compressible gas flow. Since the Tricomi equation has characteristics with cusps on the transition line $\tau = O(\omega - f(k) = 0)$ between the elliptic and hyperbolic regimes, all characteristics of (10.2.36) are tangent to the line $\tau = 0$, which is an envelope of characteristics and itself a characteristic. In particular, each characteristic given by equation (10.2.35) assumes the form

$$\tau = \pm\frac{1}{4}\left[\int_{k_0}^{k}\{\pm f''(k)\}^{\frac{1}{2}}\,dk\right]^2, \tag{10.2.37ab}$$

according to whether $f''(k) > 0$ or < 0, where $k = k_0$ is the point where the characteristic touches the transition line $\tau = 0$. Equations (10.2.37ab) do not suffer from a serious error if $f''(k)$ is replaced by a constant value $\mu_0 = f''(k_0)$ because the equations are only valid for small τ and so for small values of $|k - k_0|$.

In the hyperbolic case, the transformation

$$r = \left[\frac{4\tau}{f''(k)}\right]^{\frac{1}{2}} = 2\left[\frac{\omega - f(k)}{f''(k)}\right]^{\frac{1}{2}} \tag{10.2.38}$$

is used to reduce the dispersion equation (10.2.36) to the following axisymmetric wave equation:

$$\phi_{kk} = \phi_{rr} + \frac{1}{r}\phi_r \tag{10.2.39}$$

where terms proportional to r found in transforming ϕ_{kk} are ignored. The associated space and time variables x and t are expressed in terms of k and r in the equations

$$x = \phi_k - \frac{2f'(k)}{f''(k)}\left(\frac{1}{r}\,\phi_r\right), \qquad t = -\frac{2}{f''(k)}\left(\frac{1}{r}\,\phi_r\right). \tag{10.2.40ab}$$

Equation (10.2.39) is solved subject to the appropriate boundary conditions

$$\phi_k = x = 0, \quad \phi_\omega = -t = -t_0(k), \quad \text{on} \quad \omega = \omega_0, \tag{10.2.41}$$

where $t_0(k)$ is a decreasing function of k. These boundary conditions show that ϕ is constant on $\omega = \omega_0$, and, without loss of generality, we take this constant as zero. In terms of r and k, the initial conditions are

$$\phi = 0, \quad \phi_r = -\frac{1}{2} r f''(k) t_0(k), \quad \phi_k = -f'(k) t_0(k) \tag{10.2.42}$$

$$\text{on} \quad r = 2\left[\frac{\omega_0 - f(k)}{f''(k)}\right]^{\frac{1}{2}},$$

which become for small r, by defining k_0, u_0 and μ_0 by $\omega_0 = f(k_0)$, $u_0 = f'(k_0)$, and $\mu_0 = f''(k_0)$,

$$\phi = 0, \quad \phi_r = -\frac{1}{2} r \mu_0 t_0(k), \quad \phi_k = u_0 t_0(k) \quad \text{on} \quad k = k_0 - \frac{\mu_0 r^2}{4u_0}. \tag{10.2.43}$$

Using the known Riemann function, equation (10.2.39) can be solved with (10.2.42) by making reference to Courant and Hilbert (1962), p. 453, with x and y replaced by $r \pm k$. It turns out that the solution is

$$\phi_p = \phi(r_p, k_p) = -\frac{1}{2} u_0 \int_{r_p-(k-k_0)}^{r_p+(k-k_0)} \left(\frac{r}{r_p}\right)^{\frac{1}{2}} t_0(r)\, F\left(\frac{1}{2}, \frac{1}{2}, 1, z\right) dr, \tag{10.2.44}$$

where F is the hypergeometric function; $t_0(k)$ is rewritten as $t_0(r)$, which represents the time if the value r characterizes the waves generated at the wavemaker; and the argument z in the hypergeometric function is

$$z = \left[\frac{(k_p - k)^2 - (r_p - r)^2}{4 r r_p}\right] \tag{10.2.45}$$

which cannot be greater than $\frac{1}{2}$ or less than zero.

For example, waves are generated at the wavemaker with r as a linear function of time so that $r = \alpha t$ and the amplitude is a linear function of time t. The initial condition (10.2.43) can be simplified as

$$\phi = 0, \qquad \phi_k = -\left(\frac{u_0}{\alpha}\right) r \qquad \text{on} \quad k = k_0. \tag{10.2.46}$$

The solution of the initial value problem (10.2.39) and (10.2.46) can be obtained by integrating the elementary solution $\{g(r,k)\}^{-1} = \left\{r^2 - (k - k_0)^2\right\}^{-1/2}$ three times with respect to k, so that

$$\phi = -\left(\frac{u_0}{4\alpha}\right)\left[\left\{r^2 + 2(k - k_0)^2\right\}\sin^{-1}\left(\frac{k - k_0}{r}\right) + 3(k - k_0)\, g(r,k)\right]. \tag{10.2.47}$$

Thus, in this hyperbolic case, the amplitude at the wavemaker increases linearly with time. This conclusion seems to suggest a good test of Whitham's conjecture regarding the appearance of discontinuities.

On the other hand, when the dispersion equation is elliptic, the transformation

$$s = \left[-\frac{4\tau}{f''(k)}\right]^{1/2} = 2\left[-\frac{\omega - f(k)}{f''(k)}\right]^{1/2} \tag{10.2.48}$$

is used to reduce (10.2.36) to the axisymmetric Laplace equation in the form

$$\phi_{kk} + \phi_{ss} + \frac{1}{s}\,\phi_s = 0. \tag{10.2.49}$$

The associated space and time variables are

$$x = \phi_k + 2\frac{f'(k)}{f''(k)}\left(\frac{1}{s}\,\phi_s\right), \qquad t = \frac{2}{f''(k)}\left(\frac{1}{s}\,\phi_s\right). \tag{10.2.50ab}$$

We look for a solution of (10.2.49) subject to the boundary conditions that are obtained from the relations

$$t = -\frac{2}{\mu_0}\left(\frac{1}{s}\,\phi_s\right), \qquad x - u_0 t = \phi_k + 2k\,(k - k_0)\left(\frac{1}{s}\,\phi_s\right), \tag{10.2.51ab}$$

where k_0 is a reference wavenumber about which variations of k are small.

In general, the boundary conditions appropriate to any wave problem are of the initial value type or the hyperbolic type. This kind of elliptic problem is referred to as the *ill-posed problem*. Several authors, especially Garabedian (1964), developed the theory and the method of solution of elliptic problems by using imaginary characteristics. It can be shown that uniquely defined solutions of these problems exist, provided the boundary data are smooth in a certain sense. In general, solutions of elliptic problems determine subsequent development, as long as the boundary data remain sufficiently smooth. Often singularities appear, which simply means that after a finite time, solutions representing the amplitude and the wavenumber for nonlinear dispersive wave systems break down in the sense that they cease to vary smoothly. Evidently, these results have physical ramifications.

However, an explicit solution for the elliptic case similar to (10.2.44) for the hyperbolic dispersion equation (10.2.39) can be obtained in terms of conditions on a fixed straight line $k = k_0$. For the elliptic equation (10.2.39), the imaginary characteristics through a point (k_p, s_p), where ϕ is to be determined, are $s - s_p = \pm i\,(k_p - k)$, and these intersect the line $k = k_0$ at the points $s = s_p \pm (k_p - k_0)$, which are the new limits of integration in the modified result (10.2.44). As in the hyperbolic case, the appropriate boundary conditions are

$$\phi = 0, \qquad \phi_k = -u_0 t_0 (s), \quad \text{on} \quad k - k_0, \tag{10.2.52}$$

where $u_0 = f'(k_0)$, and $t_0(s)$ is the time at which the amplitude s assumes a given value for the waves generated at the wavemaker. With determined ϕ, equations (10.2.48) and (10.2.50ab) can be utilized in the approximate form

$$t = -\frac{2}{\mu_0}\left(\frac{1}{s}\phi_s\right), \qquad x - u_0 t = \phi_k + \frac{2}{s}(k - k_0)\,\phi_s, \tag{10.2.53ab}$$

where $f''(k_0)$ and $f'(k_0) - \mu_0$ are replaced by $-\mu_0$ and $(k - k_0)\,\mu_0$, respectively. With necessary changes in (10.2.44), the solution for this elliptic problem is given by

$$\phi(k, s) = \frac{i}{2} u_0 \int_{s-i(k-k_0)}^{s+i(k-k_0)} t_0(\alpha)\left(\frac{\alpha}{s}\right)^{1/2} F\left(\frac{1}{2}, \frac{1}{2}, 1, z'\right) d\alpha, \tag{10.2.54}$$

where

$$z' = \frac{(k - k_0)^2 + (s - \alpha)^2}{-4s\alpha}, \tag{10.2.55}$$

and $t_0(\alpha)$ is the time at which the amplitude s takes the value α at the wave maker, and admits unique analytic continuation in the complex α plane filled by the points $s \pm i(k - k_0)$. This means that $t_0(\alpha)$ is a smooth function. However, such analytic continuation is not possible whenever there are discontinuities in the first or higher derivatives of $t_0(\alpha)$. Hence, the solutions fail to exist in the elliptic case.

As an example of the general situation discussed earlier, we consider the time development of a group of waves of constant wavelength but variable amplitude so that $k = k_0$ and $s = s_0(x)$ at $t = 0$. Conditions (10.2.53ab) become

$$\phi = 0, \qquad \phi_k = x_0(s) \quad \text{on} \quad k = k_0, \tag{10.2.56}$$

where the constant involved in integration of $\phi_s = 0$ has been taken as zero and $x = x_0(s)$ is the inverse of $s = s_0(x)$. The solution of (10.2.49) under the boundary conditions (10.2.56) is

$$\phi = -\frac{i}{2}\int_{s-i(k-k_0)}^{s+i(k-k_0)} x_0(\alpha)\left(\frac{\alpha}{s}\right)^{1/2} F\left(\frac{1}{2}, \frac{1}{2}, 1; z'\right) d\alpha, \tag{10.2.57}$$

where the limits of integration are the values of α that are the points of intersection of the initial curve $k = k_0$ with the characteristics through (k, s).

We take a particular form of initial variation of amplitude in the wave group as

$$s = s_0(x) = s_m\left(1 + \frac{x^2}{b^2}\right)^{-1}, \tag{10.2.58}$$

where b is the distance in which the amplitude decreases to $\frac{1}{2}s_m$ and it is large on a scale of wavelengths. In this case, the solution does not involve any singularity in the (s, k) plane; however, after a certain critical time, $t = t_{\text{crit}} = t_c = 0.69b/\mu_0 s_m$, which is inversely proportional to the maximum amplitude s_m, a singularity in the transformation between the (x, t) and (k, s) planes develops. This means that the smoothness condition breaks down in the physical plane. For the case of water waves, the distance $u_0 t_c$ traveled by the wave group during the critical time assumes the form $(ak)^{-1}(0.49b)$. Up to the critical time $t = t_c$, the variation of amplitude in the group remains smooth and symmetrical about a point advancing with the speed u_0.

As predicted by Lighthill's theory, the total energy of the group gradually becomes concentrated within a small distance as the wave group propagates away from the wavemaker. At the same time, the maximum amplitude in the group continues to be at the center $x = u_0 t$ and is found to increase above the initial value s_m. This increase becomes more and more rapid and the rate of maximum amplitude becomes infinite. On the other hand, wavenumbers begin to change more rapidly. The tendency for the wavenumber to increase above k_0 ahead of the center of the group and to decrease below k_0 behind the center has at time $t = t_{\max}$ become so intensified as to generate an infinite gradient k_x at the center. When the critical condition is reached, the wavenumber exhibits a discontinuity at the center of the group, and the amplitude acquires a sharp cusp there. So the theory can hardly be used to investigate the solution for $t > t_c$, as the condition of smoothness has broken down. Lighthill's (1965b) Figure 4 describes this striking feature. This figure is reproduced here as Figure 10.1, which describes the change with time t in the distribution of the wavenumber k and amplitude $s \sim \sqrt{8}\,ak^2$ (for the deep water case, $f(x) = \sqrt{gk}$) in a wave group initially of constant wavenumber k_0 and with s satisfying the initial condition (10.2.58).

During the changes of form of the wavegroup as indicated earlier, the total energy in the group remains constant. To the present approximation,

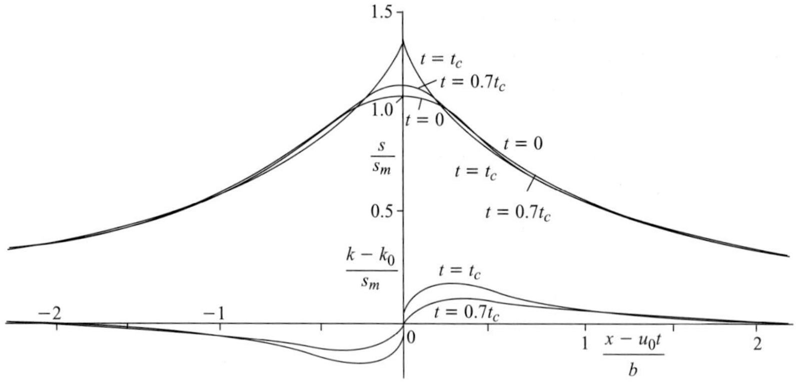

Fig. 10.1 Change with time in the distribution of the wavenumber k and the amplitude s in a wave group initially of constant wavenumber k_0. (From Lighthill, 1967).

the energy is proportional to

$$E(t) = \int_{-\infty}^{\infty} \frac{1}{2} s^2 dx = -\int (x - u_0 t)\, s\, ds = -\int [\phi_k - (k - k_0)\, \mu_0 t]\, ds. \tag{10.2.59}$$

It follows from (10.2.53ab) that

$$\frac{\partial \phi}{\partial s} = -\frac{1}{2}\mu_0 s t, \qquad \frac{\partial k}{\partial t} = -\frac{\frac{1}{2} s \mu_0}{\phi_{ks}}, \tag{10.2.60ab}$$

and hence, equation (10.2.59) gives

$$\frac{dE}{dt} = -\int \left[(\phi_{kk} - \mu_0 t) \left(-\frac{\frac{1}{2} s \mu_0}{\phi_{ks}} \right) - (k - k_0)\, \mu_0 \right] s\, ds. \tag{10.2.61}$$

We also get, from (10.2.49) and (10.2.60a) with constant value of t,

$$\frac{dE}{dt} = \int d \left[\frac{1}{2} (k - k_0)\, \mu_0 s^2 \right] = 0. \tag{10.2.62}$$

This shows that the total energy E of the wave group is constant as expected because the whole theory is based upon Hamilton's principle for a conservative system.

We now give some applications of the general theory of dispersive wave systems satisfying an elliptic or hyperbolic form of Whitham's dispersion equation. If the dispersion relation for small-amplitude waves is $\omega = f(k)$, then, for moderate amplitudes, the dispersion equation is shown to be hyperbolic or elliptic according to whether the expression

$$[\omega - f(k)]\, f''(k) \tag{10.2.63}$$

is positive or negative, respectively. In the hyperbolic case, the point (ω, k) for moderate amplitude lies on the side of the infinitesimal amplitude curve $\omega = f(k)$ towards which that curve is concave. If the expression is (10.2.63) is negative, then the point (ω, k) for moderate amplitude is in the region towards which the curve is convex.

For long-crested deep water waves, $\omega = f(k) = \sqrt{gk}$ for infinitesimal amplitude, while for moderate amplitude a, the dispersion relation is

$$\omega = \sqrt{gk}\left[1 + \frac{1}{2}(ak)^2 + \frac{1}{2}(ak)^4 + \cdots\right]. \tag{10.2.64}$$

Clearly, $\left(\omega - \sqrt{gk}\right) > 0$ for finite a, but $f''(k) < 0$; therefore, the dispersion is governed by an elliplic equation. Alternatively, the curve $\omega = f(k) = \sqrt{gk}$ is convex upwards, while for moderate amplitude ω increases above $\sqrt{gk}$. The curve is convex to the point (ω, k), and (10.2.63) is negative. This is in perfect agreement with Whitham's theory, which also predicts an elliptic equation for this case. Both Lighthill's and Whitham's theoretical predictions are found to be in good agreement with experimental findings of Benjamin and Feir (1967), which exhibit the growth of small and slow modulation to waves of finite amplitude. In the later stages of development, a change in the modulation curve was found to exhibit a cusped shape. In recent years, an additional term proportional to the square of the amplitude gradient has been included in the averaged Lagrangian to develop improved dispersion theories that are in much better agreement with this and other similar experiments.

In the case of deep water waves, gravity is the restoring force, while for water waves of shorter wavelength, surface tension T_1 plays a predominant role as an additional restoring force. In the extreme limiting case of very short waves, gravitational force can be neglected compared with surface tension T_1; an exact analytical solution for progressive capillary waves of arbitrary amplitude is given by Crapper (1957). This solution shows that $\omega - f(k) < 0$ for capillary waves of moderate amplitude

$$f(k) = \left(\frac{T_1 k^3}{\rho}\right)^{1/2} \tag{10.2.65}$$

and $f''(k) > 0$. The dispersion equation associated with this problem is elliptic. In fact, the averaged Lagrangian involved in the problem is

$$\mathcal{L} = 2T_1 - \frac{\rho\omega^2}{k^3} - \frac{T_1^2 k^3}{\rho\omega^3}. \tag{10.2.66}$$

Equation (10.2.23) can now be rewritten with $\sigma = \left(T_1^2 k^6\right)\left(\rho\omega^2\right)^{-2}$ and k and σ as new independent variables so that it becomes

$$(1 + 3\sigma) k^2 \phi_{kk} + 12\sigma(\sigma - 1)(k\phi_{k\sigma} + \sigma\phi_{\sigma\sigma}) + 6\sigma(1 + \sigma)\phi_\sigma = 0. \quad (10.2.67)$$

Without any further analysis of this equation, we recall $f(k)$ associated with the capillary-gravity waves on deep water in the form

$$f(k) = \left(gk + \frac{T_1}{\rho} k^3\right)^{1/2}. \quad (10.2.68)$$

It has been shown before that the dispersion equation is elliptic for very small wavenumbers (pure gravity waves) and for very large wavenumbers (pure capillary waves). This fact does not eliminate the possibility of a hyperbolic nature of the dispersion equation in some intermediate range of wavenumbers. With the expression (10.2.68), $f''(k)$ changes sign where $\kappa = \left(T_1 k^2 / g\rho\right)$ assumes the value $\left(\left(2\sqrt{3}\right) - 1\right) = 0.1547$. According to Wilton (1915), $\omega - f(k)$ changes sign where $\kappa = 0.5$. From a physical point of view, these two values of κ correspond to (i) minimum group velocity, and (ii) a wavenumber having the same phase speed as its first harmonic. In the intermediate range, $0.1541 < \kappa < 0.5$, the dispersion equation is hyperbolic because $\omega - f(k)$ and $f''(k)$ are both positive. Lighthill (1965) suggested to attempt the experimental test of the Witham's theory in the preceding range of κ. It is then necessary to set the fixed frequency ω in the range $0.674\gamma < \omega < 1.03\gamma$, where $\gamma^4 = \left(\rho g^3 / T_1\right)$. Such experiment seems to be suitable for comparison with theory in the hyperbolic problem.

The upshot of Lighthill's deep analysis is that, under certain conditions, very gradual variations in the properties of a wavetrain are governed by the elliptic or hyperbolic equations. In the elliptic case, solutions representing the amplitude and the wavenumber for nonlinear dispersive waves definitely become *unstable*. The most striking feature of the Lighthill theory is that there is a development of wave group whose amplitude distribution is symmetrical about its maximum. This feature is found to remain true at later times, but energy is redistributed within the group so that the maximum amplitude increases with development of a *cusp* at the center of the group when the wavenumber exhibits a singularity. Moreover, such a singularity occurs after a finite time for a wide class of elliptic initial value problems. Lighthill's theoretical predictions were found in excellent agreement with the theory and the conclusive experimental findings of Benjamin (1967) and Benjamin and Feir (1967).

Chapter 11

Nonlinear Diffraction of Water Waves by Offshore Structures

"Nonlinear hydrodynamic stability theory is really concerned, ultimately, with phenomena such as transition to turbulence. In practice, however, that phenomenon is so complex as to defy rational understanding at the present time. A more limited objective is that of gaining some understanding of nonlinear processes in fluid mechanics, perhaps with reference to the early, relatively-simple stages of the evolution of laminar flow to turbulence. Even then, the mathematical problems posed are challenging enough."

J. T. Stuart

"The transition of laminar flow, with its clean layers of flow tubes, to strongly mixed, irregular turbulent flow is one of the principal problems of modern hydrodynamics. It is certain that this fundamental change in type of motion of the fluid is traceable to an instability in the laminar flow, for laminar flows of themselves would always be possible solutions of the hydrodynamic equations."

W. Tollmien

11.1 Introduction

Diffraction of water waves by offshore solid structures is of considerable interest in ocean engineering and naval hydrodynamics. The study of wave forces on the offshore structures or natural boundaries has become more important due to the tremendous growth and need for ocean exploration and extraction of wave energy from oceans. Current methods of calculating wave forces on the offshore structures and/or harbors based upon Morison's (1950) equation seem to be useful for construction of such structures that are widely used for exploration of oil and gas from the ocean floor.

Historically, Havelock (1940) first developed the linearized diffraction theory for small-amplitude water waves in deep oceans. Based upon this work, MacCamy and Fuchs (1954) extended the theory for a fluid of finite depth. These authors used the linearized theory successfully for estimation of wave loading on a vertical cylinder extending from the ocean floor to above the free surface of the water. Subsequently, several authors including Mogridge and Jamieson (1976), Hogben *et al.* (1977), Garrison (1978, 1982) and Mei (1978) have obtained analytical solutions of the linearized diffraction problems for simple geometrical configurations. However, the linearized theory has limited applications because it is only applicable to water waves of small wave steepness. In reality, ocean waves are inherently nonlinear and often irregular in nature, and hence water waves of large amplitude are of special interest in estimating wave forces on offshore structures or harbors.

In the theory of diffraction of water waves, it is important to distinguish between a small and large body of typical dimension b compared with the characteristic wavelength $(2\pi/k)$ and the wave amplitude a. Physically, when a/b is small and kb is large $\left(b \gg k^{-1}\right)$, where k is the wavenumber, the body becomes as a generator of dipole wave radiation and the wave force on it becomes more resistive in nature. This means that flow separation becomes insignificant while diffraction is dominant. In other words, the body radiates a very large amount of scattered (or reflected) wave energy. On the other hand, when kb is very small (the characteristic dimension b of the body is small compared with k^{-1}) and (a/b) is very large $(a/b \geq 0(1))$, the body radiates a very small amount of scattered wave energy. This corresponds to the case of a rigid lid on the ocean inhibiting wave scattering, that is, diffraction is insignificant. In this limiting case, the resulting force, involving no substantial loss of energy by wave radiation, is a purely reactive force; in fact, this is an inertial force. At large distances from the body, the transient pressure distribution $-\rho\ \phi_t$ dominates the irrotational-flow pressure disturbances caused by the presence of the body. In fact, the body produces a typical dipole field.

11.2 Morison's Equation

In most practical applications of fluid dynamics, including the determination of hydrodynamic wave loading on offshore structures, it is necessary to study the complete fluid motion that consists of (i) the irrotational flow field satisfying appropriate boundary conditions and (ii) a vortex motion associated with any vorticity satisfying zero boundary conditions. When a solid body moves through externally unbounded fluid that is at rest far from the body with variable velocity $U(t)$ in the negative x-direction, it can be shown that the total kinetic energy is the sum of kinetic energies associated with components (i) and (ii). The kinetic energy of the former component is $\frac{1}{2}m_a U^2$, where m_a is the added mass of the fluid. The rate of increase of this part of the kinetic energy is $m_a U\dot{U}$. To feed this increase of energy, the body must exert the thrust $m_a\dot{U}$ to overcome the equal and opposite potential-flow drag force $m_a\dot{U}$ of the fluid on the body. At the same time, the kinetic energy associated with the vortex motion increases as more and more vorticity is shed from the body surface into the wave where the vortex lines are eventually convected and diffused. The rate of work done by the force exerted by the body on the fluid is equal to the rate of increase of the total energy of the fluid, including both kinetic energy and any thermal energy generated by viscous dissipation of kinetic energy. The rate of increase of this energy can be obtained from the rate proportional to ρAU at which the mass of the wake fluid is increasing, where A is the body's frontal surface area. Since the velocity of the vortex motion is proportional to U, the rate of increase of energy is $\frac{1}{2}\rho AU^2 C_D \cdot U$, where C_D is called the *drag coefficient*. The quantity

$$\frac{1}{2}\rho AU^2 C_D \tag{11.2.1}$$

represents the vortex-flow drage force of the fluid on the body such that the rate of work done against it is exactly equal to the rate of increase of energy.

In order to obtain results for the important case associated with offshore-structure problems, where a body is at rest in a fluid moving with variable velocity, we superimpose an additional velocity $U(t)$ in the positive x-direction on the entire system so that an additional uniform pressure gradient $-\rho\dot{U}$ exerts the so called *Froude-Krylov force*,

$$\rho\dot{U}V, \tag{11.2.2}$$

on a body of volume V. When this force is combined with the other force component, $m_a\dot{U}$, associated with irrotational motion, we obtain the total potential-flow drag force,

$$\left(\rho\dot{U}V + m_a\dot{U}\right) = \rho C_M \dot{U}V, \tag{11.2.3}$$

where C_M is called the *Morison* (or *inertial*) *coefficient*, defined by

$$C_M = 1 + \frac{m_a}{\rho V}. \tag{11.2.4}$$

Thus, the total hydrodynamic drag force can be written as a sum of its potential-flow component (11.2.3) and its vortex-flow component (11.2.1) in the form

$$D = \rho C_M V \dot{U} + \frac{1}{2}\rho C_D A U^2. \tag{11.2.5}$$

This is known as *Morison's equation.* For cases of the flow past a circular cylinder or sphere, both coefficients C_M and C_D can be determined. It is well known that $C_M = 2.0$ for a circular cylinder and $C_M = 1.5$ for a sphere.

There are several characteristic features of the Morison equation. First, the force component (11.2.3) associated the irrotational part of the flow depends linearly on U. Second, the vortex-flow drag force (11.2.1) is proportional to U^2. Third, the potential-flow component and the vortex-flow component of drag force on the solid structure in an unsteady flow are independent in the sense that there is no interaction between them.

Several studies have shown that the Morison equation is fairly satisfactory, and it is widely used by engineers to calculate the wave forces on offshore solid structures. However, several difficulties in using it in the design and construction of offshore structures have been reported in the literature. One such difficulty is concerned with the drag force, which has relatively large-scale effects. The shortage of reliable full-scale drag force data in ocean waves is another problem. There is another important question whether a linear theory of the irrotational flow response is appropriate at all to finite-amplitude water wave motions with a free surface. Despite these difficulties and shortcomings, the use of the Morison equation has extensively been documented in the past literature through plentiful data for determing the coefficients C_M and C_D.

However, for real ocean waves with solid structures, there is a significant nonlinear force associated with the irrotational component of the flow field because of the large amplitude of ocean waves. These waves are of special interest in the wave loading calculation. It seems important to include all significant nonlinear effects associated with nonlinear free surface boundary conditions in the irrotational flow component of the wave loading on the offshore structures. However, although the effects of large-amplitude waves on an offshore structure of small mean diameter may be insignificant, it is no longer true as the diameter increases compared with the wavelength of the incident wave field. Consequently, the Morison equation may no longer be applicable. However several difficulties in using it in the design and construction of offshore structures have been reported in the literature.

In 1979, Lighthill was invited to deliver a keynote address on "Waves and Hydrodynamic Loading" at the Behavior of Offshore Structure Symposium at Cranfield, England. He gave a detailed survey of this title and then presented a mathematical analysis of estimating all linear and nonlinear wave forces acting on a very large solid structure of size b comparable with the wavelength λ, where $\lambda = (2\pi/k)$. In his keynote address, Lighthill (1979) suggested a number of improvements of the Morison equation (11.2.5). Based on common principles of fluid mechanics and solid mechanics, he incorporated second-order terms in the interaction between the potential flow component of fluid motion and the solid structure.

He demonstrated that this interaction is influenced by nonlinear effects to a significant extent. Consequently, he developed a nonlinear theory that incorporates effects of large wave steepness. In particular, Lighthill pointed out that the sinusoidal wave potential

$$\phi_0 = k^{-1}e^{kz}a\omega\cos(\omega t - kx), \tag{11.2.6}$$

in deep water, produces large extensional rate-of-strain components

$$\frac{\partial^2\phi_0}{\partial x^2} = -\frac{\partial^2\phi_0}{\partial z^2} = -\left(ke^{kz}\right)a\omega\cos(\omega t - kx). \tag{11.2.7}$$

He further demonstrated that the presence of a solid body in this fluctuating rate-of-strain field generates an important potential-flow response. In fact, he showed that second-order force on the body arises from the cross terms between its response to fluctuating velocity field and to fluctuating rate-of-strain rather than from squared terms in either response separately. His analysis of potential flows with a free surface includes three major terms, one of which reduces to the expression of Taylor (1928) in the small wavenumber limit.

On the other hand, the wave loading associated with the nonlinear free surface condition is described by a new free surface equation in the form

$$\frac{\partial^2\phi}{\partial t^2} + g\frac{\partial\phi}{\partial z} = -\frac{\partial}{\partial t}\left(\nabla\phi\right)^2 \quad \text{on } z = \eta. \tag{11.2.8}$$

The right hand side of (11.2.8) is twice the value obtained by simply including the dynamic pressure term

$$-\frac{1}{2}\rho\left(\nabla\phi\right)^2, \tag{11.2.9}$$

when the water pressure is equated to atmospheric pressure at the free surface. This is due to an exactly equal contribution when the rate of change of the sum of transient and hydrostatic pressure on the free surface is represented to the second order.

Indeed, the periodic oncoming wave on deep water given by (11.2.6) satisfies the linearized free surface condition (11.2.8) without the term on the right hand side not just on $z = 0$, but for all values of z. Furthermore, $(\nabla\phi_0)^2$ assumes the value $(a\omega)^2 \exp(2kz)$ independent of time t so that the second-order equation (11.2.8) is satisfied for all z and, in particular, on $z = \eta$. Evidently, the free sinusoidal wave motion is virtually unchanged by second-order effects.

Lighthill's profound analysis reveals that nonlinear contributions to wave loading on offshore structures are made of the quadratic, dynamic, and waterline forces associated with the irrotational flow component of the wave loading. He then suggested a new nonlinear modification of the Morison equation (11.2.5) with the nonlinear contribution given by

$$F_{nl} = F_q + F_d + F_w, \tag{11.2.10}$$

where F_q, F_d and F_w are defined later by (11.5.13), (11.5.14) and (11.5.16).

11.3 Basic Equations of Water Waves in Cylindrical Polar Coordinates and Boundary Conditions

We now state the fundamental equations and the boundary conditions for the study of diffraction of finite amplitude progressive waves in deep water, incident on a vertical, surface-piercing, large rigid circular cylinder of radius b. We assume that the cylinder is acted on by a train of two-dimensional periodic progressive waves of amplitude a propagating in the positive x direction, as shown in Figure 11.1. In cylindrical polar coordinates (r, θ, z) with the z-axis vertically upwards from the origin at the mean free surface, the governing equations, free surface conditions, boundary conditions at the rigid ocean bottom and at the surface of the solid cylinder are

$$\nabla^2\phi \equiv \phi_{rr} + \frac{1}{r}\,\phi_r + \frac{1}{r^2}\,\phi_{\theta\theta} + \phi_{zz} = 0, \tag{11.3.1}$$

where $b < r < \infty$, $-\pi < \theta < \pi$, $-\infty < z \leq \eta$, $\eta \equiv \eta\,(r, \theta, t)$ is the free surface elevation, and $\phi \equiv \phi\,(r, \theta, z, t)$ is the velocity potential.

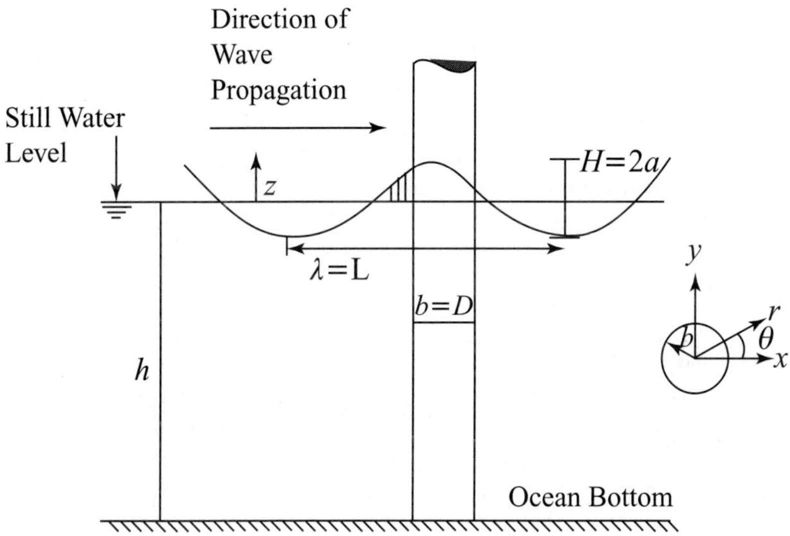

Fig. 11.1 Schematic diagram of a large cylinder in water waves.

The kinetic and the dynamic conditions at the free surface are

$$\eta_t - \phi_z + \eta_r\,\phi_r + \frac{1}{r^2}\,\eta_\theta\,\phi_\theta = 0 \quad \text{on} \quad z = \eta, \tag{11.3.2}$$

$$g\eta + \phi_t + \frac{1}{2}(\nabla\phi)^2 = 0 \quad \text{on} \quad z = \eta, \tag{11.3.3}$$

where g is the acceleration due to gravity and the velocity components are $(u_r, u_\theta, u_z) = \nabla\phi = (\phi_r, (1/r)\phi_\theta, \phi_z)$.

The boundary conditions at the rigid bottom and at the body's surface are

$$\phi_z \to 0 \quad \text{as} \quad z \to -\infty, \tag{11.3.4}$$

$$\phi_r = 0 \quad \text{at} \quad r = b. \tag{11.3.5}$$

It is noted that ϕ satisfies the three-dimensional Laplace equation (11.3.1). With a fluid of finite depth h and a vertical circular cylinder, the eigenfunctions corresponding to the real and imaginary eigenvalues are given by

$$\left\{\begin{matrix} H_m^{(1)} kr \\ H_m^{(2)}(kr) \end{matrix}\right\}\left\{\begin{matrix} \cos m\theta \\ \sin m\theta \end{matrix}\right\}\phi_0(z), \qquad m = 0, 1, 2, 3, ..., \tag{11.3.6ab}$$

$$\left\{\begin{matrix} I_m(k_n r) \\ K_m(k_n r) \end{matrix}\right\}\left\{\begin{matrix} \cos m\theta \\ \sin m\theta \end{matrix}\right\}\phi_n(z), \qquad n = 1, 2, 3, ..., \tag{11.3.7ab}$$

where $H_m^1(kr)$ and $H_m^{(2)}(kr)$ are the Hankel functions of the first and second kinds, and $I_m(kr)$ and $K_m(kr)$ are the modified Bessel functions of the first and second kinds, respectively. The most general solutions that satisfies the radiation condition at infinity is given by

$$\phi = \sum_{m=0}^{\infty} (a_{0m}\cos m\theta + b_{0m}\sin m\theta)\, H_m^{(1)}(kr)\phi_0(z)$$
$$+ \sum_{m=0}^{\infty}\sum_{n=1}^{\infty} (a_{nm}\cos m\theta + b_{nm}\sin m\theta)\, K_m(k_n r)\,\phi_n(z). \quad (11.3.8)$$

Using the asymptotic expansions for the Bessel functions for large r and noting the fact that the terms associated with $K_m(k_n r)$ represent exponentially decaying modes, the final asymptotic representation for the wave potential is

$$\phi \sim -\left(\frac{ig}{\omega}\right) \sum_{m=0}^{\infty} (a'_{0m}\cos m\theta + b'_{0m}\sin m\theta)\, H_m^{(1)}(kr) \frac{\cosh\, k(z+h)}{\cosh\, kh},$$
$$\sim -\left(\frac{ig}{\omega}\right)\left(\frac{2}{\pi kr}\right)^{1/2} A(\theta) \frac{\cosh\, k(z+h)}{\cosh\, kh} \exp\left(ikr - \frac{i\pi}{4}\right), \text{ as } kr \to \infty, \quad (11.3.9)$$

where

$$A(\theta) = \sum_{m=0}^{\infty} (a'_{0m}\cos m\theta + b'_{0m}\sin m\theta) \exp\left(-\frac{im\pi}{2}\right), \quad (11.3.10)$$

representing the angular variation of the radially spreading waves.

The asymptotic result (11.3.9) represents an outgoing wave and therefore, satisfies the radiation condition at infinity. Alternatively, this can be expressed, by differentiation, as

$$\sqrt{kr}\left(\frac{\partial\phi}{\partial r} - ik\phi\right) \to 0 \quad \text{as} \quad kr \to \infty. \quad (11.3.11)$$

In terms of small wave steepness $\varepsilon = ak \ll 1$, we write down the Stokes expansions of ϕ and η as

$$\phi = \varepsilon\phi^{(1)} + \varepsilon^2\phi^{(2)} + \cdots, \quad (11.3.12)$$
$$\eta = \varepsilon\eta^{(1)} + \varepsilon^2\eta^{(2)} + \cdots. \quad (11.3.13)$$

We must use Taylor series expansions of the nonlinear free surface conditions (11.3.2) and (11.3.3) about $z = 0$. For example

$$\phi(r, \theta, z = \eta, t) = \phi(r, \theta, 0, t) + \eta\left(\frac{\partial\phi}{\partial z}\right)_{z=0} + \frac{1}{2}\eta^2\left(\frac{\partial^2\phi}{\partial z^2}\right)_{z=0} + \cdots. \quad (11.3.14)$$

Substitution of these expansions into (11.3.2) and (11.3.3) and equating similar power of ε leads to the familiar free surface conditions on $z = 0$, valid for all r, θ, and t. To the first order, these conditions on $z = 0$ are

$$\eta_t^{(1)} - \phi_z^{(1)} = 0 \quad \text{and} \quad g\eta^{(1)} + \phi_t^{(1)} = 0, \tag{11.3.15ab}$$

which give a single equation for $\phi^{(1)}$:

$$\phi_{tt}^{(1)} + g\phi_z^{(1)} = 0 \quad \text{on} \quad z = 0. \tag{11.3.16}$$

To the second order, the free surface conditions on $z = 0$ are

$$\eta_t^{(2)} - \phi_z^{(2)} = -\left(\eta_r^{(1)}\phi_r^{(1)} + \frac{1}{r^2}\,\eta_\theta^{(1)}\phi_\theta^{(1)}\right) + \eta^{(1)}\phi_{zz}^{(1)}, \tag{11.3.17}$$

$$g\,\eta^{(2)} + \phi_t^{(2)} = -\eta^{(1)}\phi_{zt}^{(1)} - \frac{1}{2}\left(\nabla\phi^{(1)}\right)^2. \tag{11.3.18}$$

Eliminating $\eta^{(1)}$ and $\eta^{(2)}$ from these two conditions gives a single condition for the potential function on $z = 0$ as

$$\phi_{tt}^{(2)} + g\,\phi_z^{(2)} = \frac{1}{g}\,\phi_t^{(1)}\frac{\partial}{\partial z}\left(\phi_{tt}^{(1)} + g\phi_z^{(1)}\right) - \frac{\partial}{\partial t}\left(\nabla\phi^{(1)}\right)^2. \tag{11.3.19}$$

The Stokes expansions also require

$$\nabla^2\phi^{(1)} = \nabla^2\phi^{(2)} = 0, \tag{11.3.20ab}$$

$$\phi_z^{(1)} \to 0, \qquad \phi_z^{(2)} \to 0 \quad \text{as} \quad z \to -\infty, \tag{11.3.21ab}$$

$$\phi_r^{(1)} = 0, \qquad \phi_r^{(2)} = 0 \quad \text{on} \quad r = b. \tag{11.3.22ab}$$

In addition, $\phi^{(1)}$ and $\phi^{(2)}$ must satisfy the radiation condition as $r \to \infty$. The famous Bernoulli's equation

$$\frac{p}{\rho} + gz + \phi_t + \frac{1}{2}(\nabla\phi)^2 = \frac{p_a}{\rho} \tag{11.3.23}$$

gives the irrotational-flow pressure p, which consists of three components:

(i) The hydrostatic pressure, $p_a - g\rho z$ at a height z above the mean level $z = 0$, where the constant atmospheric pressure is p_a.
(ii) The transient pressure, $-\rho\phi_t$.
(iii) The dynamic pressure of magnitude $\frac{1}{2}\rho\left(U^2 - q^2\right)$ at a point where the flow speed is $q = |\nabla\phi|$.

There is a certain resultant force with which each of these pressure components acts on a body in a stream of homogenous fluid of variable speed U.

It is noted here that the hydrostatic pressure is the familiar buoyancy force. Accordingly, only the resultant of the other pressure fields gives the hydrodynamic force on the body. However, it can be shown that the dynamic pressures associated with the irrotational flow speed q have zero resultant. On the other hand, the resultant of the transient pressures is linear in U, and its component in the direction of U represents the total potential-flow drag of value (11.2.3) involved in the Morison equation (11.2.5). This is, indeed, the resultant force for rotationally symmetric bodies such as spheres and cylinders.

Finally, we evaluate (11.3.23) at $z = \eta$ and take the total derivative of (11.3.23) with respect to t, so that

$$\frac{D}{Dt}\left(\frac{p}{\rho}\right) + \frac{D}{Dt}\left[g\eta + \phi_t + \frac{1}{2}(\nabla\phi)^2\right] = 0. \tag{11.3.24}$$

This is rewritten by using the kinematic free surface condition, and the fact that $\mathbf{q}\cdot\nabla\phi_t = (\partial/\partial t)\left[\frac{1}{2}(\nabla\phi)^2\right]$ to obtain

$$\frac{D}{Dt}\left(\frac{p_a}{\rho}\right) + (\phi_{tt} + g\phi_z)_{z=\eta} + \left[\frac{\partial}{\partial t}(\nabla\phi)^2\right]_{z=\eta} + \frac{1}{2}\left[\mathbf{q}\cdot\nabla(\nabla\phi)^2\right]_{z=\eta} = 0. \tag{11.3.25}$$

If the atmospheric pressure p_a is constant, then the first term in (11.3.25) vanishes so that (11.3.25) becomes

$$(\phi_{tt} + g\phi_z) + \frac{\partial}{\partial t}(\nabla\phi)^2 + \frac{1}{2}\left[\mathbf{q}\cdot\nabla(\nabla\phi)^2\right] = 0 \quad \text{on} \quad z = \eta. \tag{11.3.26}$$

This is indeed an exact nonlinear free surface condition for ϕ with unknown free surface elevation function η.

Retaining up to the second-order term in (11.3.26), we obtain

$$\frac{\partial^2\phi}{\partial t^2} + g\frac{\partial\phi}{\partial z} = -\frac{\partial}{\partial t}(\nabla\phi)^2 \quad \text{on} \quad z = \eta. \tag{11.3.27}$$

It is noted here that a simple combination of (11.3.2) and (11.3.3) also gives the same result (11.3.27).

Equation (11.2.27) is a significant modification of the free surface condition due to Lighthill so that it becomes accurate to the second order in small quantities. The right hand side is twice the value obtained by simply taking into account the dynamic pressure term, $-\frac{1}{2}\rho(\nabla\phi)^2$ when the water pressure is equal to atmospheric pressure on the free surface.

11.4 Linear Diffraction Theory and Morison's Coefficient

The diffraction problem is usually concerned with the calculation of the irrotational-flow response of a solid body to incident sinusoidal water waves. The presence of the body in the fluid results in diffraction of the incident waves, and in the addition of a disturbance to the incident wave potential associated with the *scattering* effect of the body. Thus, the solution of the diffraction problem consists of an incident wave potential and a scattering potential, so that the linearized velocity potential is split into the form

$$\phi = \phi_I + \phi_S, \tag{11.4.1}$$

where the incident wave potential ϕ_I satisfies the required boundary and free surface conditions, and ϕ_S is the scattering potential to represent the disturbance of the incident waves by the fixed body so that ϕ_S must also satisfy those conditions. It represents a new wave field radiated by a distribution of normal motions at the surface of the body, so that ϕ_S must satisfy the boundary condition

$$\frac{\partial \phi_S}{\partial n} = -\frac{\partial \phi_I}{\partial n} \quad \text{on the body surface } S_B. \tag{11.4.2}$$

The scattered potential is also required to satisfy the standard radiation condition at infinity, that is, all scattered waves travel outwards from the body.

Havelock (1940) developed the linear diffraction-theory potential for small-amplitude deep-water waves incident on a long, vertical, surface-piercing circular cylinder of radius b. The incident wave potential appropriate to a straight-crested Stokes wave propagating in the x-direction along $\theta = 0$, is given by

$$\phi_I = \left(\frac{ga}{\omega}\right) e^{kz} \cos(\omega t - kx), \tag{11.4.3}$$

where the associated dispersion relation is

$$\omega^2 = gk, \tag{11.4.4}$$

and the corresponding free surface elevation is

$$\eta_I = a \sin(\omega t - kx). \tag{11.4.5}$$

These results assume the following forms in water of finite depth h:

$$\phi_I = \left(\frac{ga}{\omega}\right) \frac{\cosh k(z+h)}{\sinh kh} \cos(\omega t - kx), \tag{11.4.6}$$

$$\omega^2 = gk \tanh kh, \tag{11.4.7}$$

$$\eta_1 = a \sin(\omega t - kx). \tag{11.4.8}$$

According to Havelock's theory, the linear diffraction potential for a vertical cylinder in an incident wave field specified by (11.4.3) has the form

$$\phi = \mathrm{Re}\left[\left(\frac{a\omega}{k}\right)\exp\left(i\omega t + kz\right)\sum_{n=0}^{\infty}\alpha_n\phi_n(kr)\cos n\theta\right], \tag{11.4.9}$$

where

$$\alpha_0 = 1, \quad \alpha_n = 2\left(-1\right)^n, \quad (n > 0) \tag{11.4.10}$$

and

$$\phi_n(kr) = J_n(kr) - \frac{H_n^{(2)}(kr)J_n'(kb)}{H_n^{(2)'}(kb)}. \tag{11.4.11}$$

It follows from the two terms in (11.4.11) that the series containing the term $J_n(kr)$ represents ϕ_I and the series associated with the second term in (11.4.11) corresponds to ϕ_S. Since the scattered potential ϕ_S in (11.4.9) has been expanded as a series of Hankel functions $H_n^{(2)}(kr)$, it follows from the asymptotic expansion of $H_n^{(2)}(kr)$ that ϕ_S satisfies the required radiation condition at infinity.

The linear force $F_L(z,t)$ per unit height of the cylinder is

$$F_L(z,t) = \int_0^{2\pi}\left(-\rho\frac{\partial\phi_I}{\partial t}\right)(-\cos\theta)\,d\theta, \tag{11.4.12}$$

and hence, the total linear force F_L on the cylinder is

$$F_L = \int_{-\infty}^{0} F_L(z,t)dz. \tag{11.4.13}$$

A simple calculation gives the total linear force on the cylinder,

$$F_L = \rho g\pi ab^2 C_M\cos\left(\omega t + \beta\right), \tag{11.4.14}$$

where the coefficients C_M gives the cylinder's inertial reaction to a fluid acceleration advanced in phase by the phase lead β over its value on the cylinder axis $x = y = 0$, and

$$C_M\, e^{i\beta} = \left(\frac{2}{kb}\right)\phi_I(kb) = -\frac{4i}{\pi(kb)^2 H_n^{(2)'}(kb)}. \tag{11.4.15}$$

The coefficient C_M and the phase-lead β are plotted against kb, and the graphs are shown in Figure 11.2.

MacCamy and Fuchs (1954) extended Havelock's linear diffraction theory for deep water waves to waves in water of finite depth h. With the

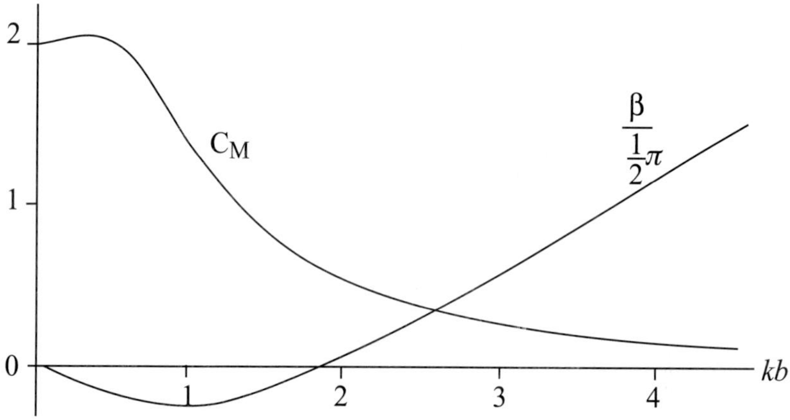

Fig. 11.2 Graphs of C_M and β plotted vs. kb.

incident wave potential (11.4.6) and with some changes of notation, their solution for the first-order potential can be expressed in the form

$$\phi = \left(\frac{ga}{\omega}\right) \frac{\cosh k(z+h)}{\cosh kh} \sum_{n=0}^{\infty} \varepsilon_n C_n(kr) X_n(kb) \cos\left(\frac{n\pi}{2} + \theta_n - \omega t\right) \cos n\theta, \tag{11.4.16}$$

where

$$C_n(kr) = J_n(kr) Y_n'(kb) - Y_n(kr) J_n'(kb), \tag{11.4.17}$$

$$X_n(kb) = \left[\{J_n'(kb)\}^2 + \{Y_n'(kb)\}^2\right]^{-1/2}, \tag{11.4.18}$$

$$\theta_n = \tan^{-1}\left[J_n'(kb)/Y_n'(kb)\right], \tag{11.4.19}$$

the coefficient ε_n denotes the Neumann's number, $\varepsilon_0 = 1$, $\varepsilon_n = 2$, $n \geq 1$, and $J_n(x)$ and $Y_n(x)$ are the Bessel functions of the first and second kinds, respectively.

The total first-order free surface elevation, $\eta = \eta_I + \eta_S$, is given by

$$\eta = a \sum_{n=0}^{\infty} \varepsilon_n C_n(kr) X_n(kb) \sin\left(\frac{n\pi}{2} + \theta_n - \omega t\right) \cos n\theta. \tag{11.4.20}$$

The total first-order force on the cylinder in the x-direction, $\varepsilon F_x^{(1)}$, and the first-order moment about $z = d$, $\varepsilon M^{(1)}(d)$, require only the term $n = 1$ in (11.4.16), since

$$F_x^{(1)} = 2b\rho \int_0^{\pi} \int_{-h}^{0} \left[\frac{\partial}{\partial t}\phi\right]_{r=b} \cos(\pi - \theta)\, d\theta\, dz, \tag{11.4.21}$$

and

$$M_x^{(1)}(d) = 2b\rho \int_0^{\pi} \int_{-h}^{0} (d-z) \left[\frac{\partial}{\partial t}\phi\right]_{r=b} \cos(\pi - \theta)\, d\theta\, dz. \quad (11.4.22)$$

In view of the Wronksian property of Bessel functions, $C_n(kb) = 2/\pi kb$, and hence, results (11.4.21) and (11.4.22) reduce to

$$F_x^{(1)} = -4\rho g a k^{-2} X_1(kb) \tanh kh \cos(\theta - \omega t), \quad (11.4.23)$$

$$M_x^{(1)}(d) = 4\rho g a k^{-3} \left[kd \tanh kh - \mathrm{sech} kh + 1\right] X_1(kb) \cos(\theta - \omega t). \quad (11.4.24)$$

Clearly, result (11.4.21) yields the moment about the ocean bottom $(d = -h)$, which is in phase with $F_x^{(1)}$.

11.5 Nonlinear Wave Forces on Offshore Structures and Nonlinear Modification of the Morison Equation

We make reference to the cylindrical model of Section 11.2 and follow Lighthill (1979) and Debnath and Rahman (1981) to write down the Stokes expansion for the velocity potential ϕ and the free surface elevation η in the form

$$\phi = \phi_1 + \phi_q + \cdots, \quad \eta = \eta_1 + \eta_q + \cdots, \quad (11.5.1ab)$$

where ϕ_1 and η_1 are solutions for a wave of frequency ω satisfying the linear diffraction problem, and ϕ_q and η_q are the associated second-order solutions of the order of the squares of the wave steepness, $\varepsilon = ak$. We assume that the linear solutions are of the order of ε, and all third- and higher-order terms are neglected. We next write the Taylor series expansions about $z = 0$ in the form

$$\begin{aligned}\phi &= \phi(r,\theta,0) + \eta\left(\frac{\partial\phi}{\partial z}\right)_{z=0} + \cdots \\ &= \phi_1 + \phi_q + \cdots + (\eta_1 + \eta_q + \cdots)\left[\frac{\partial\phi_1}{\partial z} + \frac{\partial\phi_q}{\partial z}\right]_{z=0} + \cdots. \quad (11.5.2)\end{aligned}$$

In view of the assumptions just made, (11.5.2) reduces to

$$\phi = \phi_1 + \phi_q + \eta_1\left(\frac{\partial\phi_1}{\partial z}\right)_{z=0} + \text{higher-order terms}, \quad (11.5.3)$$

where the the first term is the first-order solution and the second one represents the second-order solution. Invoking similar Taylor series expansions, we obtain

$$\frac{\partial \phi}{\partial z} = \frac{\partial \phi_1}{\partial z} + \left\{ \frac{\partial \phi_q}{\partial z} + \eta_1 \left[\frac{\partial}{\partial z} \left(\frac{\partial \phi_1}{\partial z} \right) \right]_{z=0} \right\} + \cdots, \tag{11.5.4}$$

$$\frac{\partial^2 \phi}{\partial t^2} = \frac{\partial^2 \phi_1}{\partial t^2} + \left\{ \frac{\partial^2 \phi_q}{\partial t^2} + \eta_1 \left[\frac{\partial}{\partial z} \left(\frac{\partial^2 \phi_1}{\partial t^2} \right) \right]_{z=0} \right\} + \cdots, \tag{11.5.5}$$

$$\nabla \phi = \nabla \phi_1 + \left\{ \nabla \phi_q + \eta_1 \left[\frac{\partial}{\partial z} (\nabla \phi_1) \right]_{z=0} \right\} + \cdots. \tag{11.5.6}$$

Substituting these results in (11.5.2) and (11.5.3) yields the following equations, for the case of deep water:

$$\frac{\partial^2 \phi_1}{\partial t^2} + g \frac{\partial \phi_1}{\partial z} = 0, \tag{11.5.7}$$

$$\frac{\partial^2 \phi_q}{\partial t^2} + g \frac{\partial \phi_q}{\partial z} = -\frac{\partial}{\partial t} (\nabla \phi_1)^2. \tag{11.5.8}$$

Thus, the linearized potential ϕ_1 satisfies (11.5.7) not only on $z = 0$, but everywhere in the fluid, including $z = \eta$. Equation (11.5.8), although obtained at the exact free surface, can be applied at the undisturbed free surface $z = 0$ because both sides are quadratic, so that the error involved at the free surface is of still higher order.

For the case of fluid of finite depth h, equations for ϕ_q corresponding to (11.5.8) and η_q are given by

$$\frac{\partial^2 \phi_q}{\partial t^2} + g \frac{\partial \phi_q}{\partial z} = -\frac{\partial}{\partial t} (\nabla \phi_1)^2 + \frac{1}{g} \frac{\partial \phi_1}{\partial t} \left[\frac{\partial}{\partial z} \left(\frac{\partial^2 \phi_1}{\partial t^2} + g \frac{\partial \phi_1}{\partial z} \right) \right]_{z=0}, \tag{11.5.9}$$

$$\eta_q = -\frac{1}{g} \left[\frac{\partial \phi_q}{\partial t} + \frac{1}{2} (\nabla \phi_1)^2 - \frac{1}{g} \left(\frac{\partial \phi_1}{\partial t} \right) \left(\frac{\partial^2 \phi_1}{\partial z\, \partial t} \right) \right]_{z=0}. \tag{11.5.10}$$

Several comments on these results are in order. First, the right-hand side of equation (11.5.9) contains double frequency components as functions of r, θ and z. Second, if incident waves in finite depth are considered, then (11.5.10) reveals that the terms associated with products of the linear potential will yield both mean and double frequency components. Thus, except for the case of deep water, it is necessary for the mean component of $(\partial \phi_q / \partial t)$ to have a constant term in order that the mean free surface elevation be at $z = 0$ for the undisturbed incident waves. In order to comment further, it is necessary to mention certain general results concerning irrotational fluid flows that are extremely useful for the study of both steady and unsteady surface waves in oceans. When solid boundaries are absent in irrotational fluid flows, there is no principal source of generation of vorticity. With the linearized free surface conditions at the free surface, $z = 0$, the problem of determining the irrotational flow field becomes a purely linear problem because the velocity potential satisfies the Laplace equation. For such a linear problem, the irrotational-flow response to an incident wave field of every frequency and direction in the ocean-wave spectrum can be determined explicitly. For example, the sinusoidal waves of amplitude a propagating freely in the x-direction in water of uniform depth h have the velocity potential ϕ and the free surface elevation η given by (11.4.6) and (11.4.8), respectively. The corresponding waves in deep water can be used as another example.

When incident waves are combined with the scattered waves due to the presence of a solid structure, the fluid velocity at the free surface may not remain constant. So there is a significant quadratic correction to the velocity potential of a disturbed wave, even though a perfectly periodic wave-train has none. The quadratic potential ϕ_q satisfies the equation (11.5.8) or (11.5.9) at the free surface, and zero boundary conditions everywhere else. Physically, this means that ϕ_q is uniquely determined as the potential of the linearized wave motion generated in the presence of the stationary vessel, due to a variable surface pressure P given by

$$P = \rho\left(\nabla\phi_1\right)^2 \quad \text{in deep water,} \tag{11.5.11}$$

or, in a fluid of finite depth h,

$$P = \rho\left[\left(\nabla\phi_1\right)^2 + \frac{1}{g}\left(\frac{\partial\phi_1}{\partial t}\right)\left\{\frac{\partial}{\partial z}\left(\frac{\partial^2\phi_1}{\partial t^2} + g\frac{\partial\phi_1}{\partial z}\right)\right\}_{z=0}\right]. \tag{11.5.12}$$

The *quadratic wave force* F_q generated by the second-order wave potential ϕ_q is defined by

$$F_q = -\int_{z=0} PW\, dS, \tag{11.5.13}$$

where P is given by (11.5.11) or (11.5.12) and W is the vertical velocity on the free surface associated with the unit translational oscillation of the body. The value of W at $z = 0$ can be determined by solving the linear problem. We shall find an explicit form of W for the cylindrical model.

In addition to the quadratic force F_q, another equally important contribution to the second-order loading can be obtained from the linear velocity potential ϕ_1. This second-order force may be called the *dynamic force* , F_d, which is the resultant of the dynamic pressures $-\frac{1}{2}\rho\left(\nabla\phi_1\right)^2$ associated with the linear velocity potential ϕ_1. Hence, this force is defined by an integral over the surface of the body as

$$F_d = \frac{1}{2}\int_S \rho\left(\nabla\phi_1\right)^2 n_x dS, \tag{11.5.14}$$

where n_x is the direction cosine between the outward normal and the direction of F_d being determined. It may be pointed out that the fore- and aft-symmetry makes this integral likely to vanish, provided ϕ_1 is simply the potential of how the body responds to variable velocities due to waves. However, in practice, ϕ_1 also includes a response to variable extensional motions. All the second-order terms in the irrotational-flow forces arise from the interaction between the element's response to (i) the incident wave's variable velocity and (ii) its variable extension. Neither the response to the variable velocity by itself nor the variable extension alone would produce quadratic terms in the result for the wave loading. All the nonzero second-order contributions would come from the cross-term in (11.5.14) involving products of these two kinds of responses.

Another force provides a second-order contribution to the wave loading on structures whenever a body penetrates the free surface of the liquid. Lighthill (1979) has shown that the transient and the hydrostatic pressure distributions (they are equal and opposite at the free surface) are combined to give a horizontally resolved force per unit length of the waterline where the body intersects the free surface. Thus, the combined force is

$$-\frac{1}{2}\rho g\eta^2 + g\rho\eta^2 = \frac{1}{2}g\rho\eta^2 = \frac{\rho}{2g}\left(\frac{\partial\phi_1}{\partial t}\right)^2. \tag{11.5.15}$$

Then the total additional second-order horizontal x-component of force acting at the waterline w is defined by

$$F_w = \oint_w \frac{\rho}{2g}\left(\frac{\partial\phi_1}{\partial t}\right)^2 dy, \tag{11.5.16}$$

where the integrand is the horizontally resolved force per unit length acting at the waterline, and the integral in (11.5.16) gives its resultant force in the x-direction.

In summary, the total second-order force F_{nl} acting on the offshore solid structure is the sum of three forces, F_q, F_d and F_w, given by (11.5.13), (11.5.14) and (11.5.16), respectively, so that the total nonlinear force is

$$F_{nl} = F_q + F_d + F_w, \tag{11.5.17}$$

where each term in this result can be determined in terms of computed solutions of linear wave problems.

We next calculate all linear and nonlinear wave forces acting on a very large solid structure of size b comparable with the wavelength λ so that $\lambda/2\pi = k^{-1}$. As an example, we consider a vertical cylinder of radius b with the axis $x = y = 0$ in deep water waves and with the velocity potential (11.4.3). The horizontal velocity in the undisturbed wave train on the axis of the cylinder represents the variable velocity $U(z,t)$ so that

$$U(z,t) = (a\omega \sin \omega t)\, e^{kz}. \tag{11.5.18}$$

To a first-order approximation, the irrotational response to the variable velocity (11.5.18) produced by the cylinder is the simple dipole field with the velocity potential

$$\phi_1 = U(z,t)\left(r + \frac{b^2}{r}\right)\cos\theta, \tag{11.5.19}$$

which is expressed in terms of cylindrical polar coordinates. The value of the potential, and of the horizontal velocity on the cylinder, $r = b$, are

$$(\phi_1)_{r=b} = 2Ub\cos\theta, \qquad \left(\frac{1}{r}\frac{\partial \phi_1}{\partial \theta}\right)_{r=b} = -2U\sin\theta. \tag{11.5.20ab}$$

The linear force $F_1(z,t)$ per unit length of the cylinder is the resultant of the transient pressures $(-\rho\partial\phi_1/\partial t)$ so that

$$F_1(z,t) = \int_0^{2\pi}\left(-\rho\frac{\partial\phi_1}{\partial t}\right)(-\cos\theta)\, b\, d\theta = 2\rho\pi b^2 \dot{U}, \tag{11.5.21}$$

which corresponds to the classical Morison coefficient, $C_M = 2$.

Thus, the total linear force exerted on the cylinder is

$$F_L(t) = \int_{-\infty}^{0} F_1(z,t)\, dz = 2\rho g a\pi b^2 \cos\omega t. \tag{11.5.22}$$

The total drag force involved in the Morison equation is given by

$$F_D = \frac{1}{2}\int_{-\infty}^{0} \rho A C_D U^2 dz = \frac{1}{4}\rho g A a^2 C_D \sin^2 \omega t. \tag{11.5.23}$$

In addition to the variable velocity produced near the cylinder by the incident waves, waves can also generate significant variable extensional motions, as mentioned earlier. The variable extension is given by the value of the gradient of the horizontal velocity, $(\partial/\partial x)(\partial\phi_1/\partial x)$, in the undisturbed wave train (11.4.3) on the axis of the cylinder. So the variable extension is

$$E(z,t) = \left(\frac{\partial^2 \phi_1}{\partial x^2}\right)_{x=0} = (-a\omega\cos\omega t)\, k e^{kz}. \tag{11.5.24}$$

As in solid mechanics, it is often reasonable in fluid dynamics to treat an extension as a linear combination of a pure dilatation and a dilatationless strain. The local irrotational response to the pure dilatation is a simple monopole field, whereas the corresponding response to the dilatationless strain is a quadrupole field. Thus, the combined potential takes the form

$$\phi_1 = \frac{E}{4}\left(r^2 - 2b^2 \log r\right) + \frac{E}{4}\left(r^2 + \frac{b^4}{r^2}\right)\cos 2\theta. \tag{11.5.25}$$

The horizontal velocity on the cylinder surface ($r = b$) is

$$\left(\frac{1}{r}\frac{\partial \phi_1}{\partial \theta}\right)_{r=b} = -Eb\sin 2\theta. \tag{11.5.26}$$

Thus, the square of the horizontal velocity on the cylinder surface ($r = b$) obtained from (11.5.20b) and (11.5.26) is

$$(2U\sin\theta + Eb\sin 2\theta)^2. \tag{11.5.27}$$

The vertical velocity on the cylinder surface ($r = b$) is obtained by combining the vertical velocity in the incident waves (11.4.3), which is equal to $\left(-k^{-1}E\right)$, with the vertical velocity $(\partial\phi_1/\partial z)_{r=b} = 2Ubk\cos\theta$ so that their sum squared is

$$\left(2bkU\cos\theta - k^{-1}E\right)^2. \tag{11.5.28}$$

Thus, the square of the fluid velocity on $r = b$ is

$$\left[(\nabla\phi_1)^2\right]_{r=b} = \left[\left(\frac{\partial\phi_1}{\partial r}\right)^2 + \left(\frac{1}{r}\frac{\partial\phi_1}{\partial\theta}\right)^2 + \left(\frac{\partial\phi_1}{\partial z}\right)^2\right]_{r=b}, \tag{11.5.29}$$

which is, since $(\partial\phi_1/\partial r) = 0$ on $r = b$,

$$\begin{aligned} &= \left(\frac{1}{r}\frac{\partial\phi_1}{\partial\theta}\right)^2_{r=b} + \left(\frac{\partial\phi_1}{\partial z}\right)^2_{r=b} \\ &= (2U\sin\theta + Eb\sin 2\theta)^2 + \left(2bkU\cos\theta - k^{-1}E\right)^2. \end{aligned} \tag{11.5.30}$$

Thus, the dynamic pressure per unit height of the cylinder is

$$F_d(z,t) = \frac{1}{2}\int_0^{2\pi}\left[\rho\,(\nabla\phi_1)^2_{r=b}\right](-\cos\theta)\,b\,d\theta,$$

which is, due to only non-zero contribution of the cross-term in UE and no contribution from the terms in U^2 and E^2 in (11.5.30),

$$\begin{aligned} &= -2\rho b^2 UE\int_0^{2\pi}(\cos\theta - \sin\theta\sin 2\theta)\cos\theta\, d\theta \\ &= -\rho\pi U E b^2. \end{aligned} \tag{11.5.31}$$

The total dynamic force $F_d(t)$ exerted on the cylinder is then equal to

$$F_d(t) = \int_{-\infty}^{0} -\rho\pi U E b^2 dz = \frac{1}{4}(ak)\pi g\rho a b^2 \sin 2\omega t. \tag{11.5.32}$$

A simple comparison of the total dynamic force (11.5.32) with the total linear force (11.5.22) reveals that the importance of the former increases with the wave steepness (ak). This analysis is based upon the simple periodic waves and reveals that the second-order forces depend on the product UE. So this study can be extended to a more general random wave field. In that case, U is simply the local variable velocity and E is the local variable extension in the wave field.

We next calculate the waterline force F_w defined by (11.5.16). The term $(\partial\phi_1/\partial t) = 2\dot{U}b\cos\theta$ is obtained on the cylinder surface $(r = b)$ by using (11.5.20a), and the incident wave field (11.4.3) at $x = z = 0$, so that $\partial\phi_1/\partial t = (-ga\sin\omega t)$, which is produced by the rise and fall of the water surface associated with the extension (11.5.24). Thus, the square of their sum is

$$\left(2\dot{U}b\cos\theta - ga\sin\omega t\right)^2 = (2gkab\cos\omega t\cos\theta - ga\sin\omega t)^2. \quad (11.5.33)$$

The waterline force F_w, given by (11.5.16) assumes the form

$$\begin{aligned} F_w &= \int_0^{2\pi} \frac{\rho}{2g}(2gkab\cos\omega t\cos\theta - ga\sin\omega t)^2(-\cos\theta)\, b\, d\theta \\ &= (ak)g\rho a\pi b^2 \sin 2\omega t. \end{aligned} \quad (11.5.34)$$

It is interesting to compare F_w, given by (11.5.34), with F_d, given by (11.5.32). Clearly, in the limit of small kb, $F_w = 4F_d$. The second-order correction to the Morison equation for a vertical cylinder is given by the sum of (11.5.32) and (11.5.34), so that

$$(F_d + F_w) = \frac{5}{2}(ak)g\rho a\pi b^2 \sin\omega t\cos\omega t. \quad (11.5.35)$$

The significance of the nonlinear correction term (11.5.35) is that it gives an enhanced positive peak of the total drag force, occurring when $\cos\omega t$ and $\sin\omega t$ are both positive. The importance of the correction term increases with the wave steepness ak.

The Morison equation (11.2.5) for the case of a vertical cylinder is

$$D = 2\rho g a\pi b^2 \cos\omega t + \frac{1}{2}C_D\rho g a\,(ab)\sin\omega t\,|\sin\omega t|. \quad (11.5.36)$$

We next note that the vertical velocity W involved in (11.5.13) represents the effects, at the surface, of the fluid motions generated by horizontal translational oscillations of the cylinder with unit amplitude and twice the frequency (2ω) of the incident waves. The velocity W is directly proportional to $\cos\theta$, which can also be confirmed by the exact diffraction theory.

For small structures ($kb \ll 1$), the approximate value of W near the cylinder can be given by a simple form

$$W \sim (4k)\left(-\frac{b^2}{r}\cos\theta\right), \tag{11.5.37}$$

where the first factor $4k$ is found from the linearized boundary condition, $\phi_{tt} + g\phi_z = 0$ on $z = 0$, for oscillations of frequency 2ω to relate W at the surface to a local velocity, potential. This potential is described by the second factor in (11.5.37), whose radial derivative is $\cos\theta$ on the cylinder surface, $r = b$. Since W is proportional to $\cos\theta$, it is of special interest in any $\cos\theta$ term in other factors in (11.5.13). This is the square of the fluid velocity that can be derived from the linear velocity potential given by (11.5.19) and (11.5.25). The $\cos\theta$ term in the horizontal velocity squared is obtained as the cross-term between the results derived from the velocity potentials (11.5.19) and (11.5.25). Thus, it turns out that this term is

$$\left(2r - \frac{2b^2}{r} + \frac{b^4}{r^3} + \frac{b^6}{r^5}\right) UE\cos\theta. \tag{11.5.38}$$

Similarly, the $\cos\theta$ term of comparable magnitude in the vertical velocity squared is also obtained as a cross-term between the z-derivative of the dipole potential (11.5.19) and the vertical velocity in the incident waves (11.4.3). Thus, the square of the vertical velocity is

$$\left(\frac{\partial\phi_I}{\partial z} + \frac{\partial\phi_1}{\partial z}\right)^2 = \left[Uk\left(r + \frac{b^2}{r}\right)\cos\theta - \frac{1}{k}E\right]^2, \tag{11.5.39}$$

which produces the cross-term containing $\cos\theta$ as

$$-2UE\left(r + \frac{b^2}{r}\right)\cos\theta. \tag{11.5.40}$$

We now combine (11.5.38) and (11.5.40) to obtain the resulting $\cos\theta$ term in the total fluid velocity squared at the surface, $z = 0$,

$$(\nabla\phi_1)^2 = \frac{1}{2}\left(\frac{4b^2}{r} - \frac{b^4}{r^3} - \frac{b^5}{r^5}\right)(ga^2k^2\sin 2\omega t)\cos\theta. \tag{11.5.41}$$

In the limit of small kb ($kb \ll 1$), this term is small so that the contribution to F_q, given by (11.5.13), is negligible compared with other wave forces. As a matter of fact, the first term in (11.5.41) with positive coefficient reveals that the nonlinear contribution to the Morison equation increases with kb. This is a striking contrast with the linear contribution, which decreases with kb, as indicated earlier in Fig. 11.2. Consequently, in the diffraction regime with large kb ($\gg 1$), the nonlinear contribution to the

Morison equation would be quite significant. We examine this point further by using the dipole form (11.5.37) in a range $b < r < Ck^{-1} = (2\pi)^{-1} C\lambda$ representing the near field created by a translational oscillation of the body, where C is a numerical constant to be determined later on. It is noted that beyond a limiting radius $r = (2\pi)^{-1} C\lambda$, W oscillates and has no significant contribution to the quadratic wave force. The numerical value of C would not affect the major predictions. Finally, the quadratic wave force F_q can be expressed in the form

$$\begin{aligned} F_q &= -\int_b^{Ck^{-1}} r\,dr \int_0^{2\pi} (4k)\left(-\frac{b^2}{r}\cos\theta\right)\rho\left(\nabla\phi_1\right)^2_{z=0}\,d\theta \\ &= 2\pi g\rho k\left(abk\right)^2 \sin\omega t \int_b^{Ck^{-1}} \left(\frac{4b^2}{r} - \frac{b^4}{r^3} - \frac{b^6}{r^5}\right) dr \\ &= 2\pi g\rho a b^2 (kb)^2 (ak)\left[4\left(\log\frac{C}{kb}\right) - \frac{3}{4}\right]\sin 2\omega t, \end{aligned} \tag{11.5.42}$$

where kb is assumed to be small.

Thus, the quadratic force is insignificant in the limit of very small kb and begins to put its influence as kb increases. As already mentioned, the nonlinear correction is dominated by the sum (11.5.35) of the dynamic and waterline forces. In the diffraction regime, the quadratic force contributes to the magnitude of those other forces as kb increases.

We next use a standard argument of fluid dynamics to calculate the quadratic force F_q. We assume that the variable part of the surface pressure is

$$\rho\left(\nabla\phi_1\right)^2 = \mathrm{Re}\left[Pe^{2i\omega t}\right] \tag{11.5.43}$$

and

$$\phi_q = \mathrm{Re}\left[\Phi_q e^{2i\omega t}\right], \tag{11.5.44}$$

so that (11.5.8) gives

$$\left[\frac{\partial\Phi_q}{\partial z} - \sigma\Phi_q\right]_{z=0} = -\frac{2i\omega P}{g}, \tag{11.5.45}$$

where

$$\sigma = \frac{4\omega^2}{g} = 4k. \tag{11.5.46}$$

With the potential due to a unit translational oscillation of the body in the form $\mathrm{Re}\left[\Psi e^{2i\omega t}\right]$, it follows that, on the surface S of the body,

$$\left(\frac{\partial\Psi}{\partial n}\right)_S = n_x, \qquad \left(\frac{\partial\Phi_q}{\partial n}\right)_S = 0, \tag{11.5.47ab}$$

where n_x is the x-component of a unit inward normal to S. Furthermore,

$$\left[\frac{\partial \Psi}{\partial z} - \sigma \Psi\right]_{z=0} = 0, \tag{11.5.48}$$

where Ψ also satisfies the radiation condition at infinity ($r \to \infty$). This leads us to use Green's theorem,

$$\int\int_V \left[\Phi_q \nabla^2 \Psi - \Psi \nabla^2 \Phi_q\right] dV = \int_{S \cup z=0} \left[\Phi_q \frac{\partial \Psi}{\partial n} - \Psi \frac{\partial \Phi_q}{\partial n}\right] dS, \tag{11.5.49}$$

where the surface integral is taken over the boundaries of the fluid (S and $z = 0$), and n is a normal outward from the fluid. Since both Ψ and Φ_q satisfy the Laplace equation, the left-hand side of (11.5.49) vanishes so that (11.5.49) gives

$$\int_{S \cup z=0} \left[\Phi_q \frac{\partial \Psi}{\partial n} - \Psi \frac{\partial \Phi_q}{\partial n}\right] dS = 0. \tag{11.5.50}$$

After performing indicated integration in (11.5.50) and applying the conditions (11.5.47ab) and (11.5.48), it follows that

$$\int_{z=0} \Psi \left[\frac{\partial \Phi_q}{\partial z} - \sigma \Phi_q\right] dx\, dy = \int_S \Phi_q n_x dS. \tag{11.5.51}$$

Thus, the quadratic force F_q can be written as

$$F_q = \int_S \left[-\rho \frac{\partial \Phi_q}{\partial t}\right] n_x\, dS = \text{Re}\left[-2\rho i \omega e^{2i\omega t} \int_S \Phi_q n_x dS\right]. \tag{11.5.52}$$

Invoking (11.5.51) and (11.5.45), the result (11.5.52) reduces to the form

$$\begin{aligned} F_q &= \text{Re}\left[-2i\omega\rho e^{2i\omega t}\left(-\frac{2i\omega}{g\sigma}\right)\int_{z=0} P \frac{\partial \Psi}{\partial z} dx\, dy\right] \\ &= \text{Re}\left[-\int_{z=0} P\, e^{2i\omega t} W\, dx\, dy\right], \end{aligned} \tag{11.5.53}$$

where $W = (\partial \Psi/\partial z)$ is the complex vertical velocity on the free surface at $z = 0$ associated with the unit translational oscillation of the body. This vertical velocity always contains a part that is the near-field standing wave motion generated close to the body. Also, it includes a radiating part that is significantly small for small kb.

To determine the vertical velocity W, it is necessary to find the solution for Ψ. The function Ψ satisfies the Laplace equation, the condition (11.5.47ab) on the surface $r = b$ of the cylinder, the free surface condition

(11.5.48) and the radiation condition at infinity. The solution Ψ takes the form

$$\Psi(r,\theta,z) = \left[\frac{i\sigma}{\pi}\int_{-\infty}^{\infty}\frac{K_1(|\alpha| r)}{|\alpha| K_1'(|\alpha| b)}\cdot\frac{e^{i\alpha z}d\alpha}{\alpha(\sigma - i\alpha)} + \frac{2e^{z\sigma}}{\sigma}\frac{H_1^{(2)}(\sigma r)}{H_1^{(2)'}(\sigma b)}\right]\cos\theta, \tag{11.5.54}$$

where the preceding integral is treated as the Cauchy principal value, $K_1(x)$, is the Bessel function of imaginary argument, and $H_1^{(2)}(x)$ is the Hankel function of the second kind of order one. Also, the integral in (11.5.54) is the standing-wave component: it makes the left-hand side of (11.5.48) vanish as the integral of an odd function. Its r-derivative on $r = b$ is a Fourier integral that can easily be evaluated as $1 - 2e^{z\sigma}$. The radial part added to the integral also satisfies the free surface condition on $z = 0$, and its r-derivative on the body surface ($r = b$) is $2e^{z\sigma}$. Hence, the combined result (11.5.54) satisfies the condition

$$\left(\frac{\partial\Psi}{\partial r}\right)_{r=b} = \cos\theta. \tag{11.5.55}$$

The vertical velocity $W = (\partial\Psi/\partial z)$ on $z = 0$ can be computed and assumes the following value, after some simplification:

$$W = \left[\frac{2}{\pi}\int_0^{\infty}\left\{-\frac{K_1(\sigma rm)}{mK_1'(\sigma bm)}\right\}\frac{dm}{1+m^2} + \frac{2H_1^{(2)}(\sigma r)}{H_1^{(2)'}(\sigma b)}\right]\cos\theta. \tag{11.5.56}$$

When σr and σb are very small,

$$-\frac{K_1(\sigma rm)}{mK_1'(\sigma bm)} \sim \frac{\sigma b^2}{r}. \tag{11.5.57}$$

Consequently, the first term inside the square bracket in (11.5.56) gives the value $\sigma b^2/r$, whereas the second term tends to the value $-2\sigma b^2/r$ as σr and σb are very small. Thus (11.5.56) becomes identical with (11.5.37).

As indicated before, it is necessary to find the range, $b < r < B$, within which the following result (11.5.58) holds. In order to determine the value of B, we define

$$\mathrm{Re}\int_b^{\infty} W\,dr = -\sigma b^2\left[\log\left(\frac{B}{b}\right)\right]\cos\theta. \tag{11.5.58}$$

But (11.5.56) implies that

$$\int_b^{\infty} W\,dr = \left[\frac{2}{\pi}\int_0^{\infty} -\frac{K_0(\sigma bm)}{\sigma m^2 K_1'(\sigma bm)}\cdot\frac{dm}{(1+m^2)} + \frac{2H_0^{(2)}(\sigma b)}{\sigma H_1^{(2)'}(\sigma b)}\right]\cos\theta. \tag{11.5.59}$$

When σb is very small,

$$\frac{K_0(\sigma bm)}{\sigma m^2 K_1'(\sigma bm)} \sim \sigma b^2 \left[\log\left(\frac{1}{2}\sigma bm\right) + \gamma\right], \tag{11.5.60}$$

where $\gamma = 0.577$ is Euler's constant.

We next use the result

$$\int_0^\infty \frac{\log m}{1+m^2}\, dm = 0 \tag{11.5.61}$$

to simplify (11.5.59) in the form

$$\int_b^\infty W\, dr \sim \left(-\sigma b^2\left\{\log\left(\frac{\sigma b}{2}\right) + \gamma\right\} + 2\sigma b^2\left\{\log\left(\frac{\sigma b}{2} + \gamma + \frac{i\pi}{2}\right)\right\}\right] \cos\theta. \tag{11.5.62}$$

This result agrees with (11.5.58), provided the value of B is

$$B = \frac{1}{\sigma}\left(2e^{-\gamma}\right) = \left(\frac{1}{2}e^{-\gamma}\right) k^{-1} = Ck^{-1}, \tag{11.5.63}$$

where $C = 0.28$. This value of C, approximately equal to $\frac{1}{4}$, corresponds to the fact that the wavenumber related to oscillations of frequency 2ω is quadrupled.

This analysis reveals that the nonlinear contributions to the wave loading on offshore structures are made up of the quadratic, dynamic and waterline forces associated with the irrotational flow component of the wave loading. Further, this study suggests a nonlinear modification of the classical Morison equation (11.2.5). The nonlinear contribution is given by

$$F_{nl} = F_q + F_d + F_w, \tag{11.5.64}$$

where F_q, F_d and F_w are defined by (11.5.13), (11.5.14) and (11.5.16), respectively.

The preceding analysis of nonlinear wave forces on offshore structures in deep water has been extended by Debnath and Rahman (1981) to the case of fluid of finite depth. However, it should be pointed out that the nonlinear diffraction theory for wave loading on offshore structures in a shallow ocean has been a controversial subject in recent years. More progress is needed to resolve the current controversial issues.

Finally, both the design of wave-energy extraction by submerged resonant ducts and the diffraction of water waves by offshore solid structures are of considerable interest in ocean engineering and naval hydrodynamics. Lighthill's 1979 paper entitled "Two-dimensional analysis related to wave-energy extraction by submerged resonant ducts" published in JFM represents his major interest in ocean engineering. Based on the fundamental principles of interaction between the fluid wave motion and solid structure, he also published another paper on "Fundamentals concerning wave loading on offshore structures" in JFM in 1986. This work has many important practical applications for the design of wave-energy devices consistent with the need for maximum seaworthiness. Based on two-dimensional wave hydrodynamics and linearized duct dynamics, he developed both local, models od forcing-pressure modification, and fully interactive models. His study led to inferences of many aspects of the duct's response to incident waves including forcing pressure modification, added mass, and energy absorption, transmission and reflection.

Chapter 12

Biofluid Mechanics

"Biofluiddynamics is concerned with the mechanics of the relative motions between living organisms and fluids. *External* biofluiddynamics deals with an organism's motions in relation to the dynamics of a fluid medium surrounding it, as in the study of aquatic animal locomotion or animal flight. *Internal* biofluiddynamics analyzes fluid motions inside an organism as in the study of circulation and respiration. A characteristic feature of biofluiddynamics in general is *interaction* between the motions of the fluid and of boundaries to it taking the form of highly flexible, actively motile tissue."

James Lighthill

"It is not Animal Motion performed by the Vibration of this Medium, excited in the Brain by the power of Will, and propagated from thence through the folid, pellucid and uniform Capillamenta of the Nerves into Muscles, for contracting and dilating them?"

John von Neumann

12.1 Introduction

No biographical sketch of James Lighthill's life would be complete without mention of Sir James Lighthill's life-long research interest in biofluiddynamics. The entire volume IV of his Collected Papers is devoted to his research contributions to this subject. Indeed, Sir James created this modern subject with new title biofluiddynamics and described eloquently the new scope of it in his above quotation stated in his 1975 John von Neumann Lecture on Flagellar Hydrodynamics at the SIAM Annual Meeting. In the Introduction of this Lecture Sir James states: "..., and am happy to accept the invitation to survey a "significant field" or area of science where application of mathematics can yield important results". He also told the audience: "... Those lectures were published this year by SIAM as a book, *Mathematical Biofluiddynamics.* In today's lecture, I am going to concentrate on one special field within that wide domain where the story can now be taken much further through a considerable deepening of understanding achieved during the past two years." In his introduction, Sir James emphasized the unique importance of interdisciplinary research between mathematics and other diverse areas including biology, zoology, and biophysics in collaboration with scientists of different countries, such as England and the United States. He also stated: "May I also acknowledge the 1974 Caltech Symposium,"Swimming and Flying in Nature" as a most noteworthy interdisciplinary discussion, with extensive sessions on biofluiddynamics of microorganisms, ..." All these discussions describe Sir James' philosophy and direction of future research of this new and modern field. This is followed by brief presentation of his research on both External Biofluiddynamics including studies on animal locomotion in fluid media, and Internal Biofluiddynamics including physiological fluid dynamics for the human body together with fluid flows in the microcirculation. However, my serious discussion of Biofluid Mechanics would require more space and time and competence than I have and many better references are available. So, instead, I have given a brief presentation with special reference to Sir James' research contributions to this modern subject.

12.2 External Biofluiddynamics

Sir James published at least sixteen original research, research expository and survey articles in major books and journals. In his paper "On the Squirming Motion of Nearly Spherical Deformable Bodies through Liquids at very small Reynolds Numbers" published in *Communications on Pure and Applied Mathematics*, **v** (1952) 109-118, he described that a spherical deformable body can swim, at very small Reynolds numbers, by performing small oscillations of shape. He also gave three examples of swimming motions with axial symmtery, in each of which the maximum surface strain is $\frac{1}{3}$. He also developed mathematical theory in his paper entitled, Note on the Swimming of Slender Fish published in the *Journal of Fluid Mechanics*, **9** (1960) 305-317. He worked out mathematical methods for a slender fish which makes swimming movements only in a single direction at right angles to its direction of locomotion.

In his 1975 John von Neumann Memorial Lecture on Flagellar Hydrodynamics published in *SIAM Review*, **18** (1976) 161-230, Lighthill concentrated exclusively on one particular field: the biofluiddynamics of microorganisms. According to him, a mathematical feature common to all biofluiddynamics of microorganisms is the overwhelming dominance of surface forces including pressure and viscous stresses over volume forces such as inertial effects. On the other hand, in the dynamics of swimming organisms, the inertial effects are significant which means that, at each instant, their total swimming velocity of translation through the fluid is determined by the translation motion together with undulatory motions of the organisms so that it will generate zero resultant force on the body. This is a single 3-vector condition to find a single 3-vector unknown swimming velocity.

He first gave a general overview of microorganisms possessing flagella and of related organisms, and then described undulations in a eukaryotic flagellar motions, bacterial flagellar motions and flagellar structures in some detail. In order to develop flagellar hydrodynamics, he formulated mathematical methods as follows. He stated that the fluid motion around a microorganism follows the same rule as does the microorganisms own motion so that the forces on any particle of fluid are in equilibrium. Thus, the pressure force on such aparticle is $-\nabla p$ per unit volume and the viscous force per unit volume is $\mu \nabla^2 \mathbf{u}$ which represents a diffusive transfer of momentum due to the dynamic viscosity of the fluid. Since the fluid inertia is negligible on the scale of a microorganism, the vector sum of the above two forces must be zero so that

$$-\nabla p + \mu \nabla^2 \mathbf{u} = 0. \tag{12.2.1}$$

This may be referred to the momentum equation. Combined with the equation of continuity for an incompressible fluid as

$$div\ \mathbf{u} = 0. \tag{12.2.2}$$

Consequently, any potential flow $\mathbf{u} = \nabla \phi$, where ϕ satisfies the Laplace equation $\nabla^2 \phi = 0$, with the pressure simply constant because the viscous forces vanishes. However, any such potential flow is too restricted to include that satisfies the no-slip condition at a solid boundary, such as the surface of microorganism. The velocity of the fluid at such a surface must be equal to the velocity of the solid surface itself. For example, a dipole source of vector strength $\mathbf{G}$ directed from the negative towards the positive source can be written as

$$\phi = div \left(\frac{\mathbf{G}}{4\pi r} \right), \qquad \mathbf{u} = grad\ \phi = grad \left[\left(div\, \frac{\mathbf{G}}{4\pi r} \right) \right], \tag{12.2.3ab}$$

where r is distance from the dipole source. The velocity $\mathbf{u}$ decreases rapidly as r increases, according to an inverse-cube law. On a particular spherical surface $r = a$, the normal component of $\mathbf{u}$ and of a constant vector $\mathbf{U}$ are equal if

$$\mathbf{G} = 2\pi a^3 \mathbf{U}. \tag{12.2.4}$$

It follows from the solution of the Stokes equations (12.2.1) and (12.2.2) that $\nabla^2 p = 0$. One of the fundamental solutions of special importance is called the *Stokeslet* which represents the effect of a concentrated external force $\mathbf{F}$ acting at a single point of the fluid so that a delta function distribution, $\delta(\mathbf{r})$ of force per unit volume can change the momentum equation (12.2.1) in the form

$$\mathbf{F}\delta(\mathbf{r}) - grad\, p + \mu\, \nabla^2 \mathbf{u} = 0, \tag{12.2.5}$$

where $\mathbf{r}$ is the vector displacement from the point of application of the force. We take divergence of (12.2.5) to obtain

$$\nabla^2 p = div\, [\mathbf{F}\delta(\mathbf{r})]\,. \tag{12.2.6}$$

The solution for the potential p of a dipole is given by

$$p = div\left(-\frac{\mathbf{F}}{4\pi r}\right). \tag{12.2.7}$$

The associated velocity distribution $\mathbf{u}$ can be written as

$$\mathbf{u} = \frac{\mathbf{F}}{6\pi\mu r} + \frac{1}{4} r^2 grad\left[div\left(\frac{\mathbf{F}}{6\pi\mu r}\right)\right]. \tag{12.2.8}$$

Both terms in the Stokeslet velocity field (12.2.8) satisfy the famous inverse-first-power $\left(r^{-1}\right)$ law. With the same r^{-1}-dependence in every direction, the first term represents the spherical mean of $\mathbf{u}$ (average value over a sphere of radius r). The second term changes the above fact with second order surface harmonics, that is, by a factor $1 + \frac{1}{2}P_2(\cos\theta)$ for the component parallel to $\mathbf{F}$. However, the second term in (12.2.8) generates the viscous force to balance the gradient of the pressure field (12.2.7), whereas the first term generates no viscous force, but must be included in (12.2.8) if the continuity equation is to be satisfied.

On a fixed sphere $r = a$, the second term in (12.2.8) can be completely cancelled by adding on a dipole velocity distribution (12.2.3ab) of strength

$$\mathbf{G} = -\left(\frac{a^2}{6\mu}\right)\mathbf{F}. \tag{12.2.9}$$

The velocity on the sphere $r = a$ then takes the constant Stokes value

$$\mathbf{U} = (6\pi\mu a)^{-1}\,\mathbf{F}. \tag{12.2.10}$$

Thus, the combined dipole **G** and Stokeslet **F** can represent a fluid motion for $r \geq a$ which matches exactly with the translational motion of the sphere $r = a$ at velocity **U**. The sphere acts on the fluid with a total force which satisfies the Stokes law (12.2.6). The required dipole strength (12.2.5) takes the value $-\pi a^2\mathbf{U}$ opposite in sign to the value (12.2.4) which generates the irrotational flow field with just the normal component of velocity matched.

The Stokes law (12.2.10) for the force **F** required to move a sphere with velocity **U** takes the form in terms of the volume $V = \frac{4}{3}\pi a^3$ of the sphere

$$\mathbf{F} = 11.66\mu V^{\frac{1}{3}}\mathbf{U}. \tag{12.2.11}$$

The work of Pironneau (1973) and Baurot (1974) reveals that the coefficient 11.66 in (12.2.11) is *not* far above the *minimum possible value* for resistance of a body of volume V. This minimum value is found by the above authors for axial motion of the body of revolution. This analysis shows that such drag formulas for shapes similar to cell bodies of microorganisms are of great importance in flagellar hydrodynamics because they suggest how much thurst an attached flagellum must exert to propel the organism at a given velocity.

It is worth mentioning that the classical paradox concerning the inverse-first-power Stokeslet velocity field (12.2.8), that is, the fact that in an unbounded fluid it has infinite energy, completely disappears in these applications to swimming microorganisms. Since the total force on the organism is zero, its total reaction on the fluid must be zero. Thus, the flow field is a combination of (i) Stokeslet fields involved in the flagellar thurst, and (ii) Stokeslet fields associated with equal and opposite cell body drag. Dipoles can be combined with the Stokeslet fields so that the velocity field satisfies the boundary condition for a particular flagellar motion, related to the assumed force field. If this relation can be inverted, then the force field may be known for a given flagellar motion, to a far better approximation than if Gray-Hancock resistance coefficients were utilized.

Writing the Stokeslet velocity field (12.2.8) with differentiation of r^{-1} carried out explicitly as

$$\mathbf{u} = \left(8\pi\mu r^3\right)^{-1}\left[r^2\mathbf{F} + (\mathbf{F}\cdot\mathbf{r})\mathbf{r}\right], \tag{12.2.12}$$

which is the sum of a parallel field and a radial field. Finally, the relationship of flagellar velocity distribution in terms of the flagellar force field is stated in the Lighthill Theorem as follows:

"Theorem 1. If $\mathbf{f}(\mathbf{s})$ is the force per unit length with which a flagellum of small radius a acts on a fluid, where the variable s represents length measured along the centerline of the flagellum from some given cross section, then the resulting flow field can be represented by a distribution of Stokeslet along the centerline, of strength $\mathbf{f}(s)$ per unit length, accompanied by dipoles strength

$$-\left(\frac{a^2}{4\mu}\right)\mathbf{f}_n(s) \tag{12.2.13}$$

per unit length, where $\mathbf{f}_n(s)$ is the vector normal to the centerline obtained by resolving $\mathbf{f}(s)$ onto the plane normal to the centerline. Thus, the fluid velocity field $\mathbf{u}$ closely matches flagellar motions $\mathbf{w}$ such that the entire cross section where $s = s_0$ moves with velocity

$$\mathbf{w}(s_0) = \frac{\mathbf{f}_n(s_0)}{4\pi\mu} + \int_{r_0>\delta} \left(8\pi\mu r_0^3\right)^{-1} \left\{r_0^2\mathbf{f}(s) + [\mathbf{f}(s)\cdot\mathbf{r}_0]\,\mathbf{r}_0 ds\right\}, \tag{12.2.14}$$

here $\delta = \frac{1}{2}a\sqrt{e} = 0.82a$ and $\mathbf{r}_0$ is the position vector of the point s_0 the centerline relative to the point s."

It is noted that equation (12.2.14) represents only the flagellar velocity associated with the flagellar force field; the total flagellar velocity includes a contribution from the flow field induced by the movement of the cell body. The practical use of (12.2.14) must involve some form of *inversion* so that the force distribution $\mathbf{f}(s)$ can be obtained in terms of the flagellar velocity field $\mathbf{w}(s)$. Finally, the reader is referred to Lighthill's original paper for more mathematical analysis and conclusions.

In a series of four major papers on biofluiddynamics of ballistiform and gymnotiform locomotion published in *Journal of Fluid Mechanics*, **212** (1990) 183-207; **213** (1990) 1-10; **213** (1990) 11-20; **213** (1990) 21-28, Lighthill alone and in collaboration with Robert Blake provided biological background to describe different modes of propulsion of aquatic animals which are of special hydrodynamical interest because about 10^9 years of animal evolution in an aqueous environment have produced refined means of generating fast movements at small energy costs of transport. They proposed new simplified models of both balistiform modes of swimming and gymnotiform modes of swimming. In the former model, the thickness of the fish's body is basically rigid and assumed small compared with its depth, that is, the body is treated simply as a flat plate for hydrodynamical analysis. In the latter model, fishes possess a long vertical fin but no corresponding dorsal fin, that is, a mode of balistiform locomotion in which the body of the fish is held rigid and undulations pass along ventral fin in the posterior direction. This is followed by their simplified large-amplitude elongated-body theory, especially adopted to ballistiform locomotion which allows a direct comparison of thurst and efficiency for different modes of propulsion. It is shown that thurst is dominated by the mean rate of shedding of backward momentum at the posterior end of the fish's propulsive apparatus, and this momentum is increased by a momentum enhancement factor β of around 3 or more. However, there is no such enhancement of the rate of shedding of unproductive energy into the wake, and overall efficiency is therefore improved. In their simplified model, a source of inaccuracy lies in the approach to determining the momentum vector $\mathbf{M}$ based on a locally two-dimensional flow rather than the real three-dimensional flow. In this undulatory situation, if the wavelength λ of undulations is not very large, the approximate solutions of the three-dimensional Laplace equation by solutions of the two-dimensional Laplace equation for the velocity potential $\phi(x, y)$,

$$\phi_{xx} + \phi_{yy} = 0 \tag{12.2.15}$$

may produce significant errors. These indeed are cases where a different two-dimensional equation

$$\phi_{xx} + \phi_{yy} - k^2\phi = 0, \tag{12.2.16}$$

where $k = \frac{2\pi}{\lambda}$, might be considered to describe the three-dimensional Laplace equation more accurately.

In Part 2 of his paper, Lighthill continued with a two-dimensional irrotational flow to calculate the pressure distribution due to a general symmetrical flexing motion of a flat plate and then derived a quantity needed in elongated-body theory of ballistiform locomotion. In case of ballistiform swimming, the elongated body theory produces the thurst which consists of two terms, one being the momentum associated with fin movements is significantly increased for fins of modest depth attached to a deep body, while the other term involving the area integral of excess pressure is positive, but relatively small in magnitude.

In Part 3 of his paper, Lighthill calculated the fluid momentum associated with a two-dimensional motion due to non-uniform of a fin, or of a pair of fins attached to a body of elliptic cross section. Using the conformal transformations of a complex variable, the momentum is calculated. It is found that momentum is significantly enhanced in the presence of a body of elliptic cross-section, not only in ballistiform locomotion with synchronous movement of two median fins but also in gymnotiform locomotion with movement of just a single (ventral) fin. In Part 4 of his paper, Lighthill used the three-dimensional Laplace equation (12.2.16) for fluid motions of undulatory type to examine the possibility of short-wavelength limitations on momentum enhancement predicted in his earlier papers. When fish bodies are very small width in large depth, the elongated-body theory predicted a very large momentum enhancement factor β which has now a specific upper bound $\left(\beta = 1.53\,(kb)^{-\frac{1}{2}}\right)$ based on short-wavelength consideration of the theory, where b represents the fin depth. Thus, short-wavelength limitations on momentum enhancement are of biological significance as they imply a reduction in body drag as well as of a gradual reduction of thurst enhancement.

12.3 Internal Biofluiddynamics

Lighthill published a total of five research and research expository and survey articles in internal biofluiddynamics. His interest in this subject began while he was in Imperial College and delivered a course of lectures on Blood Flow in Arteries in Spring of 1966, and established interdisciplinary research work in close and active collaboration with Dr. Colin Caro at Imperial College. His first major survey article entitled, Physiological Fluid Dynamics was published in *J. Fluid Mech.*, **52** (1972) 475-497. This survey covered a considerable amount of material including a general survey of fluid flows within the human body, both in the larger airways of the lung and in the larger arteries the Reynolds number takes values from a few hundred to several thousand. Included are both the lungs and the cardiovascular system which are extremely intricate and complex branched networks of tubes, turbulence in the blood stream. He presented the general circulation of the blood and the urinary tract, problems of microcirculation including blood flow in narrower capillaries, gas exchange with terminal airways and exchange of gas and nutrients with peripheral tissues. He explained the blood flow in vessels of diameter less than 0.1 mm (100 μm) and especially with flow in the arterioles, capillaries and venules with diameters which range from 5 μm to 50 μm. He concluded this article with a discussion of lubrication problems in very narrow capillaries, and stated that all this work on lubrication problems in the microcirculation is at a early stage and hence, more research is needed to resolve many unsolved problems.

He published a major paper on Energy Flow in the Cochlea in *J. Fluid Mech.*, **106** (1981) 149-213. He considered one-, two- or three-dimensional mathematical models of cochlear mechanics which is important because it provides a *first-filter* in the frequency-analysis function of the auditory system. Several mathematical models including traveling-wave and resonance are developed. In such a system, a traveling wave would carry energy along to the position of resonance, where it would remain in a local oscillation until it decays.

In a one-dimensional model without resonance, each cross section of the cochlea has the pressure the deviation of pressure p from its mean value P over the cross sectional area $A(x)$. The pressure p satisfies the equation

$$A\,\frac{\partial^2 p}{\partial t^2} = c^2 \frac{\partial}{\partial x}\left(A\,\frac{\partial p}{\partial x}\right), \qquad (12.3.1)$$

when $c = (2\rho D)^{-\frac{1}{2}}$ is the slow-wave velocity, ρ is the density and D is the distensibility of a single isolated tube.

Equation (12.3.1) represents a non-dispersive wave equation with the wave speed c independent of frequency so that both the wave crests and the wave energy propagate with the same speed c. The time to travel a distance x from the base $x = 0$ is

$$\int_0^x \frac{1}{c}\,dx. \qquad (12.3.2)$$

It is expected that a wave traveling in the direction x is increasing so that the pressure p has the form

$$p = f(x)\exp\left[i\omega\left(t - \int_0^x \frac{1}{c}\,dx\right)\right], \qquad (12.3.3)$$

where $f(x)$ is the real amplitude. The wave energy per unit length is

$$AD\,[f(x)]^2 \qquad (12.3.4)$$

and the wave-energy flow is equal to the energy per unit length (12.3.4) times the energy propagation velocity c so that

$$cAD\,[f(x)]^2 = A\,[f(x)]^2/2\rho c, \qquad (12.3.5)$$

where D is replaced by $\left(2\rho c^2\right)^{-1}$.

Obviously, in a traveling wave without dissipation, energy flow must be constant so that the amplitude $f(x)$ of pressure function must change with position in proportion to

$$(c/A)^{\frac{1}{2}}. \qquad (12.3.6)$$

This physical argument of deriving the amplitude is expected to give the same results as the WKB asymptotic method. In this method, a solution

$$p = f(x)\exp\left[i\omega\left(t - g(x)\right)\right] \qquad (12.3.7)$$

is substituted in the equation (12.3.1) so that f satisfies the ordinary differential equation

$$-\omega^2 Afc^{-2} = (Af')' - i\omega\left[(Ag')'f + 2Ag'f\right] - \omega^2 Afg'^2. \qquad (12.3.8)$$

In the asymptotic limit of high frequency ω, equation (12.3.8) will be closely enough satisfied if the terms in ω^2 balance so that

$$g'^2 = c^{-2} \qquad \text{and} \qquad g(x) = \int_0^x \frac{1}{c}\,dx. \qquad (12.3.9)$$

Also, the terms in ω must balance giving

$$(f'/f) = -\frac{1}{2}(Ag')'/(Ag') \qquad (12.3.10)$$

so that f is proportional to

$$(Ag')^{-\frac{1}{2}} = (c/A)^{\frac{1}{2}} \qquad (12.3.11)$$

which is exactly the same as (12.3.6). In this simple calculation, the first term on the right hand side of (12.3.8) is independent of ω and has been ignored compared with the terms in ω and ω^2 at high frequency.

Finally, Lighthill's analysis included one-, two- or three-dimensional mathematical models of cochlear mechanics. All are found to have been successful provided small damping is incorporated and if and only if they include critical-layer absorption. This resolves the paradox of why certain unrealistic one-dimensional models can provide a good prediction of cochlear response; it is due to the fact that they incorporate the one essential feature of critical layer absorption.

In order to explain the resonant phenomenon, Lighthill used the definitions of the potential and kinetic energies in vibration theory in terms of the generalized coordinates and the generalized velocity as follows:

$$\text{Potential energy} = \frac{1}{2}\, s\ (\text{generalized coordinates})^2 \qquad (12.3.12)$$

$$\text{Kinetic energy} = \frac{1}{2}\, m\ (\text{generalized velocity})^2, \qquad (12.3.13)$$

where s is called the *stiffness parameter*, m is called the *inertia parameter*, and the generalized velocity is the rate of change of generalized coordinates. In particular, for duct flows in the cochlea, both potential and kinetic energies are taken as energies per unit length of duct. The parameter s is defined as the ratio of $(2p/V)$ where $2p$ is the pressure difference required to generate the volume change V per unit length of the cochlea. Obviously, the work done by this pressure difference sV in making the volume change from 0 to V is $\frac{1}{2}sV^2$. Lighthill also expressed the stiffness parameter s in terms of distensibility parameter D by

$$s = \frac{1}{AD} = 2\rho A^{-1}c^2 \tag{12.3.14}$$

which is another form of s related to the slow-wave speed c.

The corresponding inertia parameter m depends on the shape of the bending mode of the partition in response to a constant pressure difference. If the volume change V is produced by a partition displacement

$$z = V\eta(y) \text{ across the width } 0 < y < 2l, \tag{12.3.15}$$

of the partition with thickness, $h(y)$ in $0 < y < 2l$, then the kinetic energy (K.E.) per unit length is

$$K.E. = \int_0^{2l} \frac{1}{2}\,\rho\, h(y)\left(\frac{dz}{dt}\right)^2 dy = \frac{1}{2}\int_0^{2l} \rho\, h(y)\left[\dot{V}\,\eta(y)\right]^2 dy, \tag{12.3.16}$$

where the density of cochlear partition is nearly the same density of the perilymph so that $[\rho h(y)]\,dy$ is the mass per unit length between y and $y+dy$, and $\dot{V}\,\eta(y)$ is the transverse velocity of displacement. Evidently, the definition of the kinetic energy (12.3.16) implies that the inertia coefficient m is equal to the coefficient of $\frac{1}{2}\dot{V}^2$ in (12.3.16), that is,

$$m = \rho\int_0^{2l} h(y)\eta^2(y)dy, \tag{12.3.17}$$

where $\eta(y)$ is the shape of the bounding mode can be normalized so that

$$\int_0^{2l} \eta(y)dy = 1. \tag{12.3.18}$$

In vibrations of the cochlear partition, the rate of the total energy (potential + kinetic) is equal to the rate $(2p)\dot{V}$ of working by the fluid pressure difference $2p$ so that

$$\frac{\partial}{\partial t}\left(\frac{1}{2}sV^2 + \frac{1}{2}m\dot{V}^2\right) = 2p\dot{V} \tag{12.3.19}$$

which yields the simple harmonic oscillator equation

$$m\ddot{V} + sV = 2p \tag{12.3.20}$$

where the natural resonant frequency ω_r is given by

$$\omega_r = \left(\frac{s}{m}\right)^{\frac{1}{2}}. \tag{12.3.21}$$

Using the equation $2AD\left(\frac{\partial p}{\partial t}\right) = \frac{\partial V}{\partial t}$, it follows from the fluid momentum equation

$$\rho \frac{\partial^2 V}{\partial t^2} = \frac{\partial}{\partial x}\left(A\frac{\partial p}{\partial x}\right). \tag{12.3.22}$$

For a traveling wave solution proportional to $\exp(-ikx)$ of wavenumber k, each $(\partial/\partial x)$ must be replaced $(-ik)$ so that the dispersion relation, the phase and the group velocity can be determined. Hence, as in equations (12.3.4)–(12.3.6), the amplitude variation follows from the conservation of energy flow. It also follows from equation (12.3.22) that

$$\rho \frac{\partial^2 V}{\partial t^2} = -Ak^2 p \tag{12.3.23}$$

so that (12.3.20) reduces to

$$\left(m + 2\rho A^{-1}k^{-2}\right)\frac{\partial^2 V}{\partial t^2} + sV = 0. \tag{12.3.24}$$

This clearly defines the local frequency ω as a function of k so that the local dispersion relation can be written as

$$\omega^2 = \frac{s}{(m + 2\rho A^{-1}k^{-2})} = \frac{s}{(m + m_f)}, \tag{12.3.25}$$

where m_f is the fluid inertia given by

$$m_f = 2\rho A^{-1}k^{-2}. \tag{12.3.26}$$

More interestingly, solving k in terms of ω and ω_r gives the resonant feature of k as $\omega \to \omega_r$ in the form

$$k = k_1\omega\left(\omega_r^2 - \omega^2\right)^{-\frac{1}{2}}, \tag{12.3.27}$$

where $k_1 = (2\rho/Am)^{\frac{1}{2}}$ may vary gradually with x, while the resonant frequency ω_r changes very rapidly along the cochlea.

The group velocity is equal to $\frac{\partial \omega}{\partial k}$ so that

$$U = \left(\frac{\partial k}{\partial \omega}\right)^{-1} = k_1^{-1}\omega_r^{-2}\left(\omega_r^2 - \omega^2\right)^{3/2}. \tag{12.3.28}$$

Evidently, results (12.3.27) and (12.3.28) are in complete agreement of new resonance phenomenon in a traveling-wave system which is also called *critical-layer resonance* or more appropriately known as *critical-layer absorption* because the resonance absorbs energy.

The energy per unit length still takes the form in terms of the amplitude of the pressure field $f(x)$, but it is more appropriate to express this in terms of the amplitude $|V|$ of the generalized coordinate V. The energy per unit length is the average value of sV^2 which is

$$\frac{1}{2}s|V|^2 = \frac{1}{2}m\omega_r^2|V|^2. \tag{12.3.29}$$

Consequently, the rate of energy flow is U times the result (12.3.29)

$$\frac{1}{2}mk_1^{-1}\left(\omega_r^2 - \omega^2\right)^{3/2}|V|^2. \tag{12.3.30}$$

For waves of fixed frequency ω, the constant energy flow requires $|V|$ to change proportionally as

$$(k_1/m)^{\frac{1}{2}}\left(\omega_r^2 - \omega^2\right)^{-\frac{3}{4}}. \tag{12.3.31}$$

This shows the unlimited build-up of energy at the point where $\omega_r \to \omega$.

We take a simple example of exponential form for ω_r proportional to $\exp(-\alpha x)$ so that $\frac{\partial \omega_r}{\partial x} = -\alpha\omega_r$, and also assume that A and m are almost constant. In this situation, the phase θ takes the form

$$\theta = \left(\alpha^{-1}k_1\right)\tan^{-1}\left[\frac{1}{\omega}\left(\omega_r^2 - \omega^2\right)^{\frac{1}{2}}\right] + \text{constant}. \tag{12.3.32}$$

This can be verified by differentiating (12.3.32) to show from (12.3.27) that $k = -\frac{\partial \theta}{\partial x}$. Result (12.3.32) reveals the expected decrease with increasing steepness as $\omega_r \to \omega$.

Further, for waves of fixed frequency ω, the time required for wave energy to reach a given point has the form

$$t = \left(\frac{k_1}{\alpha}\right)\frac{1}{\left(\omega_r^2 - \omega^2\right)^{1/2}} + \text{constant} \tag{12.3.33}$$

which can be verified by differentiating (12.3.33) to show from (12.3.28) that $(\partial x/dt) = U$. Result (12.3.33) shows the fundamental feature of *critical-layer absorption* as the time becomes infinite as $\omega_r \to \omega$. In other words, an unbounded time is needed for any damping, however small, to dissipate all the energy of the system. Thus, one-dimensional traveling wave models of the cochlea taking into account partition inertia exhibit critical-layer absorption. Lighthill also extended his mathematical analysis of one-dimensional models to two and three-dimensional models of cochlear mechanics. He predicted that all models are found to be successful if and only when they incorporate critical-layer absorption.

In his 1989 Rayleigh Memorial Lecture at the ASME Winter Annual Meeting in Chicago, Lighthill presented an interesting survey on the Biomechanics of Hearing Sensitivity which dealt with two major sections:

(i) macromechanics of the cochlea, and (ii) micromechanics of the cochlea. Based on experimental evidence for dispersion, he described coachlear macromechanics indicating that waves propagated along the basiliar membrane within ambient fluid medium are essentially dispersive. Using the group velocity, U of waves, the rate of energy flow must be equal to UE which is equal to input power P or less than equal to P, where there is no or some dissipation involved. In other words, $U < P/E$ only in the region where the energy is propagating very slowly, and therefore has time to suffer significant dissipation. Mathematically, the curve of U against the frequency ω comes in tangentially to zero where $\omega = \omega_r(x)$ leads to a proportionality of U to $[\omega_r(x) - \omega]^2$. The inverse function $x = x_r(\omega)$ determines the position x for each frequency ω which is the *resonant frequency* so that U is proportional to $[x_r(\omega) - x]^2$. It thus follows that the group delay for waves of frequency ω so that the travel time τ for wave energy from the base to reach a point x is given by

$$\tau = \int_0^x U^{-1} dx \tag{12.3.34}$$

which becomes unbounded as $x \to x_r(\omega)$. This is fully consistent with the idea that any damping rate, however small, has all the time it needs to dissipate the energy which is to similar to energy dissipation by viscous action in a thin Stokes boundary layer. On the other hand, the very slow velocity at which the energy is moving means that this dissipation takes place within a very small distance. Lighthill concluded this section by adding a comment which implies that some simple principles from the general theory of waves and vibrations, when applied to cochlear macromechanics, yield mathematical results which are fully consistent, not only with the existing overall knowledge of frequency selectivity in the cochlea, but also with good experimental data on basilar-membrane response at moderate sound levels.

Lighthill then described cochlear micromechanics based first on nonlinear response at low amplitudes, and second on submicroscopic studies at the cellular level. Finally, he discussed the mammalian cochlea at the microscopic and submicroscopic levels of resolution.

In one of his last major papers on Acoustic streaming in the ear itself published in *J. Fluid Mech.* **239** (1992) 551-606, Sir James described mean motions of fluid induced by sound waves, called *acoustic streaming*, and calculated the magnitude of any mean motions generated by acoustic signals after conversion into cochlear traveling waves. Such cochlear traveling waves and their role in the ear's frequency discrimination were discovered experimentally by Békésy (1960). These waves are associated with the steeply graded stiffness properties of the basilar membrane oscillating within the cochlear fluids. In regions where the wavenumber $k \geq 3\,mm^{-1}$, the fluid motions generated by the basilar membrane's wavelike vibrations are expected to penetrate only a limited distance of order k^{-1} into the cochlea. There is a close analogy between surface waves in water of depth h (or in deep water described by the relation $\omega^2 = gk$) and traveling waves in cochlear. However, there are certain dissimilarities. Since fluid motions are generated on the both sides of the basilar membrane, fluid inertia is basically doubled and it is also changed by the finite width of the membrane compared with long-crested surface water waves.

Thus, the inertia per unit length of basilar membrane must take form $A\rho k^{-1} + B$ where $A\rho k^{-1}$ is a fluid component and B is a constant representing the inertia per unit length of the membrane. Besides inertia, both stiffness and damping are also changed. The boundary condition satisfied by fluid motions at the basilar membrane leads to the existence of Stokes boundary layers on both sides of it, generating increased rate of dissipation. With the steepness $s(x)$ of the basilar membrane where x is the distance from the base, the dispersion relation can be replaced by the relation

$$\omega^2 = s(x)\left[A\rho k^{-1} + B\right]^{-1}. \tag{12.3.35}$$

The group velocity is

$$U = \frac{\partial \omega}{\partial k} = \frac{A\rho\omega^3}{2k^2 s(x)}. \tag{12.3.36}$$

At each position x, the local wavenumber k represents a good approximation determined by the given dispersion relation for waves of a prescribed frequency ω so that

$$k = \frac{A\rho}{\omega^{-2}s(x) - B} = \frac{A\rho\omega^2}{s(x) - s(x_r)}, \tag{12.3.37}$$

which tends to infinity as $x \to x_r$, where x_r is the *resonant position* $(s(x) \to s(x_r))$ and the inertia B of the basilar membrane satisfies the equation

$$\omega^2 B = s(x_r). \tag{12.3.38}$$

Combining (12.3.36)–(12.3.37) yields the group velocity

$$U = \frac{\left[s(x) - s(x_r)\right]^2}{2A\rho\,\omega\, s(x)}, \tag{12.3.39}$$

which tends to zero as $x \to x_r$. This allows wave energy an unlimited time $\int U^{-1}dx$ to reach $x = x_r$.

On the other hand, the three-dimensional wave energy outside the Stokes boundary layer is based on solutions of the Laplace equation

$$\phi_{yy} + \phi_{zz} - k^2\phi = 0, \tag{12.3.40}$$

for waves of wavenumber k, where y and z are Cartesian coordinates in a cochlear cross-section.

At a position in the cochlea where the wavenumber is k, the local wave motions outside the Stokes boundary layers are approximated by solutions of equation (12.3.40) consistent with the boundary conditions on the basilar membrane. The fundamental solution of (12.3.40) is

$$-\frac{1}{2\pi}K_0(kr), \tag{12.3.41}$$

representing for motions of wavenumber $k(>0)$ the velocity potential ϕ at a distance r from the source of unit strength, K_0 is the modified Bessel function of the second kind with the asymptotic properties

$$K_0(kr) \sim -\log\left(\frac{1}{2}kr\right) - \gamma \quad \text{as} \quad kr \to 0, \tag{12.3.42}$$

and

$$K_0(kr) \sim -\left(\frac{\pi}{2kr}\right)^{\frac{1}{2}} e^{-kr} \quad \text{as} \quad kr \to \infty, \tag{12.3.43}$$

with $\gamma = 0.577$ as the Euler constant.

In the (y, z) plane of a cochlear cross-section where the basilar membrane extends from $z = a$ to $z = b$ on the y-axis, its oscillations with frequency ω and local wavenumber k can induce a local velocity in the form

$$\frac{\partial\phi}{\partial y} = V(z) \quad \text{on} \quad y = 0, \qquad a \leq z \leq b, \tag{12.3.44}$$

where $V(a) = V(b) = 0$ and the exponential factor $\exp\left[i\left(\omega t - kx\right)\right]$ has been omitted. The corresponding potential ϕ on the side $y > 0$ can be written using the source - like solution (12.3.41) in terms of linear combination of sources over the strip $a \leq z \leq b$ with source strength $2V(z)dz$ in each small interval dz.

The kinetic energy of the fluid is

$$\frac{1}{2}\left(A\rho k^{-1}\right)\left(\frac{\partial\zeta}{\partial t}\right)^2, \tag{12.3.45}$$

where ζ is the maximum displacement of the basilar membrane so that $\zeta_t = V_{\max}$ and hence,

$$A \sim 2V_{\max}^{-2}\int_a^b V^2(z)dz. \tag{12.3.46}$$

The corresponding kinetic energy of the basilar membrane itself and of other attached material to it with the combined mass $M(z)$ per unit area is

$$\int_a^b \frac{1}{2} M(z)\, V^2(z) dz, \tag{12.3.47}$$

which assumes the form $\frac{1}{2}B\zeta_t^2 = \frac{1}{2}BV_{\max}^2$ provided that

$$B = V_{\max}^{-2} \int_a^b M(z) V^2(z) dz. \tag{12.3.48}$$

Similarly, the potential energy is of the form $\frac{1}{2}s(x)\zeta^2$ so that $s(x)$ is given by

$$s(x) = V_{\max}^{-2} \int_a^b S(x,z) V^2(z) dz, \tag{12.3.49}$$

where $S(x,z)$ is the ratio of restoring force per unit area to basilar membrane displacement.

For pure-tone forcing distribution $e^{i\omega t} f(x,z)$ in the y-direction with the amplitude $f(x,z)$ per unit area, we can write

$$f(x,z) = \int_{-\infty}^{\infty} e^{-ikx} F_1(k,z) dk, \tag{12.3.50}$$

where $F_1(k,z)$ is the Fourier transform of $f(x,z)$ with respect to x. Substituting (12.3.50) in the forcing term $e^{i\omega t} f(x,z)$ and omitting the exponential term $\exp\left[i\left(\omega t - kx\right)\right]$ gives the forcing term as $F_1(k,z)$. Using the transient pressure $-\rho\phi_t = -\rho i\omega\phi$, the transient pressure, equal and opposite to $-\rho i\omega\phi$, is acting on the side $y < 0$ of the basilar membrane, whose y-component of velocity is $V(z)$. Then the excited pressure difference is $+2i\rho\omega\phi$. In view of the asymptotic limit of $\phi \sim -k^{-1}V(z)$ as $y \to 0$. The above pressure difference is $\left(-2\rho|k|^{-1}\right)\left[i\omega V(z)\right]$. Thus, the combined forcing terms generate the motion of the basilar membrane and attached solid material, with mass $M(z)$ and total stiffness $S(x,z)$ so that

$$M(z) i\omega V(z) + S(x,z)(i\omega)^{-1} V(z) = F_1(x,z) - 2i\rho\omega|k|^{-1} V(z). \tag{12.3.51}$$

We next multiply (12.3.51) by $V(z)$, integrate with respect to z from a to b and simplify by means of results (12.3.46), (12.3.48) and (12.3.49) to obtain the forcing equation

$$V_{\max}^2 \left[\left(A\rho|k|^{-1} + B\right) i\omega + (i\omega)^{-1} s(x)\right] = \int_a^b F_1(k,z) V(z) dz. \tag{12.3.52}$$

Using the result

$$F(k) = V_{\max}^{-1} \int_a^b F_1(k,z)V(z)dz, \tag{12.3.53}$$

along with $V_{\max} = i\omega\zeta$, the forcing equation

$$\left[s(x) - \omega^2\left(A\rho|k|^{-1} + B\right)\right]\zeta = F(k). \tag{12.3.54}$$

This equation confirms that unforced waves satisfy the dispersion relation (12.3.37) and that the forcing of the principal mode $V(z)$ depends on $F(k)$ which is given by (12.3.53).

In addition to the unforced waves that propagate from the base of the cochlea towards their characteristic position, the forced waves newly induced by the forcing $e^{i\omega t}f(x,z)$ may travel in the same or in the opposite direction. Thus, for forcing $e^{i\omega t}f(x,z)$ with (12.3.50) by a continuous distribution of terms in $\exp\left[i\left(\omega t - kx\right)\right]$, the forcing equation represents the resulting basilar - membrane vibration in the form

$$\zeta = e^{i\omega t}\int_{-\infty}^{\infty} \frac{F(k)e^{-ikx}dk}{s - \omega^2\left(A\rho|k|^{-1} + B\right)}. \tag{12.3.55}$$

This integral can be evaluated in terms of polar singularities of the integrand and the resulting waves satisfy the radiation condition at infinity that the energy travels outwards from the source region.

In any two-dimensional flow the irrotational motion outside the Stokes boundary layer satisfies the modified version of (12.3.40) in the form

$$\frac{\partial^2\phi}{\partial y^2} - k^2\phi = 0. \tag{12.3.56}$$

The modified boundary condition is

$$\frac{\partial\phi}{\partial y} = V \quad \text{on} \quad y = 0, \tag{12.3.57}$$

where V is constant, which represents the local form of (12.3.44) for a particular value of z with exponential term $\exp\left[i\left(\omega t - kx\right)\right]$ again dropped.

On the side $y > 0$ of the basilar membrane, equation (12.3.56) with the boundary condition (12.3.57) admits the solution

$$\phi = -k^{-1}Ve^{-ky}, \tag{12.3.58}$$

which is found to be the asymptotic form of the corresponding three dimensional result for large k.

The x-component of the velocity is

$$u = \frac{\partial\phi}{\partial x} = iVe^{-ky}, \tag{12.3.59}$$

which satisfies the local form of the three-dimensional result on $y = 0$.

The y-component of the velocity is

$$v = \frac{\partial \phi}{\partial y} = Ve^{-ky}. \tag{12.3.60}$$

Thus, the above solutions describes the well-known circular motion of particles characteristic of waves in deep water as well as of cochlear traveling waves near their characteristic position.

Viscous dissipation takes place in the Stokes boundary layers describing that vorticity generated at the solid surface diffuses away from it with the kinematic viscosity $\nu(= \mu/\rho)$. Oscillatory results subject to such diffusion correspond to a characteristic value proportional to

$$\exp\left[-y\left(\frac{i\omega}{\nu}\right)^{\frac{1}{2}}\right], \tag{12.3.61}$$

which decreases exponentially with y, while its phase lag continually increases.

In this problem, with the x component of the velocity u given by (12.3.59) just outside the thin boundary layer given as $u = iV$, its value across the boundary layer assumes the form

$$u = iV\left\{1 - \exp\left[-y\left(\frac{i\omega}{\nu}\right)^{\frac{1}{2}}\right]\right\}, \tag{12.3.62}$$

of a rotational motion with the required form (12.3.61) of vorticity which allows the boundary condition $u = 0$ on $y = 0$ to be satisfied. The corresponding y-component of velocity v, with its boundary value V on $y = 0$, is obtained from (12.3.62) using the continuity equation

$$v = V\left[1 + \left\{1 - \exp\left[-y\left(\frac{i\omega}{\nu}\right)^{\frac{1}{2}}\right]\right\}k\left(\frac{\nu}{i\omega}\right)^{\frac{1}{2}}\right]. \tag{12.3.63}$$

The rate of energy dissipation δ per unit area of boundary layer takes the form

$$\delta = \int_0^\infty \mu\left(\frac{1}{2}|u_y|^2\right)dy = \rho V^2\left(\frac{1}{8}\omega\nu\right)^{\frac{1}{2}}. \tag{12.3.64}$$

This shows that, for cochlear traveling waves, the energy dissipation rate in either of the two boundary layers attached to the basilar membrane, and the overall rate is twice as much. The boundary-layer thickness $(\nu/\omega)^{\frac{1}{2}}$ is small compared with scale k^{-1} of the external irrotational flow. In other words the non-dimensional quantity

$$\kappa = k\left(\frac{\nu}{\omega}\right)^{\frac{1}{2}} \ll 1. \qquad (12.3.65)$$

Lighthill also calculated the mean streaming motions generated by cochlear traveling waves. He showed that there are three major contributions:

(i) The shear-stress streaming has a value

$$\frac{1}{4}V^2 k\omega^{-1} \qquad (12.3.66)$$

which is identical with that obtained in classical acoustic-streaming theory for the interaction of traveling sound waves with a rigid wall.

(ii) The normal-stress forcing gives a value

$$-\frac{1}{2}V\left(\frac{dV}{dx}\right)\omega^{-1}. \qquad (12.3.67)$$

(iii) Both for sound waves and for cochlear traveling waves, there is an extra contribution

$$-\frac{1}{4}V\,(dV/dx)\,\omega^{-1} \qquad (12.3.68)$$

which is due to the velocity amplitude V varies with x.

Thus, the three results (12.3.66)–(12.3.68) for different components of u_s, the streaming velocity forced by the Reynolds stresses acting within the Stokes boundary layer can be combined into a single equation

$$u_s = \frac{1}{4}V^2 k\omega^{-1} - \frac{3}{4}V\,(dV/dx)\,\omega^{-1}. \qquad (12.3.69)$$

This major result, representing the effective ship velocity at the boundary for the Euler mean motions external to the boundary layer, extends the classical acoustic streaming result due to Rayleigh (1896) to cochlear traveling waves.

Finally, Lighthill gave a modification of his basic result (12.3.69) for high wavenumbers κ in the form

$$u_s = \frac{1}{4}V^2 k\omega^{-1}\,\sigma(\kappa) - \frac{3}{4}V\left(\frac{dV}{dx}\right)\omega^{-1}\,\tau(\kappa), \qquad (12.3.70)$$

where two factors $\sigma(\kappa)$ and $\tau(\kappa)$ represent the modifications of the classical Stokes result. Thus, (12.3.70) describes the complete streaming motion for cochlear traveling waves.

Chapter 13

Books and Major Research Papers of Sir James Lighthill

Books and Book Chapters:

(1) *Introduction to Fourier analysis and generalized functions*, Cambridge University Press, Cambridge (1958).
(2) *Higher Approximations in Aerodynamic Theory*, Princeton University Press, Princeton (1960).
(3) *Mathematical Biofluiddynamics*, SIAM, Philadelphia (1975).
(4) *Waves in Fluids*, Cambridge University Press, Cambridge (1978).
(5) *Newer Uses of Mathematics*, Penguin Books Ltd, Hammondsworth, United Kingdom (1978).
(6) *Monsoon Dynamics* (Eds. Lighthill, J. and Pearce, R.B.), Cambridge University Press, Cambridge (1981).
(7) *An Informal Introduction to Theoretical Fluid Mechanics*, Oxford University Press, Oxford (1986).
(8) *Tropical Cyclone Disasters* (Eds. Lighthill, J., Zhang, Z., Holland, G., and Emanuel, K.), Peking University Press, Beijing (1993).
(9) *Fluid Mechanics*, A Chapter in Twentieth Century Physics (Eds. Laurie M. Brown, Abraham Pais, and Sir Brian Pippard), Institute of Physics and American Institute of Physics, New York, Volume II (1995).

Major Research Papers:

(1) Two-Dimensional Supersonic Aerofoil Theory, *Reports and Memorandum, No. 1929* (Aeronautical Research Committee), (1944) 1-19.
(2) The Conditions Behind the Trailing Edge of the Supersonic Aerofoil, *Reports and Memorandum, No. 1930* (Aeronautical Research Committee), (1944) 1-8.

(3) The Supersonic Theory of Wings of Finite Span, *Reports and Memorandum, No. 2001* (Aeronautical Research Committee), (1944) 1-15.
(4) A Note on Supersonic Biplanes, *Reports and Memorandum, No. 2002* (Aeronautical Research Committee), (1944) 1-5.
(5) Two-Dimensional Compressible Flow Past a Solid Body in Unlimited Fluid or Symmetrically Placed in a Channel (with S. Goldstein), *Philosophical Magazine*, **35** (1944) 549-568.

(6) Supersonic Flow past Bodies of Revolution, *Reports and Memorandum, No. 2003* (Aeronautical Research Committee), (1945) 1-24.
(7) A new method of two-dimensional aerodynamic design, *Reports and Memoranda, No. 2112* (Aeronautical Research Committee), (1945) 1-53.
(8) A theoretical discussion of wings with leading edge suction, *Reports and Memoranda, No. 2162* (Aeronautical Research Committee), (1945) 1-9.
(9) A mathematical method of cascade design, *Reports and Memoranda, No. 2104* (Aeronautical Research Committee), (1945) 1-17.
(10) Notes on the deflection of jets by insertion of curved surfaces and on the design of bends in wind tunnels, *Reports and Memoranda, No. 2105* (Aeronautical Research Committee), (1945) 1-7.
(11) A note on cusped cavities, *Reports and Memoranda, No. 2328* (Aeronautical Research Committee), (1945) 1-3.

(12) The hodograph transformation in trans-sonic flow. I. Symmetric Channels, *Proceedings of the Royal Society, London*, **A191** (1947) 323-369.
(13) The hodograph transformation in trans-sonic flow. II. Auxiliary theorems on the hypergeometric functions $\psi_n(\tau)$, *Proceedings of the Royal Society, London*, **A191** (1947) 341-351.
(14) The hodograph transformation in trans-sonic flow. III. *Proceedings of the Royal Society, London*, **A191** (1947) 352-369.
(15) The hodograph transformation in trans-sonic flow. IV. Tables, *Proceedings of the Royal Society, London*, **A192** (1947) 135-318.

(16) The position of the shock wave in certain aerodynamic problems, *Quarterly Journal of Mechanics and Applied Mathematics*, **1** (1948) 309-318.
(17) Supersonic flow past slender pointed bodies of revolution at yaw,

Quarterly Journal of Mechanics and Applied Mathematics, **1** (1948) 76-89.

(18) Supersonic flow past slender bodies of revolution the slope of whose meridian section is discontinuous, *Quarterly Journal of Mechanics and Applied Mathematics*, **1** (1948) 90-102.

(19) On the hodograph transformation for high-speed flow, Parts I (with S. Goldstein and J. W. Craggs) and II, *Quarterly Journal of Mechanics and Applied Mathematics*, **1** (1948) 344-357 and 442-450.

(20) Note on the ultimate form of rotary fluid motion inside a cylinder, *Quarterly Journal of Mathematics*, **19** (1948) 65-66.

(21) The flow behind a stationary shock, *Philosophical Magazine*, **40** (1949) 214-220.

(22) Methods for predicting phenomena in the high-speed flow of gases, *Journal of Aeronautical Science*, **16** (1949) 69-83.

(23) A technique for rendering approximate solutions to physical problems uniformly valid, *Philosophical Magazine*, **40** (1949) 1179-1201.

(24) The shock strength in supersonic "conical fields", *Philosophical Magazine*, **40** (1949) 1201-1223.

(25) The diffraction of blast, Part I. *Proceedings of the Royal Society, London*, **A198** (1949) 454-470.

(26) Thermal stresses in turbine blades (with F. J. Bradshaw), *Philosophical Magazine*, **40** (1949) 770-780.

(27) The drag integral in the linearised theory of compressible flow, *Quarterly Journal of Mathematics*, **20** (1949) 121-123.

(28) The diffraction of blast, Part II. *Proceedings of the Royal Society, London*, **A199** (1950) 554-565.

(29) The energy distribution behind decaying shocks I. Plane waves, *Philosophical Magazine*, **41** (1950) 1101-1128.

(30) Contributions to the theory of heat transfer through a laminar boundary layer, *Proceedings of the Royal Society, London*, **A202** (1950) 359-377.

(31) Reflection at a laminar boundary layer of a weak steady disturbance to a supersonic steam, neglecting viscosity and heat conduction, *Quarterly Journal of Mechanics and Applied Mathematics*, **3** (1950) 303-325.

(32) A note on the hodograph transformation for the two-dimensional vortex flow of an incompressible fluid (with S. Goldstein), *Quarterly*

Journal of Mechanics and Applied Mathematics, **3** (1950) 297-302.
(33) On the instability of small planetary cores II. *Monthly Notices of the Royal Astronomical Society*, **110** (1950) 339-342.

(34) A new approach to thin aerofoil theory, *Aeronautical Quarterly*, **3** (1951) 193-210.

(35) On sound generated aerodynamically, I. General theory, *Proceedings of the Royal Society, London*, **A211** (1952) 564-587.
(36) On the squirming motion of nearly spherical deformable bodies through liquids at very small Reynolds numbers, *Communications on Pure and Applied Mathematics*, **5** (1952) 109-118.

(37) Oscillating airfoils at high Mach number, *Journal of Aeronautical Science*, **20** (1953) 402-406.
(38) On boundary layers and upstream influence, I. *Proceedings of the Royal Society, London*, **A217** (1953) 344-357.
(39) On boundary layers and upstream influence, II. Supersonic flows without separation, *Proceedings of the Royal Society, London*, **A217** (1953) 478-509.
(40) Theoretical considerations on free convection in tubes, *Quarterly Journal of Mechanics and Applied Mathematics*, **6** (1953) 398-439.
(41) The hodograph transformation (chap. 7 in *High Speed Flows*), Vol. III of *Modern Developments in Fluid Dynamics* (1953).
(42) On the energy scattered from the interaction of turbulence with sound or shock waves, *Proceedings of the Cambridge Philosophy Society*, **49** (1953) 531-551.
(43) Appendix to a paper by J. C. Burns on the critical Froude number for turbulent flow over a smooth bottom, *Proceedings of the Cambridge Philosophy Society*, **49** (1953) 704-706.

(44) On sound generated aerodynamically, II. Turbulence as a source of sound, *Proceedings of the Royal Society, London*, **A222** (1954) 1-32.
(45) Mathematical methods in compressible flow theory, *Communications on Pure and Applied Mathematics*, **7** (1954) 1-10.
(46) The response of laminar skin friction and heat transfer to fluctuations in the stream velocity, *Proceedings of the Royal Society, London*, **A224** (1954) 1-23
(47) On cnoidal waves and bores (with T. Brooke Benjamin), *Proceedings*

of the Royal Society, London, **A224** (1954) 448-460.
(48) An elementary reasoned account of the general circulation of the atmosphere, *Philosophical Magazine*, **45** (1954) 1154-1162.

(49) The effect of compressibility on turbulence (with G. B. Whitham), Chapter 22 in *Gas Dynamics of Cosmic Clouds* (ed H. G. van Hulst and J. M. Burgess) North Holland Publishing Company, (1955) 121-130.
(50) On kinematic waves, I. Flood movement in long rivers (with G. B. Whitham), *Proceedings of the Royal Society, London*, **A229** (1955) 281-316.
(51) On kinematic waves, II. A theory of traffic flow on long crowded roads (with G. B. Whitham), *Proceedings of the Royal Society, London*, **A229** (1955) 317-345.
(52) The axisymmetric boundary layer on a long thin cylinder (with M. B. Glauert), *Proceedings of the Royal Society, London*, **A224** (1955) 188-203.

(53) The wave drag at zero lift of slender delta wings and similar configurations, *Journal of Fluid Mechanics*, **1**(1956) 337-348.
(54) Viscosity effects in sound waves of finite amplitude, *Surveys in Mechanics* (ed. G. K. Batchelor and R- M. Davies), Cambridge University Press, Cambridge (1956) 250-351.
(55) The image system of a vortex element in a rigid sphere, *Proceedings of the Cambridge Philosophy Society*, **52** (1956) 317-321.

(56) Drift, *Journal of Fluid Mechanics*, **1** (1956) 31-53 with corrigenda **2** (1957) 311-312.
(57) Dynamics of a dissociating gas, I. Equilibrium Flow, *Journal of Fluid Mechanics*, **2** (1957) 1-32.
(58) Contributions to the theory of the pitot-tube displacement effect, *Journal of Fluid Mechanics*, **2** (1957) 493-512.
(59) The fundamental solution for small steady three-dimensional disturbances to a two-dimensional parallel shear flow, *Journal of Fluid Mechanics*, **3** (1957) 113-144.
(60) River waves, Chapter 2 in *Proceedings of the First Symposium on Naval Hydrodynamics*, Nat. Acad. Sci./Nat. Res. Coun. Publication, **515** (1957) 17-44.
(61) The hydrodynamic analogy (contributed to discussion on "Dynamic

Instability in Road Transport"), *Operational Research Quarterly*, **8** (1957) 101-114.

(62) On displacement thickness, *Journal of Fluid Mechanics*, **4** (1958) 383-392.

(63) Shock waves (The Ramsden Memorial Lecture), *Memoirs of the Manchester Literary and Philosophical Society*, **101** (1959) 1-16.

(64) Studies in magnetohydrodynamic waves and other anisotropic wave motions, *Philosophical Transactions of the Royal Society, London*, **A252** (1960) 397-430.
(65) Dynamics of a dissociating gas, Part 2. Quasi-equilibrium transfer theory, *Journal of Fluid Mechanics*, **8** (1960) 161-182.
(66) Note on waves through gases at pressures small compared with the magnetic pressure, with applications to upper-atmosphere aerodynamics, *Journal of Fluid Mechanics*, **9** (1960) 465-472.
(67) Contributions on bodies in shear flows (ed. B. Thwaites.), *Incompressible Aerodynamics*, Oxford University Press (1960) 558-572.
(68) Note on the swimming of slender fish, *Journal of Fluid Mechanics*, **9** (1960) 305-317.
(69) Mathematics and aeronautics (The 48th Wilbur Wright Memorial Lecture), *Royal Aeronautical Society Journal*, **64** (1960) 375-394.
(70) Higher Approximations in Aerodynamic Theory, *Princeton University Press*, Princeton (1960).

(71) A technique for rendering approximate solutions to physical problems uniformly valid (The Ludwig Prandtl Memorial Lecture), *Z. Flugwissen.*, **9** (1961) 267-275.

(72) Fluid dynamics as a branch of physics, *Physics Today*, **15** (1962) 17-20.

(73) Sound generated aerodynamically (The Bakerian Lecture), *Proceedings of the Royal Society, London*, **A267** (1962) 147-181.
(74) Physical interpretation of the mathematical theory of wave generation by wind, *Journal of Fluid Mechanics*, **14** (1962) 385-398.

(75) Jet noise (The Wright Brothers Lecture), *AIAA Journal*, **1** (1963)

1507-1517.

(76) Automatic landing of aircraft independently of the weather, *Bull Inst. Phys. and Phys. Soc.*, (1963) 49-59.

(77) Introduction: Real and Ideal Fluids and Introduction: Boundary Layer Theory (Chapter 1, 1-45 and Chapter 2, 46-113 in *Laminar Boundary Layers* ed. L. Rosenhead) Oxford University Press (1963).

(78) Group velocity, *Journal of the Institute of Mathematics and Its Applications*, **1** (1965) 1-28.

(79) Contributions to the theory of waves in nonlinear dispersive systems, *Journal of the Institute of Mathematics and Its Applications*, **1** (1965) 296-306.

(80) Chapter 3 in the organisation of research establishments (ed. Sir John Cockcroft), Cambridge University Press, Cambridge (1965).

(81) Reliability, and a research and development establishment's measures to promote it, *J. Inst. Eng. Insp.*, **29** (1965) 111-123.

(82) Dynamics of rotating fluids: a survey, *Journal of Fluid Mechanics*, **26** (1966) 411-431.

(83) Initial development of diffusion in Poiseuille flow, *Journal of the Institute of Mathematics and Its Applications*, **2** (1966) 97-108.

(84) Predictions on the velocity field coming from acoustic noise and generalized turbulence in a layer overlaying a convectively unstable region, *IAU Symposium No. 28* (Aerodynamic Phenomena in Stellar Atmospheres), Academic Press (1967) 429-453.

(85) On waves generated in dispersive systems by travelling forcing effects, with applications to the dynamics of rotating fluids, *Journal of Fluid Mechanics*, **27** (1967) 725-752.

(86) Some special cases treated by the Whitham theory, *Proceedings of the Royal Society, London*, **A299** (1967) 28-53.

(87) Waves in fluids, *Communication of Pure and Applied Mathematics*, **20** (1967) 267-293.

(88) Theoretical considerations for the steady inviscid case with Rossby number smalt but not zero, Appendix to a paper 'Transverse flow past obstacles in a rotating fluid' by R. Hide and A. Ibbetson, *Journal of Fluid Mechanics*, **32** (1968) 251-272.

(89) Mathematics in control, *Journal of the Institute of Mathematics and*

Its Applications, **4** (1968) 1-19.

(90) Pressure-forcing of tightly fitting pellets along fluid-filled elastic tubes, *Journal of Fluid Mechanics*, **34** (1968) 113-144.

(91) Dynamic response of the Indian Ocean to onset of the Southwest Monsoon. *Philosophical Transactions of the Royal Society, London*, **A265** (1969) 45-92.

(92) Hydromechanics of aquatic animal propulsion, *Annual Rev. Fluid Mech.*, **1** (1969) 413-446.

(93) Unsteady wind-driven ocean currents (The Symons Memorial Lecture), *Quarterly Journal of the Royal Meteorological Society*, **95** (1969) 675-688.

(94) The outlook for a wave theory of turbulent shear flows, *Proceedings of the 1968 Conference on Computation of Turbulent Boundary Layers*, **1** 511-520. Stanford University: Thermosciences Division (1969).

(95) Motion in narrow capillaries from the stand point of lubrication theory (ed. Wolstenhotme and Knight). J. and A. Churchill, *Circulatory and Respiratory Mass Transport* (1969) 85-104.

(96) Lectures on wave mathematics, *Proceedings of the 1968 Battelle Memorial Rencontre*, Seattle: Battelle Research Institute (1969) 1-7 and 96-210.

(97) Turbulence, chapter 2 in Osborne Reynolds and Engineering Science Today (ed. D. M. McDowell and J. D. Jackson.) Manchester University Press (1970).

(98) The theory of trailing Taylor columns, *Proceedings of the Cambridge Philosophy Society*, **68** (1970) 485-492.

(99) Aquatic animal propulsion of high hydro-mechanical efficiency, *Journal of Fluid Mechanics*, **44** (1970) 265-301.

(100) How do fishes swim? *Endeavour*, **29** (1970) 77-93.

(101) Large-amplitude elongated-body theory of fish locomotion, *Proc. Roy. Soc., London*, **B179** (1971) 125-138.

(102) Time-varying currents (contribution to a R. S. discussion meeting), *Philosophical Transactions of the Royal Society, London*, **A270** (1971) 371-390.

(103) The art of teaching the art of applied mathematics, *Mathematical Gazette*, **55** (1971) 249-270.

(104) The Fourth Annual Fairey Lecture: The propagation of sound through moving fluids, *J. Sound and Vibration*, **24** (1972) 471-492.

(105) Physiological fluid dynamics: a survey, *Journal of Fluid Mechanics*, **52** (1972) 475-497.

(106) On the Weis-Fogh mechanism of lift generation, *Journal of Fluid Mechanics*, **60** (1973) 1-17.

(107) Aquatic animal locomotion, *Theoretical and Applied Mechanics* (ed. E. Becker and O. K. Mikhailov), Springer Verlag (1973) 29-46.

(108) Summing up in *Atherogenesis: Initiating Factors* (ed. R. Porter and J. Knight), CIBA Foundation Symposium, Associated Scientific Publishers (1973).

(109) Artificial intelligence: a general survey, *Artificial Intelligence*, Science Research Council (1973) 1-21.

(110) Ocean science, *The Journal of Navigation*, **27** (1974) 91-110.

(111) Aerodynamic aspects of animal flight, *Fifth Fluid Science Lecture Arranged by the British Hydromechanics Research Association* (1974) 423-491.

(112) Flagellar hydrodynamics, *SIAM Rev.*, **18** (1976) 161-230.

(113) Introduction to the scaling of aerial locomotion, *Scale Effects in Animal Locomotion* (ed. T. J. Pedley), Academic Press (1977) 365-404.

(114) Mathematical theories of fish swimming, *Fisheries Mathematics* (ed. J. H. Steele), Academic Press (1977) 131-144.

(115) The interaction between mathematics and society, *Proc. 3rd Internal. Congr. Math. Educ.*, Universität Karlsruhe (1977) 27-60.

(116) Multiple sea use, *Interdiscip. Science Reviews*, **2** (1977) 27-35.

(117) Acoustic streaming, *Journal of Sound and Vibration*, **61** (1978) 391-418.

(118) Teaching how to make and use mathematical descriptions of engineering systems, *Proc. 2nd Internal. Congr. Engin. Educ.* (ed. T. H. Darmstadt) (1978).

(119) Two-dimensional analyses related to wave-energy extraction by submerged resonant ducts. *Journal of Fluid Mechanics*, **91** (1979) 253-317.

(120) A simple fluid-flow model of ground effect on hovering, *Journal of Fluid Mechanics*, **93** (1979) 781-797.
(121) Waves and hydrodynamic loading, *Proceedings of the Second International Conference of Behaviour of Offshore Structures*, BHRA Fluid Engineering, Cranfield (1979) 1-40.

(122) Academic innovation (The Inaugural Frederic Constable Lecture), *Journ. R. Soc. Arts* (1980).

(123) Energy flow in the cochlea, *Journal of Fluid Mechanics*, **106** (1981) 149-213.
(124) A few introductory articles in *Monsoon Dynamics* (ed. Lighthill and Pearce), Cambridge University Press, Cambridge (1981).

(125) Advantages from describing cochlear mechanics in terms of energy flow (ed. E. de Boer and M. A. Viergever), *Mechanics of Hearing*, Delft University Press (1983) 63-71.
(126) The Earth's fluid envelope (Review of A. E. Gill, *Atmosphere-Ocean Dynamics*), *Nature*, **302** (1983) 733-734.

(127) The recently recognized failure of predictability in Newtonian dynamics, *Proceedings of the Royal Society, London*, **A407** (1986) 35-50.
(128) Fundamentals concerning wave loading on off-shore structures, *Journal of Fluid Mechanics*, **173** (1986) 667-681.

(129) What is mechanics? *IUTAM: A Short History 1922-1986* (ed. S. Juhasz), Springer (1988) 1-8.

(130) Emendations to a proof in the general three-dimensional theory of oscillating sources of waves, *Proceedings of the Royal Society, London*, **A427** (1990) 31-42.
(131) Biofluiddynamics of balistiform and gymnotiform locomotion, Part 1. Biological background, and analysis by elongated-body theory (with Robert Blake), *Journal of Fluid Mechanics*, **212** (1990) 183-207.
(132) Biofluiddynamics of balistiform and gymnotiform locomotion, Part 2. The pressure distribution arising in two-dimensional irrotational flow from a general symmetrical motion of a flexible flat plate normal to itself, *Journal of Fluid Mechanics*, **213** (1990) 1-10.
(133) Biofluiddynamics of balistiform and gymnotiform locomotion, Part

3. Momentum enhancement in the presence of a body of elliptic cross-section, *Journal of Fluid Mechanics*, **219** (1990) 11-20.

(134) Biofluiddynamics of balistiform and gymnotiform locomotion, Part 4. Short-wavelength limitations on momentum enhancement, *Journal of Fluid Mechanics*, **213** (1990) 21-28.

(135) Some challenging new applications for basic mathematical methods in the mechanics of fluids that were originally pursued with aeronautical aims (The Inaugural Goldstein Memorial Lecture), *Aeron. J.*, **94** (1990) 41-52.

(136) Hydrodynamic far fields, *Mathematical Approaches in Hydrodynamics* (ed. T. Miloh), SIAM (1991) 3-20.

(137) Biomechanics of hearing sensitivity (the Rayleigh Lecture), *ASME Journ. Vib. Acoust.*, **113** (1991) 1-13.

(138) A general introduction to aeroacoustics and atmospheric sound, ICASE Report No. 92-52, NASA Langley Research Center (1992).

(139) Acoustic streaming in the ear itself, *Journal of Fluid Mechanics*, **239** (1992) 551-606.

(140) La mécanique dans la biologie moderne (The Humphry Davy Lecture), *La Vie des Sciences*, **9** (1992) 117-136.

(141) Modelling of storm surge in the Bay of Bengal (with B. Johns) in *Proceedings of ICSU/WMO International Symposium on Tropical Cyclone Disasters*, (ed. James Lighthill et al.), Peking University Press, Beijing, China, (1993) 410-422.

(142) Asymptotic behaviour of anisotropic wave systems stimulated by oscillating sources, *Wave Asymptotics* (ed. Martin and Wickham), Cambridge University Press, Cambridge (1993) 1-23.

(143) Biofluiddynamics: A survey, *Fluid Dynamics in Biology*, (ed. A. Y. Cheer and C. P. van Dam), American Mathematical Society (1993) 1-32.

(144) Estimates of pressure differences across the head of a swimming clupeid fish, *Philosophical Transactions of the Royal Society, London*, **B341** (1993) 129-140.

(145) Fourier asymptotics (ed. B. D. Sleeman and R. J. Jarvis), Longman, *Ordinary and Partial Differential Equations*, **4** (1993) 210-221.

(146) Inverse Fourier asymptotics (ed. B. D. Sleeman and R. J. Jarvis), Longman, *Ordinary and Partial Differential Equations*, **4** (1993) 222-

237.

(147) Global climate change and tropical cyclones (with G. Holland, W. Gray, C. Landsea, G. Craig, J. Evans, Y. Kurihara and C. Guard), *Bull. Amer. Meteor. Soc.*, **75** (1993) 2147-2157.

(148) Some aspects of the aeroacoustics of high-speed jets (The Inaugural Theodorsen Lecture), *J. Theor. Comp. Fluid Mech.*, **6** (1994) 261-280.

(149) Ocean tides from Newton to Pekeris (Israel Academy of Sciences and Humanities) (1995).

(150) The role of the lateral line in active control of drag reduction by clupeoid fishes, *Animal Locomotion*, The Society for Experimental Biology (1995) 35-48.

(151) Fluid mechanics in *Twentieth Century Physics*, **2**, (ed. Laurie M. Brown, Abraham Pais, and Sir Brian Pippard), Institute of Physics Publishing and American Institute of Physics Press (1995) 795-912.

(152) Internal waves and related initial-value problems, *Dynamics of Atmospheres and Oceans*, **23** (1996) 3-17.

(153) Reinterpreting the basis theorem of flagellar hydrodynamics, *Journal of Engineering Mathematics*, **30** (1996) 25-34.

(154) Helical distributions of stokeslets, *Journal of Engineering Mathematics*, **30** (1996) 35-78.

Bibliography

This bibliography is not, by any means, a complete for this book. For most part, it consists of papers and books which are cited in this book. Many other selected books and papers related to material in the book have been included so that they may serve to stimulate new interest in future study and research.

Ackeret, J. (1925). Luftkräfte auf Flügel, die mit der grösserer als Schallgeschwindigkeit bewegt wrden. *Zeitschrift fur Flugtechnik und Motorluftschiffahrt*, **16**, pp. 72–95.

Arnold, V.I. (1963). *Soviet Math*, **3**, pp. 136–150.

Baines, P. G. (1998). Topographic Effects in Stratified Flows, Cambridge University Press, Cambridge.

Bekesy, G. von (1960). *Experiments in Hearing*, McGraw-Hill, New York.

Benjamin, T.B. (1967). Instability of periodic wave trains in nonlinear dispersive systems, *Proc. Roy. Soc. Lond.*, **A299**, pp. 59–75.

Benjamin, T.B. and Feir, J.E. (1967). The disintegration of wavetrains on deep water, Part 1, Theory, *J. Fluid. Mech.*, **27**, pp. 417–430.

Benney, D.J. (1979). Large amplitude Solitary Rossby Waves, *Stud. Appl. Math*, **60**, pp. 1–10.

Blandford, R.R. (1966). Mixed gravity-Rossby Waves in the Ocean, *Deep Sea Res.*, **13**, pp. 941–961.

Bourot, J. M. (1974). On the numerical computation of the optimum profile in Stokes flow *J. Fluid Mech.*, **65**, pp. 513–515.

Burgers, J.M. (1948). A mathematical model illustrating the theory of turbulence, *Adv. Appl. Mech.*, **1**, pp. 171–199.

Burton, C.V. (1893). On plane and spherical sound waves of finite amplitude, *Phil. Mag.* **35**, pp. 317–333.

Busemann, A. (1935). *Aerodynamischer Auftrieb bei Überschallgeschwindigkeit Luftfahrtforschung*, **12**, pp. 210–230.

Cap, F. (2006). Tsunamis and Hurricanes, Springer Wien New York.

Chirikov, B.V. (1979). A universal instability of many-dimensional oscillator systems, *Physics Report*, **52**, pp. 265–277.

Cole, J.D. (1951). On a quasilinear parabolic equation occurring in aero-dynamics, *Quart. Appl. Math.*, **9**, pp. 225–236.

Crapper, G.D. (1957). An exact solution for progressive capillary waves of arbitrary amplitude, *J. Fluid Mech.*, **2**, pp. 532–540.

Craik, A.D.D. (1988). Wave interactions and fluid flows, Cambridge University Press, Cambridge.

Crighton, D.G. (1981). Acoustics as a branch of fluid mechanics, *J. Fluid Mech.* **106**, pp. 261–298.

Crighton, D.G. (1999). Sir James Lighthill, *J. Fluid Mech.*, **386**, pp. 1–3.

Crighton, D.G. and Leppington, F.G. (1970). Scattering of aerodynamic noise by a semi-infinite compliant plate, *J. Fluid Mech.*, **43**, pp. 721–736.

Crighton, D.G. and Leppington, F.G. (1971). On the scattering of aerodynamic noise, *J. Fluid Mech.*, **46**, pp. 577–597.

Crighton, D.G. and Pedley, T. J. (1999). Michael James Lighthill, *Notices of the AMS* **46**, pp. 1226–1229.

Curle, N. (1955). The influence of solid boundaries on aerodynamic sound, *Proc. Roy. Soc. London* **A231**, pp. 505–514.

Debnath, L. (1967). Transient Wave Motions in Fluids, Ph. D. dissertation, University of London, England.

Debnath, L. (1976a). Lectures on Dynamics of Oceans, Center of Advanced Study in Applied Mathematics, University of Calcutta, 104 pp.

Debnath, L. (1976b). On wind-driven ocean currents in an ocean with bottom friction, *Z. Angew. Math. Mech.*, **56**, pp. 469–478.

Debnath, L. (1983). Nonlinear waves, Cambridge Unversity Press, Cambridge.

Debnath, L. (1984a). Advances in Nonlinear Waves, Vols. I & II, Pitman Publishing Company, Boston.

Debnath, L. (1984b). Poincaré, Kelvin and Rossby Waves in Oceans (a chapter of a book dedicated to H. Poincaré, edited by G.M. Rassias), in: Teubner-Texte zur Mathematik, Band 76, Teubner, Leipzig, pp. 100–124.

Debnath, L. (1992). Nonlinear Dispersive Wave Systems, World Scientific Publishing Co., Singapore.

Debnath, L. (1994). Nonlinear Water Waves, Academic Press, Boston.

Debnath, L. (1998). Wavelet transforms, fractals and turbulence in *Nonlinear Instability, Chaos and turbulence* (ed. L. Debnath and D.N. Riahi) Vol. I, Witt Press, Computational Mechanics Publications, pp. 129–196.

Debnath, L. (1999). Sir James Lighthill (1924-1998) and Modern Fluid Mechanics; A Memorial Tribute, *International Journal of Mathematics and Mathematical Sciences*, **22**, pp. 667–688.

Debnath, L. (2000). Some nonlinear evolution equations in water waves, *J. Math. Anal. Appl.*, **251**, pp. 488–503.

Debnath, L. (2005). Nonlinear Partial Differential Equations for Scientists and Engineers (Second Edition) Birkhauser, Boston.

Debnath, L., Soundalgekar, V.M., and Takhar, H.S., Recent development of the Taylor-Couette viscous flow between two rotating concentric cylinders in *Nonlinear Instability Analysis*, (Edited by L. Debnath), **Volume II**, pp. 274–348, WIT Press, Southampton, England.

Debnath, L. and Rahman, M. (1981). A theory of nonlinear wave loading on offshore structure, *Internat. J. Math and Math. Sci.*, **4**, pp. 589–613.

Debnath, L. and Rosenblat, S. (1969). The ultimate approach to the steady state in the generation of waves on a running stream, *Quart. Jour. Mech. and Appl. Math. XXII*, **22**, pp. 221–233.

Derzho, O.G. and Grimshaw, R. (1997). Solitary waves with a vortex core in a shallow layer of stratified fluid, *Phys. Fluids*, **9(11)**, pp. 3378–3385.

Derzho, O.G. and Grimshaw, R. (2005). Rossby waves on a shear flow with recirculation cores, *Stud. Appl. Math.*, **115**, pp. 387–403.

Drazin, P.G. and Reid, W.H. (1981). *Hydrodynamic Stability*, Cambridge University Press, Cambridge.

Ertekin, R.C., Webster, W.C. and Wehausen, J.V. (1984). Ship generated solitons, National Academy of Sciences, Washington, D.C. *Proc. 15th Symp. Naval Hydrodyn.*, pp. 1–15.

Ffowcs Williams, J.E. and Hall, L.H. (1970). Aerodynamic sound generation by turbulent flow in the vicinity of a scattering half plane, *J. Fluid Mech.* **40**, pp. 657–670.

Ffowcs Williams, J.E. and Hawkings, D.L. (1969). Sound generation by turbulence and surfaces in arbitrary motion. *Phil. Trans. Roy. Soc.*, **A264**, pp. 321-342.

Ford, R., McIntyre, M.E. and Norton, W.A. (2000). Balance and the slow quasimanifold: some explicit results, *J. Atmos. Sci.*, **57**, pp. 1236–1254.

Friedrichs, K.O. (1948a). Formation and decay of shock waves, *Comm. Pure Appl. Math.* **1**, pp. 211–245.

Friedrichs, K.O. (1948b). Water waves on a shallow sloping beach, *Comm. Pure Appl. Math.*, **1**, pp. 109–134.

Garabedian, P.R. (1964). Partial Differential Equations, John Wiley, New York.

Garrison, C.J. (1978). Hydrodynamic loading of large offshore structures–three-dimensional source distribution methods, *Numerical Methods in Offshore Engineering* (Ed. Ziemkiewicz, J. *et al.*), pp. 97–140.

Garrison, C.J. (1982). Forces on semi-submerged structures, *Proc. Ocean Structural Dynamics Symposium*, Oregon State University, Covallis.

Gill, A.E. (1974). The stability of planetary waves on an infinite beta-plane,*Fluid Dyn.*, **6**, pp. 29–47.

Gill, A.E. (1976). Adjustment under gravity in a rotating channel, *J. Fluid Mech.*, **77**, pp. 603–621.

Gill, A.E. (1982). Atmosphere-Ocean Dynamics, Academic Press, New York.

Goldstein, S. (1938). Modern Developments in Fluid Dynamics (2 Volumes), Oxford University Press, Oxford.

Goldstein, S. (1948). Low-Drag and suction Airofoils, *Journal of the Aeronautical Sciences*, **15**, pp. 189–200.

Goldstein, M.E. (1976). *Aeroacoustic*, McGraw-Hill, New York, 293 pp.

Gray, J. (1968). Animal Locomotion, Weidenfeld and Nicholson, London, 479 pp.

Greene, J. (1979). A method for determining a stockastic transition, *J. Math. Phys.* **20**, pp. 1183–1197.

Grimshaw, R. H. J. (2007). Solitary Waves in Fluids, WIT Press, Southampton.

Havelock, T.H. (1940). The pressure of water waves on fixed obstacle, *Proc. Roy. Soc. Lond.*, **A175**, pp. 40–421.
Haynes, P.H. (1985). Nonlinear instability of a Rossby wave critical layer, *J. Fluid Mech.*, **161**, pp. 493–511.
Hogben, N. (1977). Estimation of fluid loading on offshore structures, *Proc. Instn. Civil Engrs.*, **63**, pp. 515–562.
Hogben, N., Miller, B.L., Searle, J.W. and Ward, G. (1977). Estimation of fluid loading on offshore structures, *Nat. Maritime Inst.* Report No. 11, pp. 1–60.
Hopf, E. (1950). The partial differential equation $u_t + uu_x = \mu u_x x$, *Comm. Pure Appl. Math.*, **3**, pp. 201–230.
Howarth, L. (1953). Modern Developments in Fluid Dynamics, High Speed Flow, (Vol. 1), Oxford University Press, Oxford.
Huang, D.B., Sibul, O.J., Webster, W.C., Wehausen, J.V., Wu, D.M. and Wu, T.Y. (1982). ship moving in a transcritical range, *Proc. Conf. on Behavior of Ships in Restricted Waters* (Varna, Bulgaria), **2**, pp. 26-1–26-10.
Hussaini, M.Y. (1997). *Collected Papers of Sir James Lighthill*, (ed.) Four Volumes 1-4, Oxford University Press, Oxford.
Hussaini, M.Y. and Hardin, J.C. (1963). Proceedings of the ICASE/NASA LaRC Workshop on Computational Aeroacoustics, NASA.
Johnson, R.S. (1997). A Modern Introduction to the Mathematical Theory of Water Waves, Cambridge University Press, Cambridge.
Kamenkovich, V.M. (1977). Fundamentals of Ocean Dynamics, English translation by R. Radok, Elsevier Scientific Publishing Company, Amsterdam.
Karman, Th. von and Moore, N.B. (1932). Resistance of slender bodies with supersonic velocities with special references to projectiles, *Trans. Amer. Soc. Mech. Eng.*, **54**, pp. 303–310.
Kolmogorov, A.N. (1954). On preservation of conditionally periodic motions under a small change in Hamiltonian functions, *Dokl. Akad. Nauk, SSSR* **98**, pp. 527–530.
Kolmogorov, A.N. (1982). Newton and Contemporary Mathematical thought, *Mat. V. Shkole* **5**, pp. 58–64.
Lamb, H. (1904). On deep water waves, *Proc. London Math. Soc.* **2**, pp. 371–399.
Lamb, H. (1932). Hydrodynamcis (sixth edition), Cambridge University Press, Cambridge.
Landau, L.D. (1944). On the problem of turbulenc, *C.R. Acad. Sci. U.R.S.S.*, **44**, pp. 14–311.
Landau, L.D. and Lifshitz, E. M. (1959). Fluid Mechanics, Pergamon Press, New York.
Lassiter, L.W. and Hubbard, H.H. (1952). Experimental studies of noise from subsonic jets in still air. NACA Technical Note 2757 National Advisory Committee for Aeronautics.
LeBlond, P.H. and Mysak, L.A. (1978). Waves in the Ocean, Elsevier Scientific Publishing Company, Amsterdam.
Leibovich, S. and Seebass, A.R. (1972). Nonlinear Waves, Cornell University Press, Ithaca.
Lichtenberg, A.J. and Lieberman, M.A. (1983). Regular and Stochastic Motions,

Springer, New York.

Longuet-Higgins, M.S. (1964). Planetary Waves on a Rotating Sphere, *Proc. Roy. Soc.* **A278**, pp. 446–473.

Longuet-Higgins, M.S. (1965). Planetary Waves on a Rotating Sphere, II, *Proc. Roy. Soc.*, **A279**, pp. 40–68.

Lorenz, E.N. (1963). Deterministic non-periodic flows, *J. Atmrs. Sci.*, **20**, pp. 130–141.

Luke, J.C. (1967). A variational principle for a fluid with a free surface, *J. Fluid Mech.*, **27**, pp. 395–397.

MacCamy, R.C. and Fuchs, R.A. (1954). Wave forces on a pile: a diffraction theory, *Tech. Memo*, **69**, US. Army Board, U.S. Army Corp. of Eng.

Matais, O. and Lesieur, M. (1992). Spectral large-eddy simulation of isotropic and stably stratified turbulence, *J. Fluid Mech.* **239**, pp. 157–194.

Mei, C.C. (1978). Numerical methods in water-wave diffraction and radiation, *Ann. Rev. Fluid Mech.* **10**, pp. 393–416.

Mei, C.C., Stiassnie, M. and Yue, Dick K.-P. (2005). Theory and Applications of Ocean Surface Waves, Part I: Linear Aspects and Part 2: Nonlinear Aspects, World Scientific, New Jersey and London.

McIntyre, M.E. (1998). Balanced atmosphere-ocean dynamics, generalized Lighthill radiation, and the slow quasi-manifold, *Theor. Comput. Fluid Dyn.*, **10**, pp. 76–263.

Miche, R. (1944). Movements ondulatories de la mer en profondeur constants ou décreoissante. Forme limite de la houle lors de son déferlement. Application aux dignes maritimes. *Ann. Ponts Chaussées*, **114**, pp. 25–78, 131–164, 270–292, 369–406.

Michell, J.H. (1893). On the highest waves in water, *Phil. Mag.*, **36**, pp. 430–435.

Miles, J.H. (1962). Transient gravity wave response to an oscillating pressure, *J. Fluid. Mech.*, **13**, pp. 145–150.

Mogridge, G.R. and Jamieson, W.W. (1976). Wave loads on large circular cylinders, a design method, *National Research Council of Canada, Division of Mechanical Engineering, Rep. MH-111*, 34 pp.

Moser, J. (1962). Nachr. Akad. Wiss. Göttingen, *Math. Phys. Klasse*, pp. 1–20.

Myint-U.T. and Debnath, L. (2007). Linear Partial Differential Equations for Scientists and Engineers (Fourth Edition). Birkhauser Verlag, Boston.

Newman, J.N. (1977). Marine Hydrodynamics, M.I.T. Press, Cambridge.

Nigam, S.D. and Nigam, P.D. (1962). Wave Propagation in Rotating Liquids, *Proc. Roy. Soc. London* **A266**, pp. 247–256.

Nigam, S.D. and Nigam, P.D. (1963). Magnetohydrodynamic Waves in Rotating Liquids, *Proc. Roy. Soc. London* **A272**, pp. 529–541.

Pack, D.C. (1960). A note on the breakdown of continuity in the motion of a compressible fluid, *J. Fluid Mech.*, **8**, pp. 103–108.

Pao, Y.H. (1965). Structure of Turbulent Velocity and Scalar Fields at Large Wavenumbers *Phys. Fluids*, **8**, pp. 1063–1081.

Pedley, T.J. (1977). *Scale Effects in Animal Locomotion*, (ed) Academic Press, 545 pp.

Pedley, T.J. (1980). The fluid mechanics of large blood vessels, Cambridge University Press, Cambridge.

Pedley, T.J. (2001). Sir (Michael) James Lighthill, *Memoir of the Royal Society of London*, pp. 335–356.

Pedley, T.J. (2001). James Lighthill and his contributions to Fluid Mechanics, *Annu. Rev. Fluid Mech.*, **33**, pp. 1–41.

Pedlosky, J. (1971). Geophysical Fluid Dynamics, in: E.H. Reid (Ed.), Mathematical Problems in the Geophysical Sciences, in: Lectures in Applied Mathematics, **13**, pp. 1–60.

Pedlosky, J. (1979). Geophysical Fluid Dynamics, Springer-Verlag, Berlin.

Pironneau, O. (1973). On optimum profiles in Stokes flow, *J. Fluid Mech.*, **59**, pp. 117–128.

Rayleigh, L. (1910). Aerial plane waves of finite amplitude, *Proc. Roy. Soc. London*, **A84**, pp. 247–284.

Rayleigh, L. (1945). Theory of Sound (*2 Vols.*) second edition, Dover, New York, 362 pp.

Riemann, B. (1858). Uber die Fortpflanzung ebener Luftwellen von endlicher Schwingungsweite Göttingen Abhandlunger, Vol. viii, p. 43, (Werke, 2te Aufl., Leipzig, 1892, p. 157).

Saffman, P.G. (1993). *Vortex Dynamics*, Cambridge University Press, Cambridge.

Smart, J. and Hughes, N.F. (1972). In Insect/Plant Relationship: Sympos. R. Entomol. Soc. London, **6** pp. 143–155.

Stewartson, K. (1978). The evolution of the critical layer of a Rossby wave, *Geophysics Astrophys. Fluid Dyn.*, **9** pp. 185-200.

Stoker, J.J. (1957). Water waves, Interscience Publishers, New York.

Stuart, J.T. (1963). Unsteady boundary layers in Laminar Boundary Layers (Ed. L. Rosenhrad) Chapter 7.

Stuart, J.T. (1966). Double Boundary Layers in Oscillatory viswas flow. *Jour. Fluid Mech.*, **24**, pp. 673–687.

Taylor, G.I. (1910). The conditions necessary for discontinuous motion in gases, *Proc. Roy. Soc. London*, **A84**, pp. 371–377.

Taylor, G.I. (1928). The energy of a body moving in an infinite fluid with an application to airships, *Proc. Roy. Soc. London*, **A120**, pp. 13–21.

Taylor, G.I. (1952a). The action of waving cylindrical tails in propelling micrscopic organisms, *Proc. Roy. Soc. London* **A211**, pp. 225–239.

Taylor, G.I. (1952b). Analysis of swimming of long and narrow animals, *Proc. Roy. Soc. London*, **A214**, pp. 158–183.

Van Dyke, M. (1951). First- and Second-order theory of supersonic flow past bodies of revolution, *J. Aero. Sci.*, **18**, pp. 161–191.

von Karman, Th. and Burgers, J.M. (1934). General Aerodynamic Theory, Perfect Fluids, Vol. 2, of Aerodynamic Theory (ed. Durand, W.F.), Springer Verlag, Leipzig.

Ward, G.N. (1949). Supersonic flow past slender pointed bodies, *Quart. J. Mech. and Appl. Math.* **2**, pp. 75–97.

Warn, T. (1983). The evolution of finite solitary Rossby waves on a weak shear, *Stud. Appl. Math.* **69**, pp. 127–133.

Weis-Fogh, T. (1972). Energies of hovering in hummingbirds and in Drosophilia,

J. Exp Biol., **56**, pp. 79–104.

Weis-Fogh, T. (1973). Quick estimate of flight fitness in hovering animals, including novel mechanism for lift production, *J. Exp Biol.*, **59**, pp. 169–230.

Weis-Fogh, T. (1977). Dimensional Analysis of Hovering Flight in *Scale Effects in Animal Locomotion* (Ed. T.J. Pedley) Academic Press, London, pp. 405–420.

Weis-Fogh, T. and Jensen, M. (1956). Biology and Physics of locust flight, I-IV, *Phil. Trans. Royal Soc. London*, **B239**, pp. 415–584.

West, B.J. (1981). Deep Water Gravity Waves, Weak Interaction Theory, Lecture Notes in Physics, No. 146, Springer-Verlag, Berlin.

Whitham, G.B. (1952). The flow pattern of a supersonic projectile. *Commun. Pure Appl. Math.*, **5**, pp. 301–342.

Whitham, G.B. (1956). On the propagation of weak shock waves, *J. Fluid Mech.*, **1**, pp. 290–318.

Whitham, G.B. (1965a). A general approach to linear and nonlinear dispersive waves using a Lagrangian, *J. Fluid Mech.* **22**, pp. 273–283.

Whitham, G.B. (1965b). Nonlinear dispersive waves, *Proc. Roy. Socl Lond.*, **A283**, pp. 238–261.

Whitham, G.B. (1967). Nonlinear dispersion of water waves, *J. Fluid Mech.*, **27**, pp. 399–412.

Whitham, G.B. (1974). Linear and Nonlinear waves, John Wiley Sons, New York.

Wilton, J.R. (1913). On the highest wave in deep water, *Phil. Mag.*, **26**, pp. 1053–1058.

Wilton, J.R. (1915). On ripples, *Phil. Mag.*, **29**, pp. 688–700.

Wu, T.Y. (1961). Swimming of a waving plate, *Journal of Fluid Mechanics*, **10**, pp. 321–344.

Wu, T.Y. (1971). Hydromechanics of swimming propulsion. Part I. Swimming of a two-dimensional flexible plate at variable forward speeds in an inviscid fluid, *Journal of Fluid Mechanics*, **46**, pp. 337–355.

Wu, T.Y. (1971). Hydromechanics of swimming propulsion. Part II. Some optimum shape problems, *Journal of Fluid Mechanics*, **46**, pp. 521–544.

Wu, T.Y. (1971). Hydromechanics of swimming propulsion. Part III. Swimming and optimum movements of a slender fish with side fins, *Journal of Fluid Mechanics*, **46**, pp. 545–568.

Wu, T.Y., Brokaw, C.J. and Brennen, C. (1975). Swimming and Flying in Nature, Vol. 2, Plenum Press, New York.

Yuen, H.C. and Ferguson, W.E., Jr. (1967a). Relationship between Benjamin-Feir instability and recurrence in the nonlinear Schrödinger equation, *Phys. Fluids*, **21**, pp. 1275–1278.

Yuen, H.C. and Lake, B.M. (1980). Instabilities of waves on deep water, *Ann. Rev. Fluid Mech.*, **12**, pp. 303–334.

Yuen, H.C. and Lake, B.M. (1982). Nonlinear dynamics of deep-water gravity waves, *Advances in Appl. Mech.*, **22**, pp. 67–229.

Zabusky, N.J. and Kruskal, M.D. (1965). Interaction of solitons in a collisionless plasma and the recurrence of initial states, *Phys. Rev. Lett.*, **15**, pp. 240–243.

Zarkharov, V.E. (1968). Stability of periodic waves of finite amplitude on the surface of a deep fluid, *J. Appl. Mech. Tech. Phys.*, **9**, pp. 86–94.

Zarkharov, V.E. and Faddeev, L.D. (1971). Korteweg-de Vries equation, a completely integrable Hamiltonian system, *Func. Anal. Appl.*, **5**, pp. 280–287.

Zarkharov, V.E. and Schulman, E.I. (1982). To the integrability of the system of two coupled nonlinear Schrödinger equations, *Physica*, **4D**, pp. 270–274.

Zweig, G. (1976). Basilar Membrane Motion, Cold Spring Harbor Symposium, *Quantitative Biology*, **40**, pp. 619–633.

Index